Stability of Chemical Reactors

DANIEL D. PERLMUTTER

School of Chemical Engineering
University of Pennsylvania

PRENTICE-HALL, INC.
Englewood Cliffs, New Jersey

© 1972 by Prentice-Hall, Inc.
Englewood Cliffs, New Jersey

All rights reserved. No part of this book may be reproduced in any form or by any means without permission in writing from the publisher.

10 9 8 7 6 5 4 3 2 1

ISBN: 0–13–839969–7

Library of Congress Catalog Card Number: 70–168624

Printed in the United States of America

PRENTICE-HALL INTERNATIONAL, INC., *London*
PRENTICE-HALL OF AUSTRALIA, PTY. LTD., *Sydney*
PRENTICE-HALL OF JAPAN, INC., *Tokyo*
PRENTICE-HALL OF CANADA, LTD., *Toronto*
PRENTICE-HALL OF INDIA PRIVATE LTD., *New Delhi*

To Felice in lieu of
Formal acknowledgment giving inadequate expression,
Frank accolades getting indiscreetly effusive.

Contents

Preface

There is a children's riddle that asks, "When is a door not a door?" The apparent logical contradiction is never resolved by the riddler, for he begs the question with a play on words: "A door is not a door when it is ajar."

One might introduce the study of chemical reactor stability with a parallel conundrum: When is a steady state not a steady state? Answer: A steady state is not steady when it is unstable. This time the response is more than merely a play on words, for it rests on a thorough understanding of the words *steady* and *stable* in their technical and everyday uses. There are several related matters of more or less equal subtlety that must also be raised in a rational treatment of reactor stability. These matters have to do with systems having more than a single steady state, and with reactors that are stable to certain disturbances but not to others. Each of these subjects demands clear definitions of appropriate concepts, but because these ideas are motivated by engineering needs, their proper development requires repeated references to the study of various reactor models. The explication of these points is the major goal of this monograph.

The first chapter has a brief introduction but serves mainly to catalog the most common chemical reactor models. It will be a review for those readers who have had a course in Reactor Design. Chapter 2 is devoted to the subject of steady-state multiplicity in stirred reactors and develops uniqueness criteria for various cases that may be of interest for design. The interpretation of terms such as *steady* and *stable* constitute a major theme of Chapters 3 and 4, and Chapter 5 explores the implications of the stability concept in dealing with finite disturbances of practical magnitude.

From the viewpoint of an analyst, a major departure occurs when one moves from ordinary differential equation (lumped-parameter) models to those that require description by partial differential equation (distributed-parameter) models. The logic of the first five chapters is therefore reexamined in Chapters 6 through 8, where steady-state multiplicity, local stability, and regional stability are treated for distributed systems. Chapter 9 focuses on the effects of recycle that are so important to the practical chemical engineer.

The design-oriented engineer will wonder whether reactor stability analysis is an esoteric specialty or a useful tool. It is some of both, and it is the intent of this book to make it more of the latter. The subject is one that offers a unity in outlook that has been lacking in the field of chemical reactor design, and although it does not by any means encompass all the important material in the field, the stability viewpoint does in a sense draw together some seemingly diverse areas; for example, stirred reactors and catalyst effectiveness factors. The features in common emerge as qualitative descriptions of the behavior of the respective models, for the stability viewpoint deals with certain structural aspects present in both problems. Lastly it should be added that a stability orientation leads the designer to focus on the questions of qualitative behavior before he is immersed in details of computation, and serves as a reminder that an engineering model is one of a family of possible compromises between the usually conflicting needs of comprehensive description as opposed to numerical ease. Because of the great complexity of many practically interesting systems, the use of a simplified model is sometimes a case of making a virtue of necessity.

The mode of presentation is such that a prior course in reactor engineering, reactor modeling or kinetics, or some exposure to simple partial differential equations will be helpful as general background. As such it is intended for first-year graduate students. On the other hand, the development includes enough details that it ought to be comprehensible to readers whose backgrounds are not specifically in chemical engineering. Numerous examples with solutions are distributed through the book, chosen to make a point or to display the magnitude of the numbers involved. They are a part of the development of the subject as well as exercises for the reader.

It will be obvious from the references in each chapter that this book leans heavily on the prior contributions of many researchers, but in a most literal and personal sense it would have been impossible without the stimulation and creative efforts of the graduate students together with whom I studied reactor dynamics and stability problems. In particular I owe much to J. S. Berger, T. Z. Fahidy, I. A. Gura, A. M. Kugelman, J. G. Landis,

C. R. McGowin, W. O. Paradis, and F. S. Wang. The interest in my writing and the encouragement that came from A. E. Humphrey was certainly greater than one has a right to expect from the Director of the School of Chemical Engineering. Parts or all of the manuscript were reviewed at various times by J. S. Berger, L. C. Eagleton, Dan Luss, and M. J. Reilly. It was improved wherever their suggestions were incorporated. Needless to say, any remaining errors or shortcomings are the responsibility of the author.

For their most competent secretarial work, cheerfully offered over a span of several years, I wish to thank Miss Cora Ingram, Miss Stephanie Buividas (now Mrs. J. Sell), Miss Candice Slavin (now Mrs. K. Frankel), Mrs. Deborah Blackburn, and Miss Virginia Mathias. My children, Shira and Saul, shared enough of the proofreading chores to make the task almost painless for me.

Finally, acknowledgments should be made of the scholarly leave from the University of Pennsylvania and the Fulbright award, which in combination provided the extended period of time that was needed to organize the ideas that form the foundation of this book.

DANIEL D. PERLMUTTER

Philadelphia, Pa.

Stability of Chemical Reactors

Introduction: On Models, Stability, and Design

Modeling and Design 1

1-1. Design as a Technique

What does an engineer mean by design? The term is variously used (1) to identify a certain approach to problem solving, (2) to describe an associated technique, and (3) to give a name to the package that results from the entire enterprise. The first of these usages is more often referred to today as *systems engineering*, especially when large and complex systems are involved. This approach calls for the resolution of a problem by compromise of separately desirable but conflicting goals for the sake of an overall objective. It attempts to take account of all significant constraints on a problem, whether natural or artificial, environmental or internal. This book is not about design in the sense of systems engineering, but is concerned with aspects of the related techniques.

The technique of design has the essential character of focusing on the broadest family of problems in which a given situation can be handled, rather than the specific case. It has on occasion been called synthesis and put in opposition to analysis, but since one of these cannot exist without the other, this division is to some extent a misplaced emphasis. A more descriptive name for this procedure is *embedding*, made popular through its use in dynamic programming [Bellman (1957)].

To be sure, such a procedure has well-known difficulties. For one, the designer who attempts too broad an interpretation finds that his problem is insoluble, but this error is self-correcting and is in any case avoidable through experience. A second disadvantage can sometimes be identified as arising from the generalization process itself. It is the cost in effort that one might reasonably expect to pay for the privilege of solving more than the narrowest problem statement; however, the logic of embedding does not necessarily lead to extra work if one can take advantage of analytic relationships. A most obvious example would be an ordinary differential equation to be solved with given boundary conditions. When a closed form solution is available, it is no more trouble to solve for the family of all boundary conditions than it is for the given values, and this is of course the usual way in which such a problem is treated. Wherever possible, design is motivated by this spirit, a recurrent theme being the search for qualitative patterns.

A second major aspect of design as a technique is that it is utilitarian. The designer understands from the outset that he may be able to find an adequate result, perhaps even an excellent one, but that he is unlikely to find the abstractly best result. Just as the systems engineering approach starts out to compromise between competing objectives, the designer will compromise on the means to the end. Even at the loss of some accuracy, he will tend to favor graphical over analytic treatment of a problem, when it provides a desirable overview. He will often be willing to settle for partial solutions to a problem, accepting sufficient conditions that are not necessary (in the mathematical sense) if he believes that they are not unreasonably conservative. In summary, the technique of design is a sort of negotiation where the degree of compromise is found as part of the process, but is not known beforehand.

1-2. Mathematical Modeling

As already suggested, the ideas of embedding and utilitarian compromise are usually developed in a mathematical format. Design is based on the properties of a model, that is an equation, that represents the system under study. Implicit is the understanding that the equation is in some significant respect analogous to the system and that behavior deduced from the model will closely parallel that of the system. Simple models may be algebraic, but are more often differential equations when descriptions of dynamic change are essential. The designer chooses a model that is as simple as he can devise, without losing the essential nature of the phenomenon he is after. It will not do, for example, to formulate a first order differential equation model if he wishes to study periodic or oscillatory behavior. The degree of complexity that is justified in a given problem is not always obvious at first glance, however. There may appear to be little reason to include time dependence in a model when only steady states are of interest; nevertheless, the stability or instability of such a state is a dynamic property.

The details of a number of chemical reactor models will be presented later in this chapter. They all have a common theoretical basis in the straightforward mass and energy conservation laws. As a result they form differential equations that are each first order with respect to time derivatives, and either first or second order with respect to distance variables, depending on the geometry under consideration and the physical mechanisms assumed to be operative. Numerical solutions of such selected equations have made vital contributions to the understanding of certain reactor behavior. Although thankful for such information, the designer asks also for something other than numerical solutions for specified initial conditions. He would like to be able to describe the family of possible solutions for a range of initial conditions or a range of parameter values. In principle such results can be made available from the numerical integrations; in practice, the corresponding demands on computer time are often excessive, and it is most helpful to know something about the so-called structure of the equations. The choice of a model in a particular reactor design is therefore the result of compromise, whether the treatment is to be analytic or numerical. In the final analysis, the search for qualitative behavior is an aid to computational results, and vice versa. The two efforts are complementary, not competitive.

1-3. Stability

An intuitive idea of what is meant by stability suggests itself from the common nontechnical use of the word. It is associated with steadiness and constancy. To be of use in engineering design the meaning must be made more precise, and it should be framed in a way that lends itself to the descriptive outlook that gives the designer insight into the vital compromises of his problem solution. Thus, in the following, the question as to whether or not a system is stable will be placed in a broader context involving such subjects as multiple steady states, regions of stability, and even competing definitions of stability. The questions at the heart of the inquiry expand on the long-established steady-state design basis. They may be summarized as follows:

1. Under what conditions can a reactor system have more than one steady state? Assuming it is desirable to do so, what design conditions guarantee a unique steady state?

2. How can design assure the stability of a steady state with respect to small changes in the parameters, the instantaneous values of the dependent variables, or the inputs to the reactor?

3. How is stability to be defined for disturbances of relatively larger magnitude? Can this definition be related to bounds on the responses that

follow input changes? What design choices can insure that specifications on reactor transients will be met?

In short, through its diverse aspects, stability becomes a design concept.

That dynamic systems can be designed from a stability viewpoint is of course not a new idea. Control engineers have long recognized that an accuracy-stability trade-off is called for in feedback systems, by which they mean that it is necessary to sacrifice some degree of steady-state specification in order to achieve desired transient behavior. It is taken for granted that for a designer the stability question does not lend itself to a mere yes-or-no answer, but demands a broader view that expresses itself as a study of the degree of stability. A similar motivation is behind the approach to chemical reactor design that begins in the next chapter, but before entering that domain, it is essential to catalog the more important reactor models, noting the assumptions appropriate to each.

1-4. The CSTR

A well-stirred region uniform in composition C and temperature T is the conceptual starting point for the simplest reactor model. By noting that the accumulation of a component of interest is the algebraic sum of inflow, effluent, and loss by chemical reaction, one develops the mass balance

$$V\frac{dC}{dt} = q(C_0 - C) - VR \tag{1-1}$$

where V = the constant reaction volume
q = the flow rate in and out
C_0 = the feed concentration of the component of interest
R = the reaction rate per unit volume

In consideration of the fact that the reaction rate is ordinarily a function of C and T another relationship is called for that will account for the temperature variation with time. This is the energy balance

$$VC_p\frac{dT}{dt} = qC_p(T_0 - T) + \Delta HVR - U(T - T_w) \tag{1-2}$$

in which, except for the last, the terms parallel those in the prior equation. The exceptional term arises from the need to allow for heat transfer across the boundary of the system by means other than the material flow. Accordingly:

C_p = the heat capacity per unit volume of the flowing materials as well as the contents of the reactor
T_0 = the feed temperature

ΔH = the molar heat of reaction, taken to be positive for exothermic systems

U = the overall heat transfer coefficient, in units which include the area of transfer surface

T_w = the ambient, coolant, or heating fluid temperature

The parallelism between the two conservation equations can be made even more obvious by defining a reduced temperature $\eta = C_pT/\Delta H$. Using this substitution in equation (1-2) gives

$$V\frac{d\eta}{dt} = q(\eta_0 - \eta) + VR - \frac{U}{C_p}(\eta - \eta_w) \tag{1-3}$$

which together with (1-1) and appropriate initial conditions determines the concentration and temperature in the reactor at all later time. Since provision has been made for flow into and out of the system, the model is identified as a continuous-flow stirred tank reactor (CSTR). If $q = 0$, the flow terms are set equal to zero identically, and the result is applicable to a well-stirred constant pressure batch reactor:

$$\frac{dC}{dt} = -R \tag{1-4a}$$

$$\frac{d\eta}{dt} = +R - \frac{U}{VC_p}(\eta - \eta_w) \tag{1-4b}$$

The steady states of the CSTR are found by seeking those values of composition C_s and temperature T_s, (or reduced temperature η_s) which make the derivatives in equations (1-1) and (1-3) equal to zero:

$$0 = q(C_0 - C_s) - VR(\eta_s, C_s) \tag{1-5a}$$

$$0 = q(\eta_0 - \eta_s) + VR(\eta_s, C_s) - \frac{U}{C_p}(\eta_s - \eta_w) \tag{1-5b}$$

It should be noted that these two nonlinear algebraic equations may have more than one solution, since the reaction rate term can appear in a great variety of nonlinear forms. Typically, it is introduced in terms of a reaction order n, and Q, the ratio of an Arrhenius activation energy to the gas constant:

$$R(C, T) = k_0C^n \exp(-Q/T) \tag{1-6a}$$

The kinetics are called first-order when $n = 1$, and

$$R = k_0C \exp(-Q/T) \tag{1-6b}$$

For the batch reactor with $q = 0$, the steady state will also be the state of chemical equilibrium since the net reaction rate R_s must be zero, but this is generally not true for the CSTR.

1-5. The PFTR

For some chemical systems there is considerable advantage to be gained in using a reactor that is not at all well-mixed. Levenspiel (1963) has shown, for example, that smaller reactor volumes and better yields can in some cases be obtained from reactors that are in fact totally unmixed. Moreover, there are numerous mechanical advantages possible, related to the requirements of heat transfer, catalyst packings, and maintenance.

To formulate a model for such an unmixed situation, consider that reactant material of concentration C_0 and reduced temperature η_0 enters a tube and flows in the axial direction Z without back-mixing at a uniform linear velocity u. There are no radial gradients of either velocity, composition, or temperature, but of course these last two do change axially and their description with respect to time and position is precisely the objective of this model.

A mass balance over a differentially small tube segment, but otherwise similar to that which produced equation (1-1) leads in this case to a partial differential equation

$$\frac{\partial C}{\partial t} = -u\frac{\partial C}{\partial Z} - R \tag{1-7a}$$

The corresponding energy balance is

$$C_p\frac{\partial T}{\partial t} = -C_p u\frac{\partial T}{\partial Z} + \Delta HR - h(T - T_w)$$

where h is the wall heat transfer coefficient, in units which include the tubular reactor volume. In terms of reduced temperature

$$\frac{\partial \eta}{\partial t} = -u\frac{\partial \eta}{\partial Z} + R - \frac{h}{C_p}(\eta - \eta_w) \tag{1-7b}$$

Equations (1-7) are a reactor model usually referred to as PFTR, an acronym for plug flow tubular reactor. Because all variation occurs in a single space direction, it is as if the reacting mixture moved in an identifiable plug down the tube from entrance to exit, exchanging only heat with its surroundings. As before complete description demands that boundary conditions be specified. In this case initial condition profiles for temperature and concentration are needed, and in addition end conditions for all time. A fixed feed specification usually provides the latter.

1-6. Axial Mixing

Deviations from idealized plug flow can arise from a number of causes. When the effect is primarily the result of a large number of small scale events, such as might be expected from the repeated stream splitting and reforming that occurs in the interstices of a packed bed, the gross phenomenon is very much akin to diffusion. Indeed, the random-walk arguments that justify Fick's Law for molecular diffusion are applicable with only minor modification, and the quantitative evidence, as expected, does show a linear dependence on the second derivative with respect to distance. This leads to the following modification of the plug-flow model:

$$\frac{\partial C}{\partial t} = D\frac{\partial^2 C}{\partial Z^2} - u\frac{\partial C}{\partial Z} - R \tag{1-8a}$$

$$\frac{\partial \eta}{\partial t} = \alpha\frac{\partial^2 \eta}{\partial Z^2} - u\frac{\partial \eta}{\partial Z} + R - \frac{h}{C_p}(\eta - \eta_w) \tag{1-8b}$$

where, however, it should be emphasized that the coefficients D and α are not true diffusivities, but rather overall "dispersion" coefficients for mass and thermal energy, respectively. These equations with the foregoing qualifications will be referred to as the TRAM (tubular reactor with axial mixing).

The boundary conditions for this model are similar to those needed for the PFTR, but now two space conditions must be given. Following the work of Wehner and Wilhelm (1956) and Bischoff (1961), these are usually

$$D\frac{\partial C}{\partial Z} = u(C - C_0), \qquad Z = 0 \tag{1-9a}$$

$$\alpha\frac{\partial \eta}{\partial Z} = u(\eta - \eta_0), \qquad Z = 0 \tag{1-9b}$$

$$\frac{\partial C}{\partial Z} = \frac{\partial \eta}{\partial Z} = 0, \qquad Z = L \tag{1-9c}$$

1-7. Radial Gradients

The nonideality-of-flow arguments that justified the inclusion of dispersion terms in tubular reactor models are also applicable in the radial direction. An appreciable gradient in temperature is to be expected whenever large quantities of heat generated by reaction near the tube center line must be rejected to a coolant at the wall. Moreover, concentration gradients will result from increased reaction at the higher temperatures. Using the radial symmetry of a tubular geometry leads to a two-dimensional model in terms of radius Y and radial dispersion coefficients

$$\frac{\partial C}{\partial t} = D\left[\frac{\partial^2 C}{\partial Y^2} + \frac{1}{Y}\frac{\partial C}{\partial Y}\right] - u\frac{\partial C}{\partial Z} - R \tag{1-10a}$$

$$\frac{\partial \eta}{\partial t} = \alpha\left[\frac{\partial^2 \eta}{\partial Y^2} + \frac{1}{Y}\frac{\partial \eta}{\partial Y}\right] - u\frac{\partial \eta}{\partial Z} + R \tag{1-10b}$$

Experimental evidence as well as theoretical analysis [detailed in Kramers and Westerterp (1963)] indicate that the axial and radial dispersion coefficients are not equal in packed beds, the former being perhaps ten-fold larger. The reactor model can of course be expanded to include both effects, but this will not be done here in the interest of simplicity. Discussion of the more general model can be found in Beek (1962).

The heat removed at the wall is now handled as a boundary condition, rather than including it as a term in the balance equation

$$\alpha \frac{\partial \eta}{\partial Y} = \frac{U_r}{C_p}(\eta_w - \eta), \qquad Y = L_0 \tag{1-11a}$$

where U_r is the radial heat transfer coefficient at the wall. Since no mass can be transferred past the wall

$$\frac{\partial C}{\partial Y} = 0, \qquad Y = L_0 \tag{1-11b}$$

and assuming center-axis symmetry

$$\frac{\partial \eta}{\partial Y} = \frac{\partial C}{\partial Y} = 0, \qquad Y = 0 \tag{1-11c}$$

The abbreviated name for this model will be TRRM, short for tubular reactor with radial mixing, taking for granted the appropriate assumptions justified.

The reader should note that to this point three different overall heat transfer coefficients have been defined for as many models. To avoid confusion, their respective symbols, application, and typical units may be summarized as follows:

Symbol	*Model*	*Units*
U	CSTR (Continuous-flow Stirred Tank Reactor)	Btu/hr °F
h	{PFTR (Plug-flow Tubular Reactor) TRAM (Tubular Reactor with Axial Mixing)	Btu/ft³ hr °F
U_r	TRRM (Tubular Reactor with Radial Mixing)	Btu/ft² hr °F

1-8. Catalyst Particles

In a bed packed with catalyst particles, it is sometimes desirable to treat an individual particle as a micro-reactor. If a one-dimensional geometry is thought to be applicable, as for thin flakes or spheres active only in a thin surface layer, the following conservation equations are applicable:

$$\frac{\partial C}{\partial t} = D\frac{\partial^2 C}{\partial Z^2} - R \tag{1-12a}$$

$$\frac{\partial \eta}{\partial t} = \alpha \frac{\partial^2 \eta}{\partial Z^2} + R \tag{1-12b}$$

The terms in this model are justified by arguments very similar to those used to form equations (1-10), but because there is within the particle no possibility of convective flow, the first derivative terms are deleted. The pseudo-diffusion coefficients of this model are called "effective" diffusivities for mass and heat as a reminder that in considering a porous solid they are not the simple molecular parameters [Petersen (1965)]. When necessary, subscripts can be used to distinguish these coefficients from the axial or radial dispersion parameters of the prior models, but generally this will not be worthwhile because only one model will be under study and the meaning clear from the context.

The boundary conditions for this catalyst particle model include again initial condition profiles in concentration and temperature and two specifications in the space coordinate. These are usually obtained from known external conditions

$$C = C_0, \quad Z = L \tag{1-13a}$$

$$\eta = \eta_0, \quad Z = L \tag{1-13b}$$

and the center-line symmetry condition

$$\frac{\partial C}{\partial Z} = \frac{\partial \eta}{\partial Z} = 0, \quad Z = 0 \tag{1-13c}$$

Alternatively, if the mass and/or heat flux is known via transfer coefficients, equations (1-13a and b) are replaced by

$$D\frac{\partial C}{\partial Z} = k_G(C_0 - C), \quad Z = L \tag{1-13d}$$

$$\alpha\frac{\partial \eta}{\partial Z} = h_G(\eta_0 - \eta), \quad Z = L \tag{1-13e}$$

where the new parameters k_G and h_G are the coefficients for mass and heat transfer between the phases, defined to include the effective interfacial area of transfer.

1-9. Multi-Phase Models

Although the various models presented in the foregoing are directly applicable to many real reactors, it must be admitted that they are too simple for many others, some of great practical importance. When this is the case a closer approach to the essential characterization of a system may sometimes be obtained by combining selected simple models, using them as modules in constructing the more realistic representation.

An example of this approach can be found in the work of Luss and Amundson (1968) on a fluidized bed catalytic reactor. Assuming that the fluidizing gas phase is well-mixed allows one to view this volume as a CSTR in which chemical reaction and heat generation are replaced by mass and heat transfer from the solid phase. Thus, for the interstitial gas

$$V_1 \frac{dC_1}{dt} = q(C_0 - C_1) + k_G(C - C_1) \tag{1-14a}$$

$$V_1 C_p \frac{dT_1}{dt} = qC_p(T_0 - T_1) + h_G(T - T_1) + U(T_w - T_1) \tag{1-14b}$$

The subscript 1 refers to the volume, concentration, and temperature of the fluidizing gas as distinct from the gas held within the solid particles, whose corresponding properties are not subscripted.

Further, if the catalyst particles are small enough to be internally uniform in composition and temperature and it is assumed that the gas held within the particles is similarly homogeneous, this phase is also equivalent to a modified CSTR: the flow term of the mass balance is replaced by mass transfer between the phases, and the flow term of the energy equation is deleted entirely

$$V \frac{dC}{dt} = k_G(C_1 - C) - VR \tag{1-15a}$$

$$Vc_p \frac{dT}{dt} = h_G(T_1 - T) + \Delta HVR \tag{1-15b}$$

Because the solid provides a much greater heat reservoir then the same volume of gas, the left-side accumulation term of equation (1-15b) uses the heat capacity of the solid. For consistency with the other terms of the equation which are based on gas volumes, c_p is defined as the solid volumetric heat capacity divided by the fractional gas content of the solid phase. It can also be thought of as an "effective" heat capacity.

A packed-bed catalytic reactor is another two-phase system which has been modeled by combining subunits. Following Liu, Aris, and Amundson (1963) one can suppose that the fluid is in plug flow, but that all chemical reaction occurs within the solid and can only affect the fluid by transfer between the phases. The PFTR equations must then be modified to

$$\frac{\partial C_1}{\partial t} = -u \frac{\partial C_1}{\partial Z} + k_p(C - C_1) \tag{1-16a}$$

$$C_p \frac{\partial T_1}{\partial t} = -uC_p \frac{\partial T_1}{\partial Z} + h_p(T - T_1) - h(T_1 - T_w) \tag{1-16b}$$

where k_p and h_p are again the mass and heat transfer coefficients, subscripted differently than in equations (1-14) and (1-15) to emphasize the distinction in geometry and dimensions.

The equations of change within a particle remain the same in form as equations (1-15), but the variables now all depend on position as well as time, linked to the former through the partial differential equations (1-16). Because of this spatial variation one cannot use the volume V, but rather introduces θ, the ratio of the volumes of gas within the particles to that of the external flowing fluid, to obtain

$$\theta\frac{\partial C}{\partial t} = k_p(C_1 - C) - \theta R \tag{1-17a}$$

$$\theta c_p\frac{\partial T}{dt} = h_p(T_1 - T) + \Delta H\theta R \tag{1-17b}$$

The parameter θ may be computed from known data on void fraction external to the catalyst and pore volume internal.

Models of the form just introduced, those that include interface effects by means of added linear terms with transfer coefficients, are quite versatile. With some reasonable simplifying assumptions and redefinition of symbols, Cardoso and Luss (1969) used equations (1-15) for example to represent the behavior of a gas-phase reaction catalyzed over a heated wire. The reader who is interested in furthering this pursuit of multi-phase models will find discussion of adiabatic packed beds in Liu and Amundson (1962), and axial dispersion included by Liu and Amundson (1963). The complexities of two liquid phases were explored most thoroughly by Schmitz and Amundson (1963).

1-10. Other Balance Equations

In all the foregoing it was assumed that the specification of a single concentration C would adequately describe the compositions of interest at that time. Aris (1965) has shown that this will certainly be the case for a single reaction occurring in a CSTR, provided that the feed composition is constant and has a simple relation to the composition of material initially in the vessel, such that the latter may be thought of as being the product of partial reaction of the feed. If, for example, following an extended period of steady state operation, the dynamic behavior of a reactor is to be studied in response to a step change in feed composition, it is only necessary to write a single material balance equation, provided that the step input change maintains the stoichiometry; that is to say, the new feed may only be at a different conversion level. Otherwise, more than a single mass balance will be needed in modeling a chemical system.

Supposing that thermal effects are insignificant in a given system, an isothermal CSTR adequately described by two compositions (say one reactant C_A and one product C_B) would present the equations

$$V\frac{dC_A}{dt} = q(C_{0A} - C_A) - VR(C_A, C_B) \tag{1-18a}$$

$$V\frac{dC_B}{dt} = q(C_{0B} - C_B) + \gamma VR(C_A, C_B) \tag{1-18b}$$

where the change in sign and the added coefficient in the last term of (1-18b) arise from the reaction stoichiometry, here assumed to yield γ moles of product for each mole of reactant consumed. Of course, if the reactor cannot reasonably be held to be isothermal, the rate will depend on three state variables $R(T, C_A, C_B)$, and a third equation of the form (1-3) must be added to the set (1-18). An example of such a model may be found in a paper by Leathrum, Johnson, and Lapidus (1964). The study of nonisothermal consecutive reactions also called for a set of three equations in Sabo and Dranoff's (1970) work. Many isothermal systems of more complex biological relevance are described by Higgins (1967).

Yet another variation on this theme comes from a study of a polymerization system [Hoftyzer and Zwietering (1961)], in which the critical mass balance involves the kinetics of initiator decomposition

$$R_i = k_{0i} C \exp(-Q_i/T) \tag{1-19}$$

whereas the thermal balance depends instead on the heat released by the very exothermic polymerization of monomer

$$R = k_0 C^{1/2} \exp(-Q/T) \tag{1-20}$$

Although the kinetic terms use different reaction orders and activation energies, the simplification of writing only two equations may be retained by assuming a constant monomer level. If in turn this restriction is relaxed, a set of three equations is needed in terms of initiator concentration, temperature, and either monomer concentration or (what is equivalent) pressure.

EXAMPLE 1-1. A study of Matsuura and Kato (1967), based on earlier experimental work of Kunugi, Matsuura, and Shiro (1965), suggests that the liquid phase isopropanol oxidation proceeds as a chain reaction involving:

Initiation:

$$A \xrightarrow[\text{second-order}]{k_2} B$$

Propagation:

$$B \xrightarrow[\text{first-order}]{k_5} E + F \quad \text{(fast reaction with dissolved oxygen)}$$

$$E + G \xrightarrow[\text{second-order}]{k_4} A + B$$

Termination:

$$2E \xrightarrow[\text{second-order}]{k_3} \text{Inerts}$$

where A = peroxide $[H_2O_2]$
B = free radical $[(CH_3)_2COH]^{\cdot}$
E = free radical $[HOO]^{\cdot}$
F = acetone
G = isopropanol

Use this information to formulate an isothermal CSTR model in terms of the concentrations of the peroxide product C_A and the free radical intermediate C_B, assuming isopropanol to be in excess. Use the so-called pseudo steady-state hypothesis [Aris (1965)] and treat the concentration of free radical E as being constant. Is the result a special case of equations (1-18)?

Solution: The peroxide A is formed by second-order reaction of free radical E with alcohol G, and is simultaneously being used up by the second-order initiation. As a result the net peroxide formation is at the rate

$$R_1 = k_4 C_G C_E - k_2 C_A^2$$

Similar arguments for the components B and E lead to the net rates of free radical formation

$$R_2 = k_2 C_A^2 - k_5 C_B + k_4 C_G C_E$$

$$R_3 = -k_3 C_E^2 - k_4 C_G C_E + k_5 C_B$$

With isopropanol in excess, $k_4 C_G$ will be effectively constant and may be designated by k_1. Then

$$R_1 = k_1 C_E - k_2 C_A^2$$

$$R_2 = k_2 C_A^2 - k_5 C_B + k_1 C_E$$

$$R_3 = -k_3 C_E^2 - k_1 C_E + k_5 C_B$$

The pseudo steady-state hypothesis implies that $R_3 = 0$, and

$$-k_5 C_B + k_4 C_G C_E = -k_3 C_E^2$$

As a result

$$R_2 = k_2 C_A^2 - k_3 C_E^2$$

Finally, because E is formed by fast reaction from B, the concentration C_E may be approximated by C_B, and

$$R_1 = k_1 C_B - k_2 C_A^2 \tag{1-21a}$$

$$R_2 = k_2 C_A^2 - k_3 C_B^2 \tag{1-21b}$$

Since the CSTR is ordinarily operated with neither peroxide nor free radicals in the feed, the isothermal balance equations are

$$V\frac{dC_A}{dt} = -qC_A + VR_1 \tag{1-22a}$$

$$V\frac{dC_B}{dt} = -qC_B + VR_2 \tag{1-22b}$$

The result does resemble equations (1-18), but differs in the essential aspect that the equations are not in terms of the same reaction rate.

REFERENCES

Aris, R., *Introduction to the Analysis of Chemical Reactors*, Englewood Cliffs, N.J.: Prentice-Hall (1965).

BEEK, J., *Advances in Chemical Engineering*, **3**, 203 (1962).

BELLMAN, R., *Dynamic Programming*, Princeton University Press (1957).

BISCHOFF, K. B., *Chem. Eng. Sci.*, **16**, 131 (1961).

CARDOSO, M. A. A. and LUSS, D., *Chem. Eng. Sci.*, **24**, 1599 (1969).

HIGGINS, J., *Ind. Eng. Chem.*, **59**, No. 5, 19 (1967).

HOFTYZER, P. J. and ZWIETERING, T. N., *Chem. Eng. Sci.*, **14**, 241 (1961).

KRAMERS, H. and WESTERTERP, K. R., *Elements of Chemical Reactor Design and Operation*, New York: Academic Press (1963).

KUNUGI, T., MATSUURA, T. and SHIRO, O., *Hydrocarb. Proc. Pet. Refining*, **44**, 116 (July, 1965).

LEATHRUM, J. R., JOHNSON, E. F. and LAPIDUS, L. *AIChE Journal*, **10**, 16 (1964).

LEVENSPIEL, O., *Chemical Reaction Engineering*, New York: Wiley (1963) p. 125ff.

LIU, S. L. and AMUNDSON, N. R., *Ind. Eng. Chem. Fund.*, **1**, 200 (1962).

LIU, S. L. and AMUNDSON, N. R., *Ibid*, **2**, 183 (1963).

LIU, S. L., ARIS, R., and AMUNDSON, N. R., *Ind. Eng. Chem. Fund.*, **2**, 12 (1963).

LUSS, D. and AMUNDSON, N. R., *AIChE Journal*, **14**, 211 (1968).

MATSUURA, T. and KATO, M., *Chem. Eng. Sci.*, **22**, 171 (1967).

PETERSEN, E. E., *Chemical Reaction Analysis*, Englewood Cliffs, N. J.: Prentice-Hall (1965).

SABO, D. S. and DRANOFF, J. S., *AIChE Journal*, **16**, 211 (1970).

SCHMITZ, R. A. and AMUNDSON, N. R., *Chem. Eng. Sci.*, **18**, 265, 391 (1963).

WEHNER, J. R. and WILHELM, R. H., *Chem. Eng. Sci.*, **6**, 89 (1956).

Part I:

Lumped-Parameter Systems Modeled by Ordinary Differential Equations

Steady States of CSTR *Models* 2

2-1. Steady-State Multiplicity

It is common practice to design continuous-flow reactors and indeed most chemical engineering equipment by steady-state analysis. The mass and energy conservation laws are applied together with appropriate rate expressions with the object of attaining what is sometimes called a *balance* between input and output for a given system; that is to say, all time dependence is considered to be sufficiently small in magnitude or so far removed in the past as to be irrelevant. But the idea of a mass or energy balance does not in itself guarantee a unique set of conditions, and a designer should recognize that for some choices of system parameters mass and energy balances at various levels may exist for reactor conditions quite removed from those intended. The basis for this argument emerges quite naturally from a study of the system models presented in Chapter 1.

Consider for example an isothermal CSTR described by equation (1-1). Since the concept of a balance requires only that the time derivative be zero, any solution to

$$0 = F(C) = q(C_0 - C) - VR(C) \tag{2-1}$$

is a steady state. In general the reaction rate is not linear in concentration and multiple solutions are possible. A reasonable certainty that all the steady states are found will require successive trials or approximations over the full range of interest of the concentration variable. Appreciable tedious computation can be avoided, however, by establishing ranges of the system parameters over which unique solution of equation (2-1) is assured. To accomplish this, recall that a continuous single-valued function $F(C)$ can only be equal to zero at two different values of C, if it has a zero derivative somewhere between (a special case of the mean value theorem of elementary calculus). It follows that any criterion that makes it impossible for (dF/dC) to be zero is sufficient to guarantee a unique solution of equation (1-1). Consider the derivative of $F(C)$:

$$\frac{dF}{dC} = -q - V\frac{dR}{dC} \tag{2-2}$$

and note that the left-hand side can never be zero if either

$$\frac{dR}{dC} \geq 0 \tag{2-3a}$$

or

$$\frac{dR}{dC} > -\frac{q}{V} \tag{2-3b}$$

The latter inequality can also be combined with equation (2-1) to give

$$\frac{(C - C_0)}{R}\frac{dR}{dC} = (C - C_0)\frac{d \ln R}{dC} < 1 \tag{2-3c}$$

The utility of these simple results may be most immediately demonstrated by examining kinetics of the form

$$R = kC^n \tag{2-4}$$

for which

$$\frac{dR}{dC} = nkC^{n-1} \geq 0 \tag{2-5}$$

Comparison with the uniqueness criterion (2-3a) establishes that such an isothermal system would have a unique steady state, regardless of the flow and kinetic parameters chosen. Evidently, multiple steady states can only arise for more complicated rate equations. One such was pointed out by Matsuura and Kato (1967):

$$R = \frac{kC}{(1 + KC)^2} \tag{2-6}$$

If for example one selects the parameter values

$$(q/Vk) = 0.019$$

$$C_0 = 1.32 \text{ moles/ft}^3$$

$$K = 10 \text{ ft}^3\text{/mole}$$

three steady states may be found at

$$C = 0.065 \text{ moles/ft}^3$$
$$C = 0.25$$
$$C = 0.80$$

It is evident that these three conditions cannot exist simultaneously within a given stirred reactor. Which of them will in fact emerge depends not on the steady-state analysis, but on the system dynamics, a subject which we delay to a subsequent chapter. Even for this system, however, it is possible to limit the range of parameter choices over which multiplicity can occur.

If one's interest is primarily in the feed condition as compared to the kinetics, substitution of rate equation (2-6) into criterion (2-3c) produces

$$2KC^2 - KC_0C + C_0 > 0$$

Since the function is quadratic in the concentration variable, its sign can be guaranteed by requiring complex roots in the corresponding equality. It is therefore sufficient that the discriminant be negative, which reduces in this problem to $KC_0 < 8$. If on the other hand, the designer's interest is focused on the size of the system, it is more useful to observe that the derivative of equation (2-6) goes through a minimum at the inflection point $C = (2/K)$:

$$\text{Min}\left[\frac{dR}{dC}\right] = \text{Min}\left[\frac{k(1 - KC)}{(1 + KC)^3}\right] = -\frac{k}{27} \tag{2-7}$$

Combining equations (2-3b) and (2-7) produces a guarantee of uniqueness when

$$\frac{q}{Vk} > \frac{1}{27} \tag{2-8}$$

EXAMPLE 2-1. Two kinetic expressions that commonly arise in enzyme reactions are the Michaelis–Menten form

$$R = \frac{kC}{K + C}$$

and the substrate inhibition form

$$R = \frac{kC}{K + C + K'C^2}$$

Can either system show multiple steady states in an isothermal CSTR? Typical experimental results give $KK' \sim 0.04$.

Solution: The steady state of a system that follows Michaelis–Menten kinetics is always unique because

$$\frac{dR}{dC} = \frac{kK}{(K + C)^2} \geq 0$$

which satisfies equation (2-3a). However, when substrate inhibition exists, multiple solutions are not excluded, since

$$\frac{dR}{dC} = \frac{k(K - K'C^2)}{(K + C + K'C^2)^2}$$

will be negative in the concentration range

$$C > +\sqrt{K/K'}$$

Since typical constants are related as $KK' \sim 0.04$, steady-state multiplicity can be excluded only for approximately $C < 5K$.

2-2. Analysis of a Two-Equation Model

It was noted in Chapter 1 that some systems must be modeled by more than a single equation. Retaining the isothermal assumption, consider the system modeled by equations (1-22), which have steady states where

$$0 = -qC_A + VR_1 \tag{2-9a}$$

$$0 = -qC_B + VR_2 \tag{2-9b}$$

If the second of these relationships is used to solve for C_B, this variable can be eliminated from equation (2-9a) to give a single equation in one variable

$$F(C_A) = 0 = -qC_A + VR_1(C_A) \tag{2-10}$$

which may be thought of as a balance between mass generated by reaction

$$M_G = VR_1(C_A) \tag{2-11}$$

and mass removed by flow

$$M_R = qC_A \tag{2-12}$$

Assuming the kinetics of equations (1-21), these functions are sketched in Figure 2-1 for several choices of flow rate. The points of intersection may be

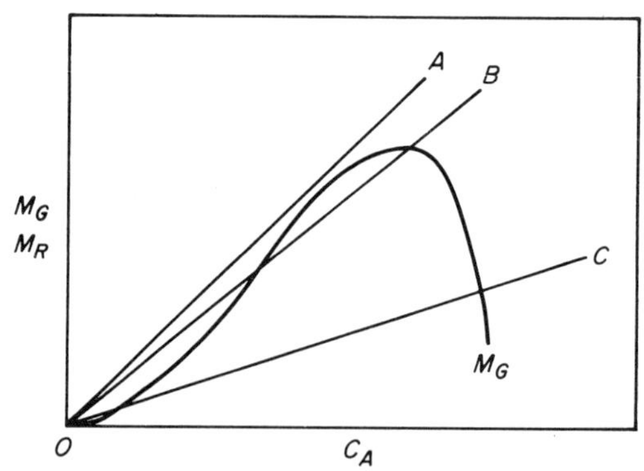

Figure 2-1. Typical mass-generated curve and three mass-removal lines for different flow rates [refer to equations (2-11) and (2-12)].

identified with values of C_A for which $M_G = M_R$; they are the steady states. Multiple solutions are possible only for sufficiently small flows. In particular, it may be observed that under conditions such as those giving line B, a relatively small change in the flow rate q could move the steady state to the origin and entirely quench the chain reaction.

With regard to the development of uniqueness criteria, the fact that equation (2-10) is in terms of only a single variable suggests that the previous approach should be applicable. By differentiation of equation (2-10)

$$\frac{dF}{dC_A} = -q + V\frac{dR_1}{dC_A} \tag{2-13}$$

from which it is clear that a sufficient uniqueness condition may be written as either

$$\frac{dR_1}{dC_A} < \frac{q}{V} \tag{2-14}$$

or

$$\text{Max}\left[\frac{dR_1}{dC_A}\right] < \frac{q}{V} \tag{2-15}$$

or, by recalling equations (2-11) and (2-12), the same statement can be related to the slopes

$$\frac{dM_G}{dC_A} < \frac{dM_R}{dC_A} \tag{2-16}$$

Adopting the kinetics of equations (1-21), and eliminating C_B via equation (2-9b):

$$R_1(C_A) = \frac{k_1}{2Vk_3}\left(-q + \sqrt{q^2 + 4V^2k_2k_3C_A^2}\right) - k_2C_A^2 \tag{2-17}$$

Differentiating:

$$\frac{dR_1}{dC_A} = 2k_2C_A\left[\frac{k_1V}{\sqrt{q^2 + 4V^2k_2k_3C_A^2}} - 1\right] \tag{2-18}$$

This derivative function has a maximum value when

$$C_A = \frac{q}{2V}\sqrt{\frac{1}{k_2k_3}}\left[\left(\frac{k_1V}{q}\right)^{2/3} - 1\right]^{1/2} \tag{2-19}$$

at which point

$$\text{Max}\left[\frac{dR_1}{dC_A}\right] = \frac{q}{V}\sqrt{\frac{k_2}{k_3}}\left[\left(\frac{k_1V}{q}\right)^{2/3} - 1\right]^{3/2} \tag{2-20}$$

Combining (2-15) with (2-20) produces the uniqueness condition

$$\sqrt{\frac{k_2}{k_3}}\left[\left(\frac{k_1 V}{q}\right)^{2/3} - 1\right]^{3/2} < 1 \tag{2-21}$$

To this point it was tacitly assumed that the design is directed to a unique steady state. In this isothermal problem this is decidedly not the case, since as Figure 2-1 showed, only the origin can be a unique solution, and such a steady state is a totally quenched system. Accordingly, a sensible design must violate condition (2-21), although the reversal of this inequality does not necessarily guarantee a good design. This idea will be further developed in Section 2-6 on design considerations.

EXAMPLE 2-2. What is a simple uniqueness criterion for the system modeled by equations (1-18)? Apply your result to the kinetics

$$R = k_1 C_A^n - k_2 C_B^m \tag{2-22}$$

Solution: From the equations referred to, a steady state is defined by a solution of

$$0 = q(C_{0A} - C_A) - VR \tag{2-23}$$

$$0 = q(C_{0B} - C_B) + \gamma VR \tag{2-24}$$

Multiplying the first of these by γ and adding:

$$C_B = C_{0B} + \gamma(C_{0A} - C_A) \tag{2-25}$$

showing that (2-23) can be written as a function of one variable:

$$F(C_A) = 0 = q(C_{0A} - C_A) - VR(C_A) \tag{2-26}$$

Then

$$\frac{dF}{dC_A} = -q - V\frac{dR}{dC_A} \tag{2-27}$$

and a sufficient condition for uniqueness is that

$$\frac{dR}{dC_A} > -\frac{q}{V} \tag{2-28}$$

Applying this criterion to the kinetics of equation (2-22) one finds from relationship (2-25) that

$$R = k_1 C_A^n - k_2[C_{0B} + \gamma(C_{0A} - C_A)]^m \tag{2-29}$$

$$\frac{dR}{dC_A} = nk_1\frac{C_A^n}{C_A} + mk_2\gamma\frac{C_B^m}{C_B} > 0 \tag{2-30}$$

Hence, inequality (2-28) is satisfied and the solution is unique for any positive values of the kinetic constants.

2-3. The Temperature-Dependent Reactor

Even in its simplest form, a CSTR analysis that includes energy considerations requires two equations. As a consequence the search for all the steady states of such a reactor would require iterative numerical computation on the simultaneous algebraic equations (1-5), were it not possible to simplify by reducing the problem to a form in which one of the equations is explicit in only one of the variables. This can be done from equations (1-5) by using the fact that their sum[1]

$$0 = q(C_0 - C) + q(\eta_0 - \eta) - \frac{U}{C_p}(\eta - \eta_w) \tag{2-31a}$$

is linear in concentration and temperature and can be rearranged to yield

$$C = C_0 + (\eta_0 - \eta) + \frac{U}{qC_p}(\eta_w - \eta) \tag{2-31b}$$

Since this is so, equation (1-5b) can be written as

$$0 = F(\eta) = q(\eta_0 - \eta) + VR(\eta) - \frac{U}{C_p}(\eta - \eta_w) \tag{2-32}$$

and the search for solutions can be a uni-variable search on equation (2-32) with results checked for consistency with equation (2-31) and the physical limitations of the problem. Among these are for example that for an irreversible reaction

$$C \leq C_0 \tag{2-33}$$

Furthermore, when the reaction is exothermic and heat loss can only occur by mixing with cold feed or transfer to coolant, either

$$T \geq T_0 \qquad \text{if } T_0 < T_w \tag{2-34a}$$

or

$$T \geq T_w \qquad \text{if } T_0 > T_w. \tag{2-34b}$$

Equation (2-32) is of course one of the balances referred to above. Using the reduced temperature scale, it specifies that the heat generated by chemical reaction

$$Q_G = \Delta HVR(\eta) \tag{2-35}$$

just equals the total heat removed by flow and heat transfer

$$\begin{aligned} Q_R &= qC_p(T - T_0) + U(T - T_w) \\ &= \Delta H\left[q(\eta - \eta_0) + \frac{U}{C_p}(\eta - \eta_w)\right] \end{aligned} \tag{2-36}$$

by requiring that their difference be zero. Considering, for example, kinetics of the form of equation (1-6a), substitution of equation (2-31b) gives:

$$Q_G = \Delta HVk_0\left[C_0 + (\eta_0 - \eta) + \frac{U}{qC_p}(\eta_w - \eta)\right]^n \exp\left(-\frac{QC_p}{\Delta H\eta}\right) \tag{2-37}$$

This function rises sharply with temperature, goes through a maximum, and

[1] The subscript s has been omitted, since it is clear in the context that steady-state values are being considered.

then falls to zero as the concentration falls at high temperature. It must therefore have a sigmoidal shape as in Figure 2-2. When the linear function Q_R is superimposed on the same diagram, it is evident that $Q_R = Q_G$ can define one or more steady states at the points of intersection, depending on the system parameters that determine the position of the heat removal line by its slope

$$\frac{dQ_R}{dT} = qC_p + U \tag{2-38a}$$

and intercept (on the T-axis)

$$\left[\frac{UT_w + qC_pT_0}{U + qC_p}\right] \tag{2-38b}$$

These observations were first made by Liljenroth (1918), but were not widely recognized at the time. The concepts were well understood by Frank-Kamenetskii (1947) and his colleagues in combustion research; however because the classic text was not translated until 1955, the basic ideas were independently rediscovered by Van Heerden (1953) for whom the diagram is sometimes named.

Examination of the Van Heerden diagram shows clearly that a reactor steady state can be very sensitive to small changes in parameters, if conditions are close to those that produce the triple intersection. Note for example the radical shift in temperature and composition that would follow the relatively small movement from line A to C in Figure 2-2a or from line E to G in Figure 2-2b. When such a jump in temperature occurs suddenly, it is referred to as *ignition*, or *extinction* if it is in the downward direction. Either phenomenon is a manifestation of extreme sensitivity to some parameter. The terms arise of course from their natural association with the field of combustion,

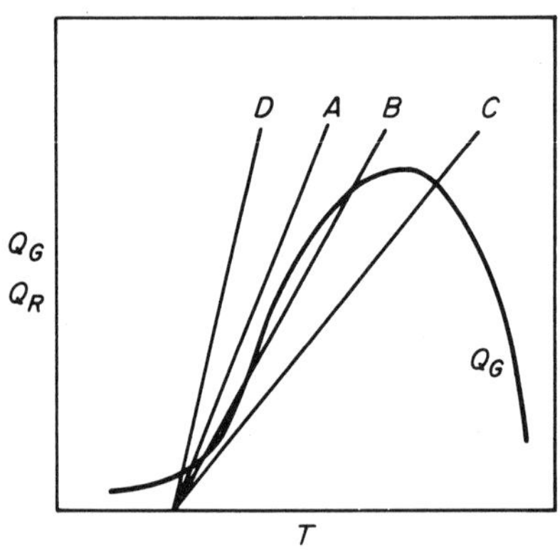

Figure 2-2a. Typical heat-generated curve and three heat removal lines of different slope.

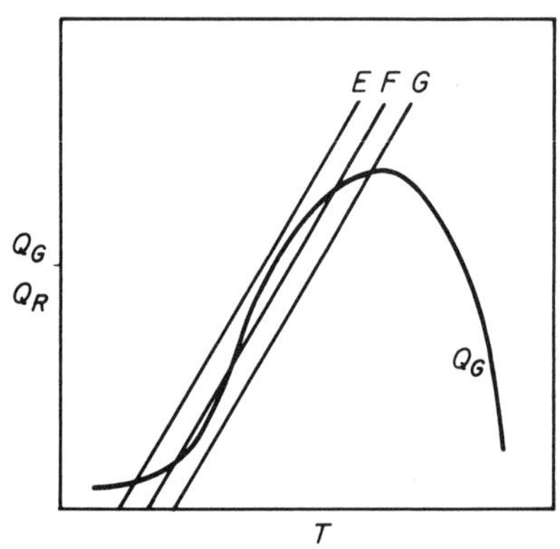

Figure 2-2b. Typical heat-generated curve and three heat removal lines of different intercept.

where the behavior was first observed, but such transitions between steady states are by no means restricted to systems with open flames. Recent supporting experimental evidence is the CSTR work of Furusawa, Nashimura, and Miyauchi (1969), and Vajtasa and Schmitz (1970).

One would perhaps expect that this relatively high sensitivity associated with multiple steady states ought to be avoided, and indeed this is a reasonable basis for design so long as the unique states that are attainable are satisfactory as regards conversion. The steady state that is determined by line A of Figure 2-2a, for example, corresponds to a quite low conversion. A more acceptable level of reaction is found at the single steady state fixed by line C, but this design presupposes that the needed slope and intercept parameters are economically available. The matter of multiple steady states is a case in point that illustrates the need for judgement to reconcile conflicting goals.

Assuming that the desirable flexibility is to be had, it is worthwhile to explore the effects of parameter choices to determine conditions which always produce a unique steady state. The analysis follows the pattern of prior developments by differentiating equation (2-32) to obtain

$$\frac{dF}{d\eta} = -q + V\frac{dR}{d\eta} - \frac{U}{C_p} \tag{2-39}$$

and noting that the left-hand side can never be zero, if

$$\frac{dR}{d\eta} \leq 0 \qquad \text{or} \qquad \frac{dR}{dT} \leq 0 \tag{2-40}$$

A more specific application can be derived from an examination of a particular rate expression. Differentiating rate equation (1-6a) one obtains

$$\frac{dR}{dT} = R\left[\frac{Q}{T^2} + \frac{n}{C}\frac{dC}{dT}\right] \tag{2-41}$$

Since the rate R can only be positive, condition (2-40) is satisfied if

$$\frac{Q}{T^2} + \frac{n}{C}\frac{dC}{dT} \leq 0 \tag{2-42}$$

But from equation (2-31) and the definition of η

$$\frac{dC}{dT} = \frac{C_p}{\Delta H}\frac{dC}{d\eta} = -\frac{C_p}{\Delta H}\left(1 + \frac{U}{qC_p}\right) \tag{2-43}$$

Substitution of equation (2-43) into (2-42) and some rearrangement of the inequality gives

$$\frac{Q\Delta HC}{nC_pT^2} \leq 1 + \frac{U}{qC_p} \tag{2-44}$$

Finally, making use of inequalities (2-33) and (2-34) a uniqueness criterion may be written in terms of only the system parameters:

$$\frac{Q\Delta HC_0}{nC_pT_0^2} \leq 1 + \frac{U}{qC_p} \tag{2-45}$$

or in terms of T_w if it is smaller than T_0.

It should be noted that criterion (2-45) is safe but quite conservative, sufficient but not necessary. This is the accumulated result of a string of simplifying approximations in the foregoing derivation, each of which introduced some degree of departure from the necessary condition. It is worthwhile to list them.

First of all, it is significant that although multiple solutions require $(dF/d\eta)$ to be zero at some value of η, the converse is not true: a function is shown, for example, as curve C in Figure 2-3 which has only a single solution in

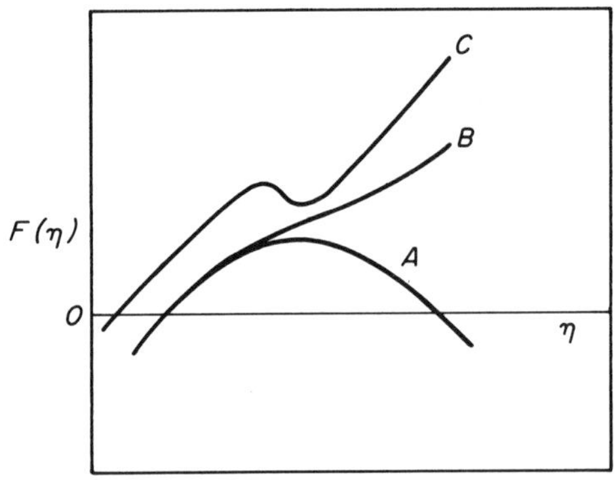

Figure 2-3. Hypothetical $F(\eta)$ functions: curve A has two solutions to $F(\eta) = 0$; curve B shows only a single solution; curve C has a local zero slope, yet only a single solution.

spite of several locations of zero slope. Second, having accepted the $(dF/d\eta)$ < 0 inequality, it follows exactly that

$$\frac{dR}{d\eta} < \frac{q}{V} + \frac{U}{VC_p} \tag{2-46a}$$

or

$$\frac{dR}{dT} < \beta\frac{q}{V} \tag{2-46b}$$

where

$$\beta = \frac{C_p}{\Delta H} + \frac{U}{q\Delta H} \tag{2-46c}$$

rather than the correct but simplified equation (2-40). Finally, yet a third estimate is involved in the step from (2-44) to (2-45), since the steady state is in general not found at the extreme limit of possible conditions. Although the relative contributions of these three factors will depend on the particular problem, it appears that the second and third are more likely to matter than the first, since curves such as C of Figure 2-3 are unlikely for simple kinetic functions.

An interesting physical interpretation may be made of inequality (2-46a) by rearranging it in the form

$$\Delta HV\frac{dR}{dT} < qC_p + U \tag{2-47}$$

and comparing it to the derivatives of equations (2-35) and (2-36):

$$\frac{dQ_G}{dT} = \Delta HV\frac{dR}{dT} \tag{2-48a}$$

and

$$\frac{dQ_R}{dT} = qC_p + U \tag{2-48b}$$

Evidently, inequality (2-47) is another form of

$$\frac{dQ_G}{dT} < \frac{dQ_R}{dT} \tag{2-49}$$

In other words, a CSTR will have a unique steady state if the slope of its heat removal line on Figure 2-2 is larger than the slope of the heat generation line over the entire temperature range of interest, as in the case of line *D*. One would expect a safe but very conservative result from such a design.

This question may be explored in further detail by returning to inequality (2-46) to recognize that it can be satisfied by a less conservative relationship than (2-40). The derivative $(dR/d\eta)$ need not necessarily be negative, so long as its largest value over the temperature range of interest is less than a known positive number:

$$\text{Max}\left[\frac{dR}{dT}\right] < \beta\frac{q}{V} \tag{2-50}$$

In exploiting this more general inequality, the relative size of the reactor is included for the first time. By combining equations (2-41) and (2-43) the derivative function may be written as

$$\frac{dR}{dT} = k_0\left[\frac{QC^n}{T^2} - n\beta C^{n-1}\right]\exp\left(-\frac{Q}{T}\right) \tag{2-51}$$

which for first-order kinetics reduces to

$$\frac{dR}{dT} = k_0\left[\frac{QC}{T^2} - \beta\right]\exp\left(-\frac{Q}{T}\right) \tag{2-52}$$

Again, taking advantage of the linear equation (2-31)

$$\frac{dR}{dT} = k_0\left[\frac{Q}{T^2}\left(C_0 - \beta T + \frac{C_pT_0}{\Delta H} + \frac{UT_w}{q\Delta H}\right) - \beta\right]\exp\left(-\frac{Q}{T}\right) \tag{2-53}$$

where for consistency all the temperatures are to be taken in degrees absolute. Now because the derivative is explicitly available as a function of a single variable, it is a simple matter to find its maximum at the temperature where the second derivative equals zero. In preparation for this step it is convenient to combine the various parameters by defining

$$\psi = \frac{\beta Q}{C_0 + C_p T_0/\Delta H + U T_w/q\Delta H} \tag{2-54a}$$

which reduces, if $T_w = T_0$, to

$$\psi = \frac{\beta Q}{C_0 + \beta T_0} \tag{2-54b}$$

With the substitution of (2-54a), equation (2-53) becomes

$$\frac{dR}{dT} = \beta k_0 \left[\frac{Q}{T^2}\left(\frac{Q}{\psi} - T\right) - 1\right] \exp\left(-\frac{Q}{T}\right) \tag{2-55}$$

The second derivative

$$\frac{d^2R}{dT^2} = \beta k_0 \left[\frac{Q^2}{T^3}\right]\left[\frac{Q}{T\psi} - \frac{2}{\psi} - 1\right] \exp\left(-\frac{Q}{T}\right) \tag{2-56}$$

will evidently be zero at the inflection point determined by

$$T = \frac{Q}{2 + \psi} \tag{2-57}$$

where

$$\text{Max}\left[\frac{dR}{dT}\right] = \beta k_0 \left[\frac{\psi + 4}{\psi}\right] \exp(-\psi - 2) \tag{2-58}$$

The most significant behavior of the derivative equation (2-55) may be shown pictorially as Figure 2-4. The range of interest of the temperature variable is always to the right of either T_0 or T_w, whichever is smaller. When the parameters of a problem place this lowest temperature at T_w for example,

$$T_0 \text{ or } T_w \leq \frac{Q}{2 + \psi} \tag{2-59}$$

and the largest possible (dR/dT) will occur at the peak of the curve as dictated by equations (2-57) and (2-58). However, if the temperature range of interest were located to the right of the peak (say $T_0 = T_w = T_B$), the largest possible (dR/dT) would be found at T_0, the boundary of the range. Thus, if

$$T_0 = T_w \geq \frac{Q}{2 + \psi} \tag{2-60}$$

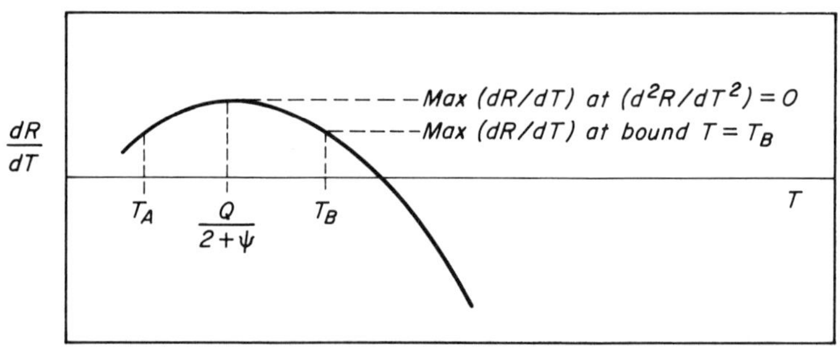

Figure 2-4. The derivative function versus temperature for equation (2-55).

then

$$\text{Max}\left[\frac{dR}{dT}\right] = \beta k_0\left[\frac{Q}{T_0^2}\left(\frac{Q}{\psi} - T_0\right) - 1\right]\exp\left(-\frac{Q}{T_0}\right) \tag{2-61}$$

or if T_w is smaller than T_0, the maximum should be found at this lower value.

In consideration of the foregoing, it is evident that two different sufficient conditions for uniqueness may be obtained from (2-50), depending on whether the local maximum or the boundary maximum are appropriate. For the former, equations (2-50) and (2-58) give

$$\frac{q}{k_0 V} > \frac{\psi + 4}{\psi}\exp(-\psi - 2) = G(\psi) \tag{2-62}$$

whereas for the latter the substitution of (2-61) leads to

$$\frac{q}{k_0 V} > \left[\frac{Q}{T_0^2}\left(\frac{Q}{\psi} - T_0\right) - 1\right]\exp\left(-\frac{Q}{T_0}\right) \tag{2-63}$$

The decision on which of these to use in a particular problem rests with the determining inequalities (2-59) and (2-60), but it turns out that (2-59) is more often satisfied than its opposite. Accordingly the right side of inequality (2-62) is graphed as Figure 2-5a for convenience in numerical calculations.

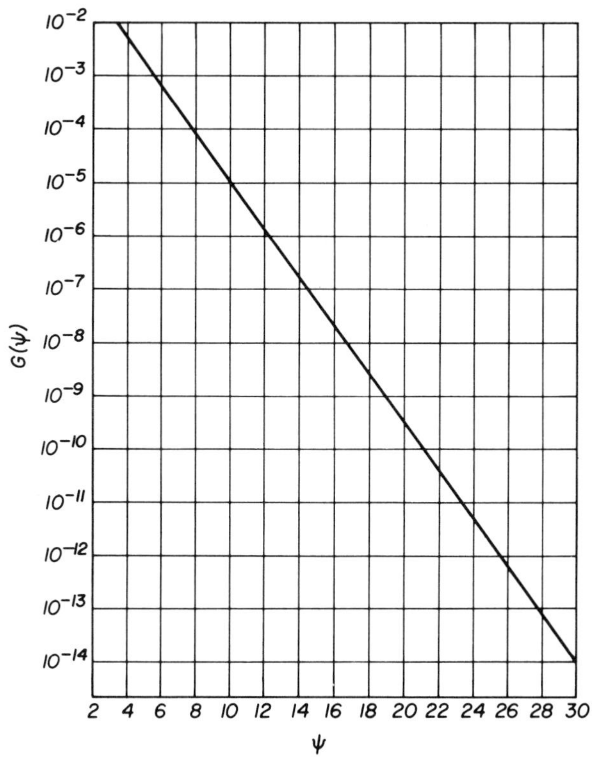

Figure 2-5a. The function $G(\psi)$ of equation (2-62).

Further discussion of these and other uniqueness criteria may be found in the survey paper by Aris (1969).

The same sequence of arguments may also be used to develop a design graph that will insure uniqueness for other than first-order kinetics; however, the analysis for $n \neq 1$ no longer gives a simple closed-form algebraic criterion. Instead, numerical studies of equation (2-51) are needed over a range of the parameters of interest. The results of a series of such computations [Chow (1970)] are summarized in Figure 2-5b, where each line is a safe upper bound on individual points obtained for the range of values:

$$0.001 < \beta < 0.02 \text{ lb-mol/ft}^3 \text{ °R}$$

$$7{,}000 < Q < 17{,}000 \text{ °R}$$

$$0.05 < C_0 < 0.75 \text{ lb-mol/ft}^3$$

$$T_0 = T_w = 530 \text{ °R}$$

The line for $n = 1$ was also obtained numerically, but coincides with the analytic solution equation (2-62).

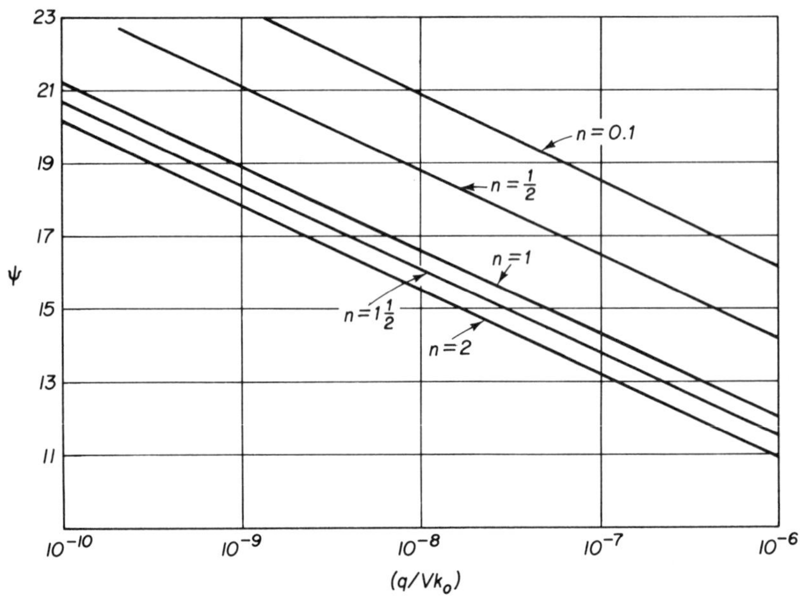

Figure 2-5b. Uniqueness criteria for different orders of reaction.

EXAMPLE 2-3. Given the necessary parameter values, find the heat transfer required to assure a unique steady state for a CSTR with first-order, Arrhenius-form kinetics, and estimate the degree of conservatism associated with inequality (2-45). The following information is available from particular CSTR examples discussed in the literature:

Aris and Amundson (1958)	Berger and Perlmutter (1964)
$Q = 10^4$°K	$Q = 10^4$°R
$n = 1$	$n = 1$
$C_0 = 1$ g-mol/l	$C_0 = 0.270$ lb-mol/ft^3
$T_0 = T_w = 350$°K	$T_0 = T_w = 530$°R
$\frac{\Delta H}{C_p} = 200$°K/(g-mol/l)	$\Delta H = 10^4$ Btu/lb-mol
	$C_p = 50$ Btu/ft^3°F
$\frac{U}{VC_p} = 1$ min^{-1}	$U = 500$ Btu/hr°F
	$V = 100$ ft^3
$\frac{q}{V} = 1$ min^{-1}	$q = 200$ ft^3/hr
$k_0 = e^{25}$ min^{-1}	$k_0 = 10^8$hr^{-1}
Steady states = 3	Steady states = 1

Solution: To compare Berger's choice of U with that estimated by inequality (2-45), it is most convenient to substitute his parameter values into the rearranged form:

$$U \geq \frac{qQC_0\Delta H}{nT_0^2} - C_p q = 9{,}220 \text{ Btu/hr°F}$$

Evidently, the estimate is safe and conservative to over an order of magnitude. The comparison with Aris and Amundson's numbers is facilitated by writing (2-45) in the form:

$$\left(\frac{U}{VC_p}\right) > \left(\frac{q}{V}\right)\left(\frac{\Delta H}{C_p}\right)\left(\frac{QC_0}{nT_0^2}\right) - \left(\frac{q}{V}\right) = 15.3 \text{ min}^{-1}$$

Again, the estimate is considerably greater than the tabulated value, but of course in this case the latter number corresponds to a multiple steady-state example and the true limiting value must lie somewhere between the two.

For a less conservative estimate of the heat transfer required to insure a unique steady state, equations (2-62) and (2-63) are appropriate. To decide which is applicable, find the value of ψ that is transitional between inequalities (2-59) and (2-60). From Berger's parameters

$$\psi = \frac{Q}{T_0} - 2 = 16.9$$

Equation (2-62) will be the proper choice if it yields a smaller value than this. Substituting:

$$G(\psi) < \frac{q}{k_0 V} = 2 \times 10^{-8}$$

which from Figure 2-5a requires that $\psi > 16$. Since this number is in the range sought, it may be used to compute U via equation (2-54):

$$U > \frac{16q(C_0\Delta H + C_p T_0) - QqC_p}{Q - 16T_w} = -4{,}340 \text{ Btu/hr°F}$$

Not only does this result confirm that a unique steady state exists for $U = 500$, it establishes also that the same would be true for any positive heat transfer, and in fact also for an adiabatic system or one that had heat transfer *into* the CSTR if it did not numerically exceed 4,340 Btu/hr°F (negative U).

Turning to the numbers chosen by Aris and Amundson, the transitional ψ is

$$\psi = \frac{Q}{T_0} - 2 = 26.6$$

Once again, equation (2-62) is applicable since it gives

$$G(\psi) < \frac{q}{k_0 V} = e^{-25} = 1.32 \times 10^{-11}$$

which corresponds (Figure 2-5a) to $\psi > 23$. As a result, rearranged equation (2-54a) leads to

$$\frac{U}{VC_p} > \frac{23q/V[(C_0 \Delta H/C_p) + T_0] - (q/V)Q}{Q - 23T_w} = 1.36 \text{ min}^{-1}$$

Evidently this condition has produced an estimate that is only very slightly conservative: a heat transfer coefficient that gives $(U/VC_p) = 1$ would already have crossed the boundary between unique and multiple solution.

EXAMPLE 2-4. Show that for the special case of $T_w = T_0$ a result analogous to inequality (2-3c) can be derived for the temperature-dependent CSTR. Substitute the kinetics of equation (1-6b) to derive the uniqueness condition

$$\frac{C_0}{T_0}\left[\frac{Q}{4T_0} - 1\right] < \beta$$

Solution: For the given special case, $\eta_w = \eta_0$ and equation (1-5b) may be rearranged to

$$\frac{q}{V} + \frac{U}{VC_p} = \frac{R}{\eta - \eta_0}$$

which in combination with inequality (2-46a) yields the desired analog:

$$(\eta - \eta_0)\frac{1}{R}\frac{dR}{d\eta} = (T - T_0)\frac{d \ln R}{dT} < 1$$

For first-order Arrhenius kinetics substitute equations (2-41) and (2-43) to obtain

$$(T - T_0)\left[\frac{Q}{T^2} - \frac{\beta}{C}\right] < 1$$

Using once again the fact that the feed stream and coolant temperatures are equal simplifies equation (2-31b) to

$$C = C_0 + \beta(T_0 - T)$$

and restates the inequality as

$$(T - T_0)\left[\frac{Q}{T^2} - \frac{\beta}{C_0 + \beta(T_0 - T)}\right] < 1$$

or rearranged as the quadratic

$$(C_0 + \beta Q)T^2 - Q(C_0 + 2\beta T_0)T + QT_0(C_0 + \beta T_0) > 0$$

This requirement will always be met if the discriminant has proper sign

$$Q^2(C_0 + 2\beta T_0)^2 < 4QT_0(C_0 + \beta T_0)(C_0 + \beta Q)$$

An explicit rearrangement for β will produce the uniqueness condition sought. It might be noted in passing that the same criterion can also be written in the form

$$\frac{Q\Delta HC_0}{C_pT_0^2} < 4\left[1 + \frac{U}{qC_p} + \frac{C_0\Delta H}{T_0C_p}\right]$$

to show that it is a considerably sharper bound than inequality (2-45) for this case.

2-4. An Alternative Approach

The key to the detailed development of the prior section was in the simplification that reduced the problem to terms of a single variable. But the method used for this step is not unique and correspondingly the results can differ, each being sufficient to its purpose but conservative over a different range of the system parameters. This section compares two methods by highlighting the similarities and differences between them.

If the kinetics equation (1-6b) is substituted into the steady-state balance equation (1-5a), the result may be rearranged to

$$C = \frac{qC_0}{q + Vk_0 \exp(-Q/T)} \tag{2-64}$$

This relationship can now be used instead of (2-31b) to develop the single-variable equation (2-32), and leads also in combination with equation (1-6b) to the modifications

$$R = \left[\frac{k_0qC_0 \exp(-Q/T)}{q + Vk_0 \exp(-Q/T)}\right] \tag{2-65}$$

and

$$\frac{dR}{dT} = \frac{k_0q^2C_0Q \exp(-Q/T)}{T^2[q + Vk_0 \exp(-Q/T)]^2} = R\frac{Q}{T^2}\frac{C}{C_0} \tag{2-66}$$

Since this derivative is positive at all temperatures and concentrations the $R(T)$ curve obtained by this method may still be sigmoid in form, but cannot show a local maximum as appeared in Figure 2-2. The inflection point of the $R(T)$ curve occurs at the temperature T_i satisfying

$$\left[\frac{\dfrac{Q}{2T_i} + 1}{\dfrac{Q}{2T_i} - 1}\right] \exp\left(-\frac{Q}{T_i}\right) = \frac{q}{Vk_0} \tag{2-67}$$

at which point

$$\text{Max}\left[\frac{dR}{dT}\right] = \frac{C_0q}{QV}\left[\left(\frac{Q}{2T_i}\right)^2 - 1\right] \tag{2-68}$$

Combining this result with inequality (2-50) gives the uniqueness criterion analogous to (2-62):

$$\frac{\beta Q}{C_0} > \left(\frac{Q}{2T_i}\right)^2 - 1 \tag{2-69}$$

In order to facilitate a comparison of the criteria (2-62) and (2-69) it is convenient to express the latter in terms of the parameter ψ. By substitution of equation (2-54b), this criterion becomes

$$\left[\frac{\psi\left(\frac{Q}{T_0}\right)}{\left(\frac{Q}{T_0}\right) - \psi}\right] > \left(\frac{Q}{2T_i}\right)^2 - 1 \tag{2-70}$$

The called-for comparison is shown in Figure 2-6, where the family of solid lines at various (Q/T_0) are solutions of (2-70) with (2-67) for (Q/T_i), and the dashed line is from (2-62). The less conservative uniqueness criterion is the one that gives the smaller value of ψ. Evidently this will depend on the dimensionless parameters (q/Vk_0) and (Q/T_0).

Westerterp (1962) has extended this approach to encompass systems of first-order parallel and consecutive reactions, and found for the latter that the heat generation curve was "double S-shaped" for high selectivity of intermediate product. Under these circumstances as many as five steady states are possible. The analysis was applied in a study of phthalic anhydride production from naphthalene, by modeling a fluid-bed reactor as an equiva-

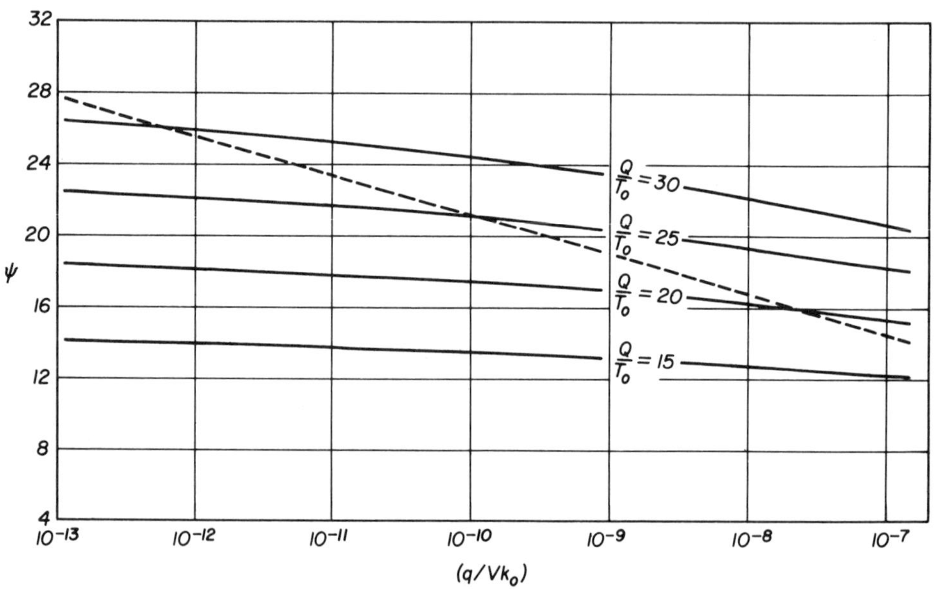

Figure 2-6. Comparison of two uniqueness criteria.

lent CSTR. A justification for such an approximation is included in Section 2-7.

EXAMPLE 2-5. Reconsider Example 2-3 in light of the alternative criterion.

Solution: To decide which of the criteria is less conservative, a given problem must be located on the coordinates of Figure 2-6. These are

$$\frac{q}{Vk_0} = 2 \times 10^{-8}, \qquad \frac{Q}{T_0} = 18.9$$

and

$$\frac{q}{Vk_0} = 1.32 \times 10^{-11}, \qquad \frac{Q}{T_0} = 28.6$$

for the Berger, and Aris and Amundson data, respectively. The less conservative estimates of ψ for the two cases are $\psi \geq 15.1$ [from equations (2-67) and (2-70)] and $\psi \geq 23.0$ [from equation (2-62)]. It may be concluded that for the Aris and Amundson data, the prior estimate of heat transfer required for uniqueness cannot be improved upon by the alternative method. Some less conservatism can be introduced however by applying the alternative method to the parameters chosen by Berger. With $\psi \geq 15.1$ equation (2-54b) rearranges to

$$\beta \geq \frac{\psi C_0}{Q - \psi T_0} = 0.00206$$

Then from the defining equation (2-46c):

$$U \geq \beta q \Delta H - C_p q = -5{,}880 \text{ Btu/hr°F}$$

As expected, this result is less conservative than $U \geq -4{,}340$ which was the comparable result of prior Example 2-3.

2-5. Multi-Phase Systems

Having found that the Van Heerden diagram is useful in the study of steady-state multiplicity in isothermal as well as temperature-dependent CSTR models, one might wonder whether the same ideas can be extended to multi-phase models of higher than second-order. Brief reflection shows that it is not the order of the starting model that is critical, but that the essential element is the ability to reduce the steady-state relations to a single algebraic equation in one variable.

An application to a fourth-order model was demonstrated by Luss and Amundson (1968) who studied the model of a catalytic fluidized bed given as equations (1-14) and (1-15). For steady state:

$$0 = q(C_0 - C_1) + k_G(C - C_1) \tag{2-71a}$$

$$0 = qC_p(T_0 - T_1) + h_G(T - T_1) + U(T_w - T_1) \tag{2-71b}$$

$$0 = k_G(C_1 - C) - VR(C, T) \tag{2-71c}$$

$$0 = h_G(T_1 - T) + \Delta HVR \tag{2-71d}$$

The heat balance on catalyst, equation (2-71d), may be used as the basis of the usual argument by defining

$$Q_G = \Delta HVR \tag{2-72}$$

and

$$Q_R = h_G(T - T_1) \tag{2-73}$$

These must in turn be reduced to single-variable dependence on T, the catalyst temperature. This may be accomplished by eliminating T_1, via rearranged equation (2-71b)

$$T_1 = \frac{qC_pT_0 + UT_w + h_GT}{qC_p + U + h_G} \tag{2-74}$$

and eliminating C by using the simultaneous solution of equations (2-71a and c):

$$\frac{k_G}{G_1V}(C_0 - C) = R \tag{2-75a}$$

where

$$G_1 = 1 + (k_G/q) \tag{2-75b}$$

An explicit solution for C can be obtained from equation (2-75) only for certain kinetic forms. If equation (1-6b) is applicable

$$C = \frac{C_0}{1 + G_1(k_0V/k_G)\exp(-Q/T)} \tag{2-76}$$

Uniqueness conditions should follow from the by-now-familiar inequality (2-49) which may also be written as

$$\frac{dR}{dT} < \frac{h_G}{\Delta HV}\left(1 - \frac{dT_1}{dT}\right) \tag{2-77}$$

From equation (2-74)

$$\frac{dT_1}{dT} = \frac{h_G}{qC_p + U + h_G}$$

which by substitution into (2-77) gives

$$\frac{dR}{dT} < \frac{h_G(qC_p + U)}{\Delta HV(qC_p + U + h_G)} \tag{2-78}$$

Invoking (1-6b):

$$\frac{dR}{dT} = R\left[\frac{Q}{T^2} + \frac{1}{C}\frac{dC}{dT}\right] \tag{2-79}$$

which in combination with (2-76) and its derivative gives

$$\frac{dR}{dT} = R\frac{Q}{T^2}\frac{C}{C_0} \tag{2-80}$$

By substituting equation (2-75) to eliminate R:

$$\frac{dR}{dT} = \frac{k_G Q C(C_0 - C)}{G_1 V C_0 T^2} \tag{2-81}$$

An upper bound may be placed on this derivative by either of the two methods described in the prior Section 2-3. First by observing that inequalities (2-3) and (2-4) apply for irreversible exothermic reaction, the bound is

$$\text{Max}\left[\frac{dR}{dT}\right] < \frac{k_G C_0 Q}{V G_1 T_0^2} \tag{2-82}$$

and by substitution in inequality (2-78) a sufficient uniqueness condition is

$$\frac{\Delta H C_0 Q}{C_p T_0^2} < \frac{G_1 G_2}{(G_1 - 1) + G_2 G_3} \tag{2-83a}$$

where the dimensionless groups are

$$G_2 = 1 + (U/qC_p) \tag{2-83b}$$

$$G_3 = k_G C_p / h_G \tag{2-83c}$$

and G_1 is as previously defined in equation (2-75b). The similarity to condition (2-45) is striking, and the numerical experience of Example 2-5 suggests that this estimate is probably quite conservative.

The second method of finding a bound for (dR/dT) involves analysis for its maximum. By the usual procedures of elementary calculus

$$\text{Max}\left[\frac{dR}{dT}\right] = \frac{k_G C_0}{G_1 Q V}\left[\left(\frac{Q}{2T}\right)^2 - 1\right] \tag{2-84}$$

where the ratio of (Q/T) is evaluated by solution of

$$\left[\frac{\frac{Q}{2T} + 1}{\frac{Q}{2T} - 1}\right] \exp\left(-\frac{Q}{T}\right) = \frac{k_G}{G_1 V k_0} \tag{2-85}$$

That these equations bear close resemblance to equations (2-67) and (2-68) is not unexpected, in view of the similarity between equations (2-66) and (2-80). In the usual range of numerical values, $(Q/T) > 10$, and the bracketed term on the left side of equation (2-85) is not greatly different from unity. It may be neglected and the equation explicitly solved:

$$\frac{Q}{T} = -\ln\left[k_G / k_0 G_1 V\right] \tag{2-86}$$

For the same reason the numerator of equation (2-84) may be simplified by dropping the 1 in comparison to $(Q/T)^2$. By now substituting (2-86) into (2-84)

$$\text{Max}\left[\frac{dR}{dT}\right] = \frac{k_G C_0}{4 Q G_1 V}\left[\ln\left(\frac{k_G}{k_0 G_1 V}\right)\right]^2 \tag{2-87}$$

Using this upper bound in place of inequality (2-82) produces a design equation differing from (2-83) only in its left side:

$$\frac{\Delta H C_0}{4QC_p}\left[\ln\left(\frac{k_G}{k_0 G_1 V}\right)\right]^2 < \frac{G_1 G_2}{(G_1 - 1) + G_2 G_3} \tag{2-88}$$

EXAMPLE 2-6. Use the following data (adapted from the Luss and Amundson (1968) paper) to evaluate the minimum heat transfer that will assure uniqueness according to the inequalities (2-83) and (2-88). How conservative are these estimates in view of the fact that the system is reported to have three steady states for $(U/qC_p) = 1.60$?

$h_G = 5.0 \times 10^6$ Btu/hr°F $\qquad Q = 15{,}000$°R

$k_G = 2.4 \times 10^8$ ft³/hr $\qquad C_p = 0.025$ Btu/ft³°F

$C_0 = 2.32 \times 10^{-4}$ mol/ft³ $\qquad \Delta H = 8 \times 10^4$ Btu/mol

$T_0 = 600$°R $\qquad k_0 = 1.20 \times 10^{12}$ hr⁻¹

$V = 120$ ft³ (gas held within solid)

$V_1 = 700$ ft³ (interstitial fluidizing gas)

$q = 7.5 \times 10^5$ ft³/hr

Solution: Part (a) based on inequality (2-83):
From the data given

$$\frac{\Delta H C_0 Q}{C_p T_0^2} = 27.7$$

$$G_1 = 1 + \frac{k_G}{q} = 321$$

$$G_3 = 1.19$$

and as a result

$$G_2 = \left(1 + \frac{U}{qC_p}\right) > 30.8$$

and

$$\frac{U}{qC_p} > 29.8$$

Part (b) based on inequality (2-88):

$$\frac{\Delta H C_0}{4QC_p}\left[\ln\left(\frac{k_G}{k_0 G_1 V}\right)\right]^2 = 4.52$$

$$G_2 > 4.58$$

and

$$\frac{U}{qC_p} > 3.04$$

In comparison with the given value of (U/qC_p) for which multiple solutions result, it appears that the part (a) estimate is very crude, but the part (b) estimate is quite reasonable. The transition from unique to multiple steady states probably occurs at about $(U/qC_p) = 2$. The high quality of the part (b) design estimate is due to the fact that the reactor steady-state temperature is in fact close to the value which corresponds to maximum (dR/dT). A supporting observation may be made by computing T from equation (2-86):

$$T = \left[Q/\ln\left(\frac{k_0 G_1 V}{k_G}\right)\right] = 785°\text{R}$$

and comparing this number to the given

$$T_0 = 600°\text{R}$$
$$T_w = 720°\text{R}$$

2-6. Design Considerations

It has already been noted that practical design does not always favor a choice of parameters that produces a unique steady state. The circumstance may be so extreme as to entirely remove this possibility from consideration, as was the case in the isothermal CSTR for oxidation of isopropanol, or more commonly, the design based on a uniqueness condition may just not be economic. When preliminary estimates indicate that this is the case, the focus of design shifts perceptively. While conditions for uniqueness may still be used as guides for limiting cases, other bases are needed which may or may not produce reactors with several steady states.

The essential ideas can best be developed with respect to the model for the temperature-dependent CSTR, for which regardless of parameters chosen the steady states must lie on the locus of equation (2-31). The designer seeks additional constraints on this linear relation in order to reduce the seemingly limitless choice of parameters and steady states to manageable proportions. Short of repeatedly plotting the entire Van Heerden diagram several simple alternatives are available, all having in common that they position the heat removal line by specifying one of its intersection points with the heat generation curve. They differ by being based either on a given conversion, or on a property of the Q_G-curve such as its maximum or inflection point. For convenience in discussion they may be numbered as follows:

Method	*Basis*	*Quantitative Condition*
I	Fractional conversion	C is given
II	Maximum rate	$(dR/dT) = 0$
III	Inflection in rate	$(d^2R/dT^2) = 0$

By Method I the steady-state concentration is picked a priori. This fixes the reaction rate according to equation (1-5) as

$$R = \frac{q}{V}(C_0 - C) \tag{2-89}$$

The steady-state temperature may then be evaluated from specific kinetics. Using rate equation (1-6b):

$$T = \frac{Q}{\ln\left[\frac{k_0 V}{q} \frac{C}{(C_0 - C)}\right]} \tag{2-90}$$

With both steady-state temperature and concentration determined, equation (2-31) can be thought of as a design equation for finding acceptable combinations of the various parameters. A simple case arises for example when $T_0 = T_w$ and equation (2-31b) reduces to

$$C = C_0 - \beta(T - T_0) \tag{2-91}$$

but the more general equation must be used if the feed temperature and coolant temperature are not the same. The results of a typical Method I analysis are conveniently summarized in a family of steady-state *operating curves* such as those given in Figure 2-7. These results of McGowin (1970) are for first-order kinetics and the parameter specifications:

$$\frac{k_0 V}{q} = 1.0 \times 10^{-11}$$

$$\frac{C_p Q}{\Delta H C_0} = 75$$

$$\frac{C_p T_w}{\Delta H C_0} = 2.5$$

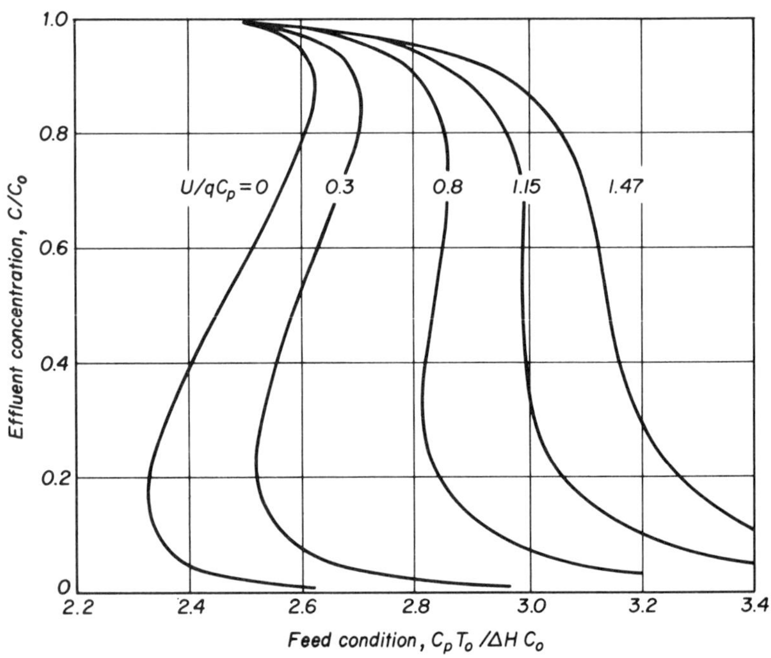

Figure 2-7. Steady-state operating curves for the temperature-dependent CSTR.

The multiple steady states may be identified from such a diagram when a vertical line for a fixed-feed condition intersects the curve at a given (U/qC_p) at more than one concentration level. In this particular example there are only unique steady states for $(U/qC_p) > 1.15$.

The variation called Method II is similarly developed, but specifies that the steady-state temperature must be such as to produce a maximum in the reaction rate. From equation (2-55) this must occur at

$$T = \frac{Q}{2}\left[-1 \pm \sqrt{\frac{\psi + 4}{\psi}}\right] \tag{2-92}$$

and by rearrangement of equation (2-90) the corresponding concentration is

$$C = \frac{C_0}{\left[1 + \frac{k_0 V}{q} \exp\left(-\frac{Q}{T}\right)\right]} \tag{2-93}$$

The design equation (2-91) is still applicable, but because the temperature appears as a function of ψ, it is convenient to combine equations (2-91) and (2-54b) to obtain

$$\psi = \frac{Q(C_0 - C)}{C_0 T - T_0 C} \tag{2-94}$$

Method III differs from II only in the criterion for determining the steady-state temperature. To place this temperature at the inflection point of the Q_G-curve, equation (2-57) gives

$$T = \frac{Q}{2 + \psi} \tag{2-95}$$

Both Methods II and III rely on equation (2-94); they do not supply explicit solutions for ψ because the steady-state temperature and concentration are also dependent on this variable. Trial solutions are relatively simple, however, and converge rapidly to the design estimate.

It should also be mentioned at this point that the alternative approach described in Section 2-4 can also be used to estimate an inflection temperature; however, it will be based on equation (2-66) which is of course a different function than (2-53). The alternative inflection point is given by equation (2-67). Comparison with equation (2-90) shows that Method I for the special case of $(C/C_0) = 0.5$ and $(Q/T_i) \gg 1$ is indistinguishable from Alternative Method III. There is no Alternative Method II, since the rate given by equation (2-65) never goes through a maximum.

EXAMPLE 2-7. Develop Method II for the isopropanol oxidation problem, assuming the CSTR model of Matsuura and Kato with their suggested value of $(k_3/k_2) = 10^6$. If $k_1 = 100\ \text{min}^{-1}$, estimate a reactor holding time that will produce multiple steady states.

Solution: The uniqueness condition (2-21) gives

$$\left(\frac{k_1 V}{q}\right) < \left[1 + \left(\frac{k_3}{k_2}\right)^{1/3}\right]^{3/2} = 1{,}000$$

Reversing this inequality does not yet guarantee multiple intersections between the M_G and M_R curves. Such a condition will certainly exist, however, if the M_R line is made to pass through the maximum point of the M_G curve, which occurs [see equation (2-49)] where

$$C_A = \frac{q}{2V}\sqrt{\frac{1}{k_2 k_3}}\left[\left(\frac{k_1 V}{q}\right)^2 - 1\right]^{1/2}$$

Since it is known that $(k_1 V/q) > 1{,}000 \gg 1$, this equation simplifies to

$$C_A = \frac{q}{2V}\sqrt{\frac{1}{k_2 k_3}}\left[\frac{k_1 V}{q}\right]$$

This is a constraint that can be imposed on the balance equation (2-17) to ensure a design with multiple solutions. By substitution and rearrangement

$$\left(\frac{k_1 V}{q}\right) = 2\sqrt{\frac{k_3}{k_2}} = 2{,}000$$

Any larger value than this will also be acceptable from the point of view of avoiding washout of all the radicals in the reactor, and perhaps some smaller values are possible. Much of a reduction below this level is probably too close to the point of transition to a unique steady state at $C_A = 0$. With $k_1 = 100\ \text{min}^{-1}$ the required holding time is $(V/q) \geq 20$ minutes.

2-7. Autothermal Reactor Systems

From the viewpoint of energy costs, it is wasteful to remove the heat liberated by an exothermic reaction by rejecting it to cooling water. By operating instead in an adiabatic manner this energy is absorbed as sensible heat in the effluent and can be partially recovered by heat exchange with the cooler feed stream, as shown schematically in Figure 2-8. The enthalpy of the reactor effluent relative to the fresh feed is

$$qC_p(T - T_F)$$

Of this quantity only part is recovered:

$$qC_p(T_0 - T_F)$$

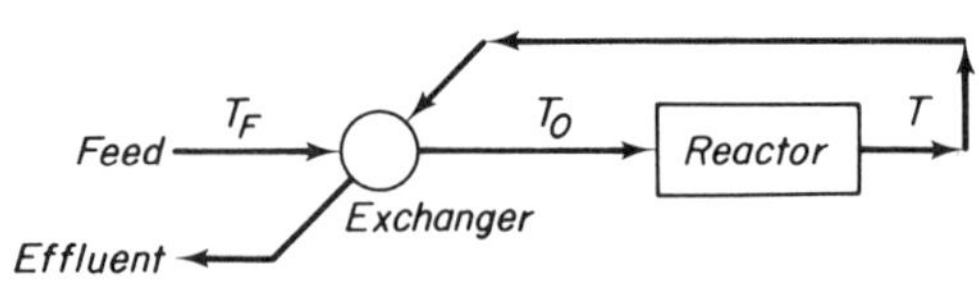

Figure 2-8. Flowsheet for heat recovery.

the remainder being lost in the effluent of the heat exchanger:

$$qC_p(T - T_0)$$

A convenient measure of this heat-exchanging capacity has been defined by Van Heerden (1958) as

$$H = \frac{T - T_F}{T - T_0} \tag{2-96}$$

which may be thought of as the reciprocal of the fractional heat loss. Because the feed pre-heat in such a system comes entirely from reaction heat, it is sometimes called autothermal (self-heating).

The steady-state mass balance of an autothermal reactor is not different from the usual CSTR equation, but the heat balance is altered in two respects. First, the heat transfer term is dropped for adiabatic reaction, and secondly, the sensible heat is in terms of the pre-heated feed temperature:

$$0 = qC_p(T_0 - T) + V\Delta HR \tag{2-97}$$

Combining with equation (2-96) reestablishes the relationship in terms of the fresh feed:

$$0 = \frac{qC_p}{H}(T_F - T) + V\Delta HR \tag{2-98}$$

Elimination of the rate term that is common to equations (2-98) and (1-5a) produces

$$C = C_0 - \frac{C_p}{H\Delta H}(T - T_F) \tag{2-99}$$

which is identical to equation (2-91), if

$$\beta = \frac{C_p}{H\Delta H} \tag{2-100}$$

Since the mass balance and kinetics are unchanged, the uniqueness criteria and design approaches of the prior sections are directly applicable to the autothermal system. In effect the only change of consequence is that the heat transfer parameter is now H instead of U reflecting merely the new flow arrangement for removing the reaction heat. If equation (2-99) is rearranged to solve for H,

$$H = \frac{C_p}{\Delta H}\frac{(T - T_F)}{(C_0 - C)} \tag{2-101}$$

It has been suggested by Van Heerden (1958) that this same equation may be applicable to an autothermal fluidized bed system. This idea may be examined in terms of the Luss and Amundson model introduced in Chapter 1. Of the four steady-state equations (2-71), only (2-71b) needs to be modified for autothermal operation to

$$0 = \frac{qC_p}{H}(T_F - T_1) + h_G(T - T_1) \tag{2-102}$$

The elimination of R from equations (2-71c and d) gives

$$k_G \Delta H(C - C_1) = h_G(T_1 - T) \tag{2-103}$$

Elimination of C_1 and T_1 by use of equations (2-71a and b) respectively, yields

$$C = C_0 - \left[\frac{h_G C_p(q + k_G)}{k_G \Delta H(Hh_G + qC_p)}\right](T - T_F) \tag{2-104}$$

which is yet again of the form of equation (2-91) with

$$\beta = \frac{C_p G_1}{\Delta H[(G_1 - 1)H + G_3]} \tag{2-105}$$

The temperature and concentration are related through the fluidized bed equation (2-76).

These relationships may be simplified by noting from the numerical work in Example 2-6 that typical parameter values for catalytic fluidized beds show $G_1 \sim (G_1 - 1)$. In addition the term G_3 in equation (2-105) is negligible in comparison with the first term. These simplifications lead to exactly equations (2-90) and (2-101) originally derived for the single-phase CSTR, and justify the approach taken by Van Heerden.

EXAMPLE 2-8. Use the following data reported by Van Heerden to assess the possibility of autothermal design for four well-known catalytic processes. Note that the column headed T contains steady-state temperatures for $(C/C_0) = 0.37$.

	In Units of degrees Kelvin		
Process	$\left(\frac{\Delta H C_0}{C_p}\right)$	T_F	T
NH_3-synthesis	270	320	670
CO-conversion	200	300	570
SO_2-oxidation	180	300	700
naphthalene-oxidation	600	300	700

Solution: Substituting the given values into equation (2-101) and using $(C/C_0) = 0.37$, the results are:

	H	*Per cent Heat Recovery* $1 - (1/H)$
NH_3-synthesis	2.06	52%
CO-conversion	2.14	53%
SO_2-oxidation	3.54	72%
naphthalene-oxidation	1.05	5%

All these processes would appear to be possible candidates for autothermal operation. The most promising is the last, since only 5 per cent of the effluent heat needs to be recovered. Allowing for some uncertainty in the data used (or the model) it is reasonable, as Van Heerden proposes, to consider autothermal operation without any heat exchange. Among the others, the SO_2-oxidation would be the most difficult to run autothermally. Values of H as high as 3 or 4 are not very difficult to achieve, but may require unacceptably high capital investment.

2-8. Feedback Control

There are many ways in which a feedback loop may be used to control a chemical reactor, depending among other things on the variables that have specifications to be met, and on the choice of one or more manipulated variables. The considerations which apply to such decisions are given elsewhere [Perlmutter (1965), Harriot (1964), Coughanowr and Koppel (1965)] and are not in the province of this book. Nevertheless, it needs to be shown how feedback effects arise, and how the analyst can account for them.

An interesting example is found in a paper by Luyben (1966) on a CSTR of the kind used in alkylation reactions or in the manufacture of tetraethyl lead. These reactors are cooled by boiling solvent which is subsequently condensed and refluxed. The reactor temperature is controlled by a feedback loop which adjusts the reflux flow rate, as shown schematically in Figure 2-9. The system dynamics may be modeled by retaining equation (1-1) but modifying the energy equation (1-2) to

$$VC_p\frac{dT}{dt} = qC_p(T_0 - T) + \Delta HVR - \Delta H_v q_c \tag{2-106a}$$

where

ΔH_v = the latent heat of vaporization of solvent

q_c = the solvent flow rate

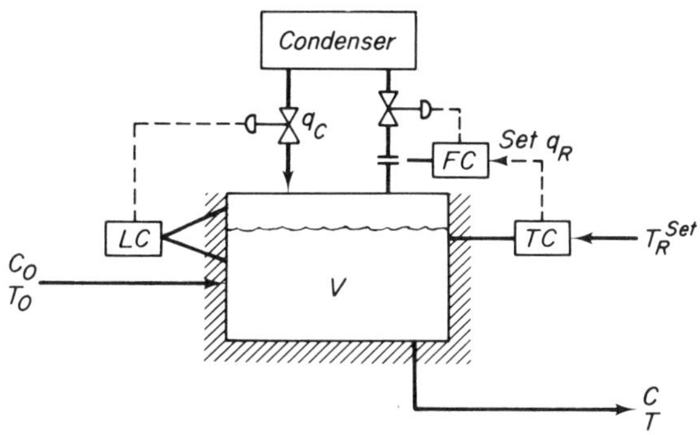

Figure 2-9. Schematic diagram of temperature-control feedback loop based on reflux manipulation (after Luyben, 1966).

The simplest control function is a proportionality between a measured temperature error and a corrective solvent flow deviation

$$q_c - q_R = K_p(T - T_R) \tag{2-106b}$$

where

T_R = the reference value of reactor temperature, the *set point*

q_R = the reference value of solvent flow rate, the value of q_c when $T = T_R$

K_p = an adjustable control constant called the *proportional gain.*

Combining (2-106a) and (2-106b)

$$VC_p\frac{dT}{dt} = qC_p(T_0 - T) + \Delta HVR - \Delta H_v[q_R + K_p(T - T_R)] \tag{2-107a}$$

which may also be written as

$$VC_p\frac{dT}{dt} = qC_p(T_0 - T) + \Delta HVR - \Delta H_vK_p\left[T - \left(T_R - \frac{q_R}{K_p}\right)\right] \tag{2-107b}$$

A term by term comparison of this equation with equation (1-2) shows them to be identical if

$$U = \Delta H_vK_p \tag{2-108a}$$

$$T_w = T_R - \frac{q_R}{K_p} \tag{2-108b}$$

That is to say, when ΔH_v is constant equation (1-2) can be used to describe the Luyben model with the understanding that the constants in the heat transfer term are to be interpreted by equations (2-108).

Turning to the question of steady-state multiplicity, the uniqueness criteria (2-45), (2-62) and (2-63) remain valid, but modifications are needed to:

$$\beta = \frac{C_p}{\Delta H} + \frac{\Delta H_vK_p}{q\Delta H} \tag{2-109a}$$

$$\psi = \frac{\beta Q}{C_0 + \dfrac{C_pT_0}{\Delta H} + \dfrac{\Delta H_vT_RK_p}{q\Delta H} - \dfrac{\Delta H_vq_R}{q\Delta H}} \tag{2-109b}$$

EXAMPLE 2-9. Use the parameter values specified by Luyben:

$q = 2{,}000$ lb/hr	$Q = 15{,}000°$R
$\Delta H = 600$ Btu/lb	$\Delta H_v = 274$ Btu/lb.
$C_0 = 0.50$ lb/lb	$q_R = 729$ lb/hr
$C_p = 0.75$ Btu/lb°F	$T_R = 600°$R
$T_0 = 530°$R	$k_0 = 7.08 \times 10^{10}$ hr^{-1}
$V = 2{,}400$ lb	

to estimate a range of K_p for which a unique steady-state solution is guaranteed for the model presented above.

Solution: Compute the value of the dimensionless group

$$\frac{q}{k_0 V} = 1.18 \times 10^{-11}$$

Assume that inequality (2-62) is applicable (to be checked later), and use Figure 2-5 to obtain

$$G(\psi) < 1.18 \times 10^{-11}$$
$$\psi > 23.2$$

Combine equations (2-109) and rearrange to

$$K_p = \frac{\psi[q(\Delta H C_0 + C_p T_0) - \Delta H_v q_R] - q C_p Q}{\Delta H_v (Q - \psi T_R)}$$

which reduces numerically to

$$K_p > 79.4 \text{ (lb/hr)/°F}$$

The ratio

$$\frac{Q}{2 + \psi} > \frac{15{,}000}{25.2} = 593°\text{R}$$

should be compared with

$$T_0 = 530°\text{R}$$
$$T_w \geq T_R - \frac{q_R}{K_p} = 590.8°\text{R}$$
$$T_w < T_R = 600°\text{R}$$

to justify the use of inequality (2-62). Moreover, the relative closeness of 593°R to the set point $T_R = 600°$R suggests that the K_p is not overly conservative, since the maximum (dR/dT) used must be a rather close estimate of the actual value at the steady state. Independent confirmation may be found in Luyben's paper where it is shown that unique solution requires a K_p of at least 30.

It should be noted that the inclusion of feedback control led to a very simple solution in this problem because the control term was chosen linear in temperature. The same would be true if a model called for heating by condensing steam, but not in the usual forms of sensible heat transfer without phase change [Grethlein and Lapidus (1963), and Aris and Amundson (1958)]. Such nonlinear models can also be treated by this approach, but the computations are much more tedious.

REFERENCES

Aris, R., *Chem. Eng. Sci.*, **24**, 149 (1969).

Aris, R. and N. R. Amundson, *Chem. Eng. Sci.*, **7**, 121 (1958).

BERGER, J. S. and D. D. PERLMUTTER, *AIChE Journal*, **10**, 233 (1964).

CHOW, P., Report on NSF sponsored undergraduate summer research, Univ. of Pa. (1970).

COUGHANOWR, D. R. and L. B. KOPPEL, *Process Systems Analysis and Control*, New York: McGraw-Hill, Inc. (1965).

FRANK-KAMENETSKII, D. A., *Diffusion and Heat Exchange in Chemical Kinetics*, Princeton Univ. Press (1955) p. 288 ff. Translated by H. Thom from the Russian edition of 1947, Publishing House of the U.S.S.R.

FURUSAWA, T., H. NASHIMURA, and T. MIYAUCHI, *J. Chem. Eng. Japan*, **2**, 95 (1969).

GRETHLEIN, H. E. and L. LAPIDUS, *AIChE Journal*, **9**, 230 (1963).

HARRIOT, P., *Process Control*, New York: McGraw-Hill, Inc. (1964).

LILJENROTH, F. G., *Chem. Met. Eng.*, **19**, No. 6, 287 (1918).

LUSS, D. and N. R. AMUNDSON, *AIChE Journal*, **14**, 211 (1968).

LUYBEN, W. L., *AIChE Journal*, **12**, 662 (1966).

MATSUURA, T. and M. KATO, *Chem. Eng. Sci.*, **22**, 171, (1967).

MCGOWIN, C. R. and D. D. PERLMUTTER, *AIChE Journal*, **17**, 831 (1971).

PERLMUTTER, D. D., *Introduction to Chemical Process Control*, New York: Wiley (1965).

VAN HEERDEN, C., *Ind. Eng. Chem.*, **45**, 1242 (1953).

VAN HEERDEN, C., *Chem. Eng. Sci.*, **8**, 133 (1958).

VEJTASA, S. A. and R. A. SCHMITZ, *AIChE Journal*, **16**, 410 (1970).

WESTERTERP, K. R., *Chem. Eng. Sci.*, **17** 423 (1962).

Geometry of Dynamics 3

3-1. Steady States and Stability

In Chapter 2 it was pointed out that the term steady state can be misleading unless one allows for the possibility of nonuniqueness, since there is in general more than one possible input-output balance point in a CSTR with a given set of parameter values. Similarly, confusion can arise from a second notion which is not always explicitly clarified: the idea that once such a state is obtained, it will remain unchanged barring unforeseen disturbances. This stability is suggested by the word steady, but it is essential to understand that the condition of time-derivatives equal to zero does not in and of itself guarantee such steadiness of behavior.

This observation can most readily be supported by an analog from mechanics. Figures 3-1 and 3-2 illustrate the movement of a ball along a surface. In each case the origin of coordinates is a point which will satisfy

$$\frac{dx_1}{dt} = \frac{dx_2}{dt} = 0$$

but whereas the ball will find a stable rest position at the bottom of the cup, it will not rest but rather move past the origin of Figure 3-2 and slide further down the saddle shape. If we are still to retain the name of steady state for this point, as has in fact become customary among chemical engineers, it will have to be called an *unstable steady state*. There are other names that have

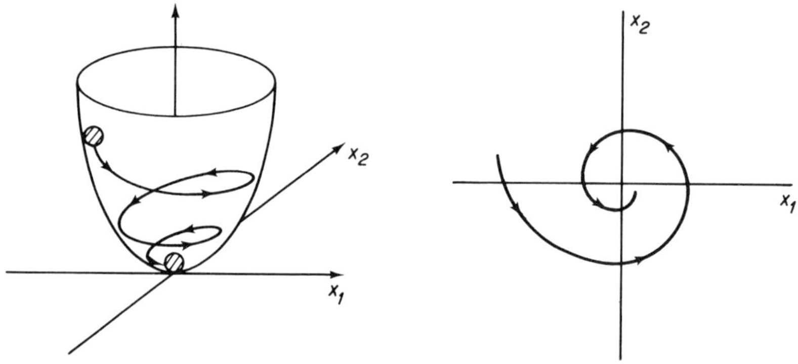

Figure 3-1. (a) Cup in three dimensions; (b) projected trajectory.

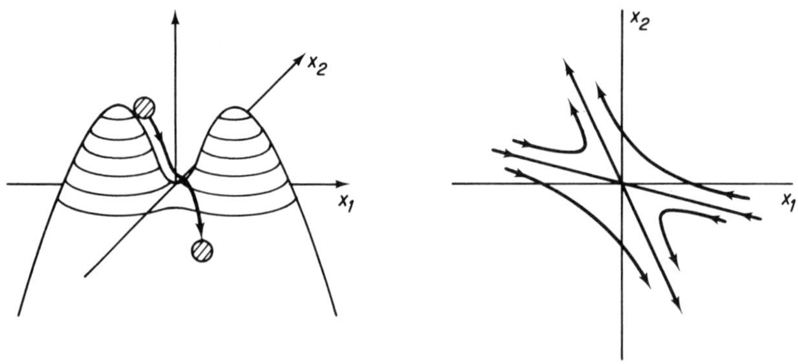

Figure 3-2. (a) Saddle in three dimensions; (b) projected trajectories.

been applied to these critical points. The alternative favored in mechanics is *equilibrium point*, but chemical engineers have shied away from the possible confusion with chemical equilibrium. The term *stationary point* is more satisfactory because of its familiarity from optimization terminology, although the word *stationary* suffers in part from the same disadvantage as *steady*. In mathematically oriented discussion, these points are *singular*, or *critical*.

The essential differences between the stable cup and unstable saddle can also be presented by projecting some typical lines of movement on the (x_1, x_2) plane. These paths are especially easy to visualize if one assumes the ball to start on the symmetrical surfaces without initial velocity. The resulting *trajectories* show the respectively convergent and divergent natures of the steady states, as for example in Figures 3-1b and 3-2b. For simplicity, oscillatory movements that repeatedly cover the same line in the (x_1, x_2) plane are not shown in this diagram.

There is one other aspect of the rolling-ball arguments that deserves attention because it will have a significant analog in the reactor problems to follow. Suppose as must be the case for any practical problem, that the stable cup has a finite size. Then it might for example end in a lip that folds back to produce the volcano-like shape in Figure 3-3a. With this modification the pattern of trajectories on the (x_1, x_2) plane will define a finite *region of asymptotic stability* from within which all paths lead to the stable steady state, and beyond which all trajectories diverge (Figure 3-3b).

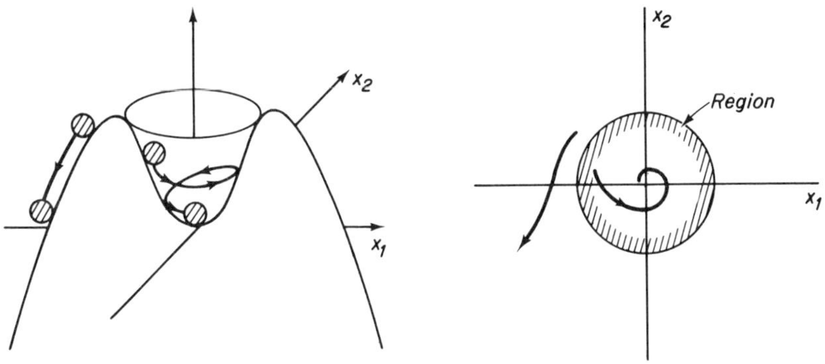

Figure 3-3. (a) Volcano in three dimensions; (b) projected trajectories and region of asymptotic stability.

The assumptions of symmetry and restricted initial velocity were made in order to present these arguments in as simple a manner as possible. The reader should be reassured that these are not limitations in any sense nor do they even arise in the reactor analyses that follow.

3-2. Phase-Plane For The CSTR

In spite of some exceptions noted, one of the more striking aspects of the catalog of reactor models presented in Chapter 1 is its emphasis on second-order systems consisting of pairs of simultaneous first-order equations. The CSTR models are typically of the form

$$\frac{d\eta}{dt} = \dot{\eta} = f_1(\eta, C) \tag{3-1a}$$

$$\frac{dC}{dt} = \dot{C} = f_2(\eta, C) \tag{3-1b}$$

Such models lend themselves to a graphical analysis in which the dependent variables (*state variables*) are plotted against one another on what is called a *phase-plane*.

Referring to equation (1-1) and (1-3) for the CSTR, for example, it is possible for fixed values of the various parameters to identify each point in the (η, C) plane with a state of the system. As the system evolves in time through a series of states from its initial to its final condition, its mapping on the (η, C) plane will trace a trajectory. The instantaneous slope of this curve can be obtained from equations (3-1) as

$$\frac{dC}{d\eta} = \frac{f_2(\eta, C)}{f_1(\eta, C)} = m \tag{3-2}$$

and the entire trajectory can be drawn as a sequence of small steps using the computed slope at each step to move on to the next. Alternatively, some of the tedium of such a process may be avoided by drawing a set of auxiliary *isoclines* in the (η, C) plane:

$$f_2 = mf_1 \tag{3-3}$$

which are the loci of points where the trajectories have slope m. By considering in turn various starting points of interest, the plane can be covered with trajectories to whatever degree of density the analyst may desire. Long-established theorems [Bellman (1953) and Struble (1962)] assure us that for systems of the form (3-1) the trajectory from any point will be unique when the f_1 and f_2 functions have continuous first partial derivatives. Since this is virtually always true for chemical reactor models, no trajectories can ever cross, except at singular points where the derivatives (dC/dt) and $(d\eta/dt)$ are both zero. These are the one or more reactor steady states determined by equations (1-5). More detailed procedures for constructing phase-plane portraits are given elsewhere [e.g., Truxal (1955) and Perlmutter (1965)].

Some CSTR results selected for specific features are shown in Figures 3-4 to 3-6. The first of these is notable because it represents a reactor system with three steady states; moreover, the behavior of trajectories close to these respective critical points illustrates with great clarity the three types of local behavior that are found: (1) nonoscillatory convergence, a *node;* (2) oscillatory spiraling convergence, a *focus;* and (3) a mixed initially convergent, ultimately divergent trajectory that depends in detail on the initial condition that is chosen, a *saddle*. The apparently diverging trajectories going down to the right will bend and return to steady state C provided that the equations still apply at the elevated temperatures. Each of these three behaviors can be thought of as analogous to a case of the rolling-ball problem. The saddle point is most obviously similar to Figure 3-2. A node would be expected on the (x_1, x_2) plane, if there were enough friction on the surface of a cup to prevent the ball from overshooting the bottom, and a focus would require that the ball have enough tangential velocity to miss the bottom on its first downward pass.

Figure 3-5, showing only a single steady state at $T = 140°\text{F}$, $C = 0.25$ is from the model equations (2-106) formulated by Luyben to include feedback control, modified to allow for a solvent latent heat that is temperature

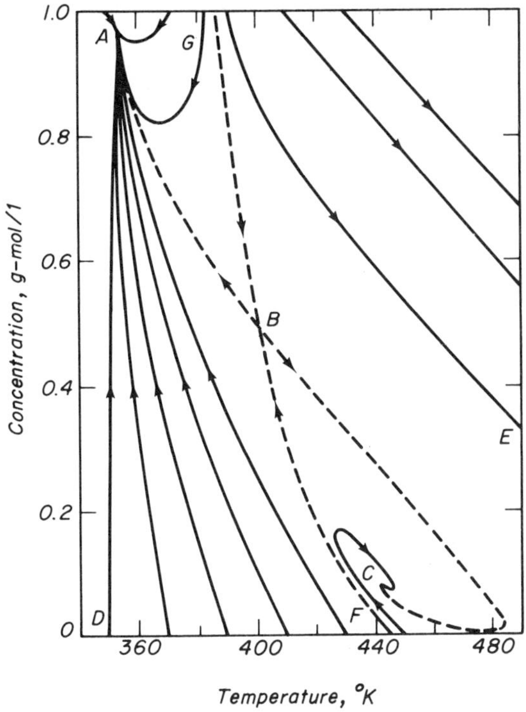

Figure 3-4. Phase-plane with multiple steady states (after Aris and Amundson, 1958).

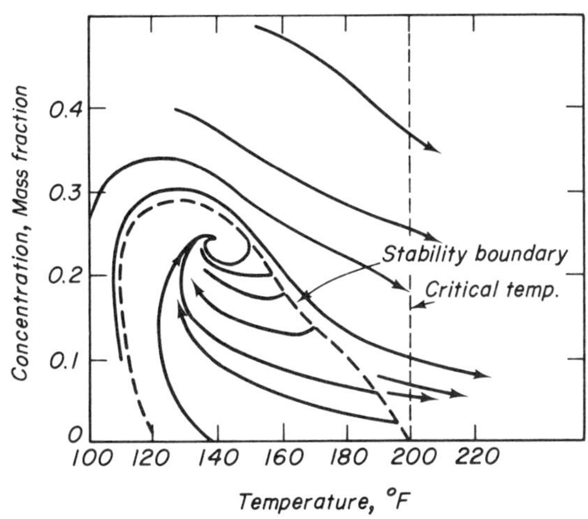

Figure 3-5. Phase-plane with region of stability (after Luyben, 1965).

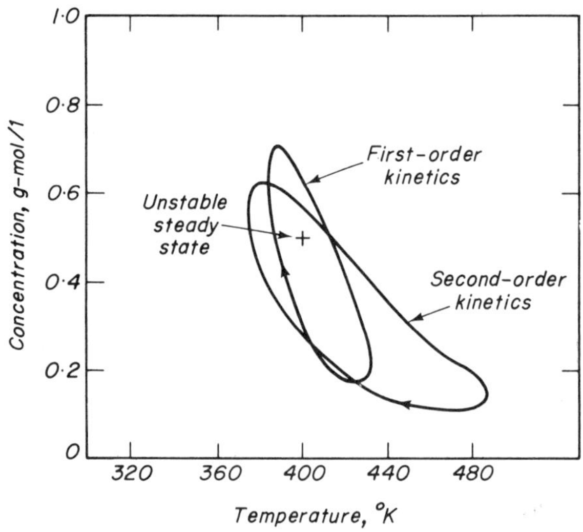

Figure 3-6. Phase-plane with limit cycles (after Gurel and Lapidus, 1965).

dependent. Above the critical temperature of the solvent the latent heat is zero and this cooling mechanism is inoperable, as the trajectories indicate. Two other steady states exist for this system but are not shown because they are beyond the range of practical interest. The distinctive character of this example is in its well-defined region of stability having a boundary rather close to the desired steady state. Comparatively small disturbances in the direction of increased concentration and temperature would cause reactor runaway at this level of control. Luyben (1966) has shown that the situation can be improved with better choice of the control constant K_p. The similarity of this diagram to Figure 3-3 should be noted.

The trajectories chosen for Figure 3-6 are *limit cycles.* They close upon themselves and are characteristic of periodic, nonlinear behavior, possessing in particular an unstable focus at the steady state within the loop: all trajectories starting within the enclosed area will spiral out until they merge with the limit cycle. These limit cycles are called stable because trajectories starting outside the cycle will also decay to the same periodic pattern. It is worth noting at this point that the term *stability* has already been used in two contexts to describe quite different characteristics, in the first case for a steady state and in the second instance for the series of states that form a limit cycle. Evidently this is a term that demands more precise definition and elaboration; this will be a major goal of Chapter 4.

Because trajectories are often most conveniently described by their relation to a particular steady state of interest, it is often useful to make that steady state the origin of the phase-plane. This may be accomplished by using as state variables the deviations from the respective steady-state values:

$$r = R(\eta, C) - R(\eta_s, C_s) \tag{3-4a}$$

$$x_1 = \eta - \eta_s \tag{3-4b}$$

$$x_2 = C - C_s \tag{3-4c}$$

These substitutions transform the CSTR equations (1-3) and (1-1) to

$$V\frac{dx_1}{dt} = -qx_1 + Vr - \frac{U}{C_p}x_1 + \left[q(\eta_0 - \eta_s) + VR_s + \frac{U}{C_p}(\eta_w - \eta_s)\right] \tag{3-5a}$$

$$V\frac{dx_2}{dt} = -qx_2 - Vr + [q(C_0 - C_s) - VR_s] \tag{3-5b}$$

which in combination with relations (1-5) yields

$$V\frac{dx_1}{dt} = -qx_1 + Vr - \frac{U}{C_p}x_1 \tag{3-6a}$$

$$V\frac{dx_2}{dt} = -qx_2 - Vr \tag{3-6b}$$

In more general notation this can be written

$$\frac{dx_1}{dt} = \dot{x}_1 = f_1(x_1, x_2) \tag{3-7a}$$

$$\frac{dx_2}{dt} = \dot{x}_2 = f_2(x_1, x_2) \tag{3-7b}$$

with steady state as $x_{1s} = x_{2s} = 0$.

3-3. Vector Notation and Measures of Distance

The position of a point in the phase-plane can be fixed by reporting its coordinates, (x_1, x_2) or equivalently it can be identified as the head of a vector $\mathbf{x}$ from the origin. The two ideas are hardly distinguishable at all, when the vector is in turn described by its components in the x_1 and x_2 directions:

$$\mathbf{x} = \begin{bmatrix} x_1 \\ x_2 \end{bmatrix}$$

The significant advantage that favors the vector notation is conciseness, especially when general theorems are to be developed applicable to systems of multiple dimensions in $\mathbf{x}$. Equations (3-7), for example, may be written as:

$$\dot{\mathbf{x}} = \mathbf{f}(\mathbf{x}) \tag{3-8}$$

with steady state $\mathbf{x}_s = \mathbf{0}$, where the vector

$$\mathbf{f} = \begin{bmatrix} f_1(x_1, x_2) \\ f_2(x_1, x_2) \end{bmatrix}$$

In dealing with graphical arguments, it might be expected that a recurrent yardstick for comparisons of several kinds will be the distance of a given state from the steady state of interest; expressed in terms of deviation variables, this will be distance from the origin.

The most immediately intuitive measure of distance from the origin in the (x_1, x_2) plane is the Euclidian measure

$$\sqrt{x_1^2 + x_2^2}$$

but this is by no means the only definition that will suit. An alternative, for example, may be developed from a recognition that the Euclidian measure is in essence an identification of the point (x_1, x_2) by means of a family of concentric circles. What we have called Euclidian distance is the radius of that circle which passes through the point in question. Could we not then modify our definition to use a family of concentric squares? or diamonds? or in general, any convenient set of contours? To facilitate this generalization, such a measure is called a *norm*, a function of $\mathbf{x}$ designated by the symbol $||\mathbf{x}||$.

To meet the geometric specifications we have outlined, the norm (distance) must be positive for any $\mathbf{x}$ except the trivial $\mathbf{x} = \mathbf{0}$. Thus:

$$||\mathbf{x}|| = 0 \qquad \text{when } \mathbf{x} = \mathbf{0} \tag{3-9a}$$

$$||\mathbf{x}|| > 0 \qquad \text{when } \mathbf{x} \neq \mathbf{0} \tag{3-9b}$$

Further, since multiplication of a vector by a constant increases its length proportionally:

$$||c\mathbf{x}|| = |c|\,||\mathbf{x}|| \tag{3-9c}$$

and, because the sum of the lengths of two sides of a triangle is at least as large as that of the third side:

$$||\mathbf{x}_A + \mathbf{x}_B|| \leq ||\mathbf{x}_A|| + ||\mathbf{x}_B|| \qquad \text{(triangle inequality)} \tag{3-9d}$$

One other property of a vector norm which we will need is

$$\left\| \int_0^t \mathbf{x}\, dt \right\| \leq \int_0^t ||\mathbf{x}||\, dt \tag{3-9e}$$

This is easily proved from its definition for some norms, but requires the subtler arguments of Beckenbach and Bellman (1961) or Hardy, Littlewood, and Polya (1952) for some others.

By putting our requirements in vector notation, we have made the statements more concise, but even more important, have stated relationships valid for a space of any number of dimensions, not merely (x_1, x_2). In this new notation the Euclidian norm is:

$$\|\mathbf{x}\|_E = \sqrt{x_1^2 + x_2^2 + \ldots + x_n^2} \tag{3-10}$$

and the diamond norm alluded to above is:

$$\|\mathbf{x}\|_D = |x_1| + |x_2| + \ldots + |x_n| \tag{3-11}$$

but now it is pertinent to note that contours of constant distance from the origin are surfaces for three-dimensional $\mathbf{x}$ (spheres, polyhedra) and corresponding hypersurfaces for higher-dimensional systems. Second-order reactor models will not evoke these difficulties.

By analogy to the norm of a vector, a measure of the magnitude of a matrix may also be defined. A norm of the matrix:

$$\mathbf{X} = \begin{bmatrix} x_{11} & x_{12} & \cdots & x_{1n} \\ x_{21} & x_{22} & \cdots & \\ \cdots & & & \\ \cdots & & & \\ x_{m1} & x_{m2} & \cdots & x_{mn} \end{bmatrix} = [\mathbf{x}_1 \quad \mathbf{x}_2 \quad \cdots \quad \mathbf{x}_n] \tag{3-12}$$

is a function $\|\mathbf{x}\|$ that satisfies:

$$\|\mathbf{X}\| = 0 \qquad \text{when } \mathbf{X} = \mathbf{0} \tag{3-13a}$$

$$\|\mathbf{X}\| > 0 \qquad \text{when } \mathbf{X} \neq \mathbf{0} \tag{3-13b}$$

$$\|c\mathbf{X}\| \leq |c|\,\|\mathbf{X}\| \tag{3-13c}$$

$$\|\mathbf{X}_A + \mathbf{X}_B\| \leq \|\mathbf{X}_A\| + \|\mathbf{X}_B\| \qquad \text{(triangle inequality)} \tag{3-13d}$$

$$\left\| \int_0^t \mathbf{X}\, dt \right\| \leq \int_0^t \|\mathbf{X}\|\, dt \tag{3-13e}$$

and in addition, we will find use for the Schwartz inequality:

$$\|\mathbf{X}_A \mathbf{X}_B\| \leq \|\mathbf{X}_A\|\,\|\mathbf{X}_B\| \tag{3-13f}$$

applicable to the two norms of interest, the Euclidian

$$\|\mathbf{X}\|_E = \sqrt{\sum_{i=1}^{m} \sum_{j=1}^{n} x_{ij}^2} = \sqrt{\|\mathbf{x}_1\|_E^2 + \|\mathbf{x}_2\|_E^2 + \ldots + \|\mathbf{x}_n\|_E^2} \tag{3-14}$$

and the diamond norm

$$\|\mathbf{X}\|_D = \sum_{i=1}^{m} \sum_{j=1}^{n} |x_{ij}| = \|\mathbf{x}_1\|_D + \|\mathbf{x}_2\|_D + \ldots + \|\mathbf{x}_n\|_D \tag{3-15}$$

3-4. Inequalities

Geometrically, a norm equation produces a contour in the phase-plane. The most elementary example is the circle $\|\mathbf{x}\|_E = \epsilon$. As equations represent lines in phase-space, so inequalities represent areas, as can be deduced from the interpretation of $\|\mathbf{x}\|_E < \epsilon$ as the set of points *within* the circle of radius ϵ, that is, points closer to the origin than distance ϵ.

Recalling that $\mathbf{x}$ is implicitly associated with a time scale as it changes along a trajectory, calls attention to the norm as a scaler function $\|\mathbf{x}(t)\|$, having an initial condition $\|\mathbf{x}(0)\|$. When the time history of the norm is of particular interest, its scaler character may be used to advantage in a graph such as Figure 3-7, where as a further example of a norm inequality, the geometric meaning of

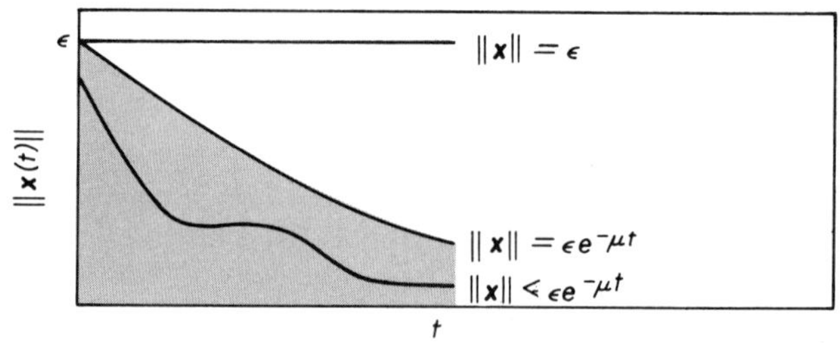

Figure 3-7. Examples of a norm as a scaler function.

$$\|\mathbf{x}(t)\| \leq \epsilon e^{-\mu t} \tag{3-16}$$

is that the norm is restricted to be within the darkened region for all t. A sample of a curve which satisfies this condition is also shown.

The extension to a matrix-norm inequality is straightforward, as may be seen in the following numerical example:

EXAMPLE 3-1. Given the matrix function

$$\mathbf{X}(t) = \frac{1}{4}\begin{bmatrix}(-e^{-5t} + 5e^{-t}) & (-e^{-5t} + e^{-t}) \\ (5e^{-5t} - 5e^{-t}) & (5e^{-5t} - e^{-t})\end{bmatrix}$$

find values of c and μ that will satisfy the bound

$$\|\mathbf{X}(t)\| \leq ce^{-\mu t} \tag{3-17}$$

Is there a largest (μ/c) ratio beyond which the inequality cannot be satisfied for all time?

Solution: The answer will depend on the norm chosen. Using the diamond norm

$$\|\mathbf{X}(t)\|_D = \tfrac{1}{4}[|-e^{-5t} + 5e^{-t}| + |-e^{-5t} + e^{-t}| \\ + |5e^{-5t} - 5e^{-t}| + |5e^{-5t} - e^{-t}|]$$

which may be simplified to

$$\|\mathbf{X}(t)\|_D = \begin{cases} 2.5e^{-t} - 0.5e^{-5t}, & t \leq 0.403 \\ 3e^{-t} - 3e^{-5t}, & t \geq 0.403 \end{cases}$$

This function is shown in Figure 3-8, together with several possible upper bounds. The semilog coordinates are used to show equation (3-17) as a straight line of slope $-\mu$ and intercept c. The following sets of values are read from the graph:

μ	c	μ/c
1.0	3.0	0.33
0.9	2.7	0.33
0.8	2.5	0.32
0.7	2.3	0.30
0.5	2.1	0.24

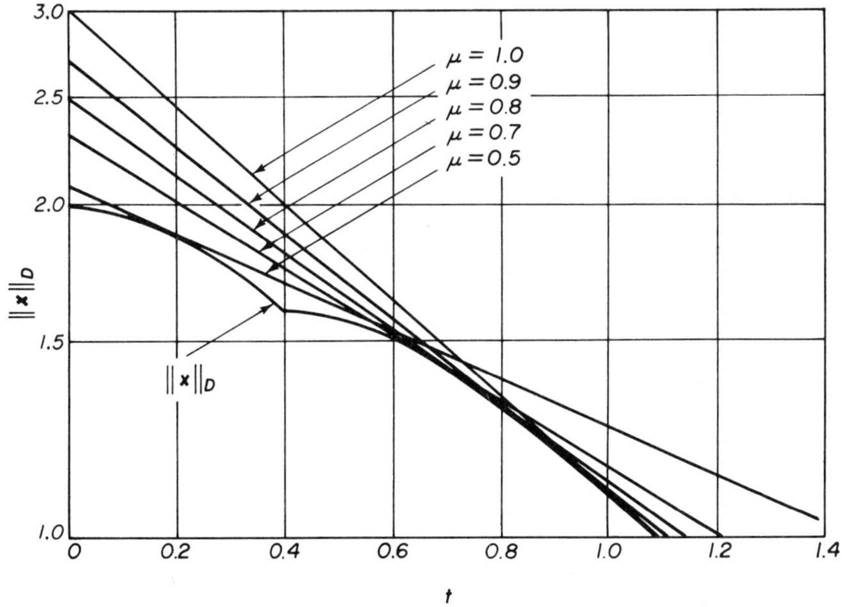

Figure 3-8. Diamond norm versus time and several bounds.

Considering now the Euclidian norm:

$$\|\mathbf{X}(t)\|_E = \tfrac{1}{4}\sqrt{52e^{-10t} + 52e^{-2t} - 72e^{-6t}}$$

the data from Figure 3-9 are:

μ	c	(μ/c)
1.00	1.80	0.56
0.90	1.60	0.56
0.63	1.41	0.45

and as before the largest (μ/c) is at one end of the feasible range. This is often but not always the case.

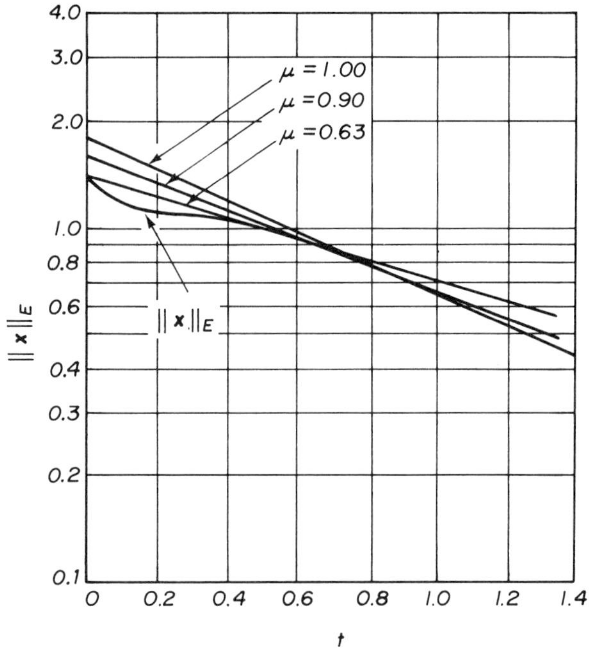

Figure 3-9. Euclidean norm versus time and several bounds.

3-5. The Linearized CSTR

The differential equations of the CSTR model are nonlinear only by virtue of the reaction rate term. Because of the relative ease with which linear equations can be manipulated, this function is sometimes approximated in a region of interest by a truncated Taylor-series expansion. Such a linearized approximation can be more or less satisfactory, improving as the expansion variables are more confined about the point of expansion. Assuming that the concentration and temperature are always only slightly removed from their steady-state values C_s and T_s:

$$R(T, C) = R(T_s, C_s) + \left(\frac{\partial R}{\partial T}\right)_{T_s, C_s}(T - T_s) + \left(\frac{\partial R}{\partial C}\right)_{T_s, C_s}(C - C_s) \qquad \text{(3-18a)}$$

or

$$R(\eta, C) = R(\eta_s, C_s) + \left(\frac{\partial R}{\partial \eta}\right)_s(\eta - \eta_s) + \left(\frac{\partial R}{\partial C}\right)_s(C - C_s) \qquad \text{(3-18b)}$$

The same statement can also be framed in the more concise notation of deviation variables (3-4) as

$$r(x_1, x_2) = \frac{\partial r}{\partial x_1}x_1 + \frac{\partial r}{\partial x_2}x_2 \qquad \text{(3-18c)}$$

where the partials are to be evaluated at the steady state $x_1 = x_2 = 0$.

The linearized CSTR equations are obtained by using (3-18) to eliminate r from equations (3-6) and combining the constants to:

$$a_{11} = -\frac{q}{V} - \frac{U}{C_p V} + \frac{\partial r}{\partial x_1} \tag{3-19a}$$

$$a_{12} = \frac{\partial r}{\partial x_2} \tag{3-19b}$$

$$a_{21} = -\frac{\partial r}{\partial x_1} \tag{3-19c}$$

$$a_{22} = -\frac{q}{V} - \frac{\partial r}{\partial x_2} \tag{3-19d}$$

The result is

$$\frac{dx_1}{dt} = a_{11}x_1 + a_{12}x_2 \tag{3-20a}$$

$$\frac{dx_2}{dt} = a_{21}x_1 + a_{22}x_2 \tag{3-20b}$$

which it may be noted is a form general enough to represent *any* linear second-order system. Indeed, equations (3-20) could as well have been written to start this section by merely noting that the system under study is modeled by two first-order equations. The detailed development was offered instead to show how the four a_{ij} constants of (3-20) are found.

Further reduction is possible if the kinetics are specified. If equation (1-6a) applies

$$\frac{\partial r}{\partial x_1} = \left(\frac{\partial r}{\partial \eta}\right)_s = \left(\frac{\Delta H}{C_p}\right)\left(\frac{\partial R}{\partial T}\right)_s = \frac{\Delta H}{C_p}\frac{Q}{T_s^2}R_s \tag{3-21a}$$

$$\frac{\partial r}{\partial x_2} = \left(\frac{\partial r}{\partial C}\right)_s = \left(\frac{\partial R}{\partial C}\right)_s = \frac{n}{C_s}R_s \tag{3-21b}$$

and replacing the rate term by its equivalent from equation (1-5a) gives

$$\frac{\partial r}{\partial x_1} = \frac{\Delta H}{C_p}\frac{Q}{T_s^2}\frac{q}{V}(C_0 - C_s) \tag{3-22a}$$

$$\frac{\partial r}{\partial x_2} = \frac{nq}{C_s V}(C_0 - C_s) \tag{3-22b}$$

Then by substitution into (3-19):

$$a_{11} = -\frac{q}{V} - \frac{U}{C_p V} + \frac{\Delta H Q q(C_0 - C_s)}{C_p T_s^2 V} \tag{3-23a}$$

$$a_{12} = \frac{nq}{VC_s}(C_0 - C_s) \tag{3-23b}$$

$$a_{21} = -\frac{\Delta H Q q(C_0 - C_s)}{C_p T_s^2 V} \tag{3-23c}$$

$$a_{22} = -\frac{q}{V}\left[1 + \frac{n(C_0 - C_s)}{C_s}\right] \tag{3-23d}$$

In vector-matrix notation equations (3-20) become

$$\dot{\mathbf{x}} = \mathbf{A}\mathbf{x} \tag{3-24}$$

with

$$\mathbf{A} = \begin{bmatrix} a_{11} & a_{12} \\ a_{21} & a_{22} \end{bmatrix} \tag{3-25a}$$

A result of identical form (but different a_{ij} elements) would have been obtained from a higher-order model. For a system of n ordinary differential equations

$$\mathbf{A} = \begin{bmatrix} a_{11} & a_{12} & \cdots & a_{1n} \\ a_{21} & a_{22} & \cdots & a_{2n} \\ \cdots & & & \\ \cdots & & & \\ a_{n1} & a_{n2} & \cdots & a_{nn} \end{bmatrix} \tag{3-25b}$$

3-6. Closed-Form Solution

Among the great variety of possible reactor models the form of equation (3-24) is unique for several reasons. Above all, this is virtually the only multi-variable ordinary differential equation that can be solved analytically. Furthermore, the details of solution can be reduced to manipulations in matrix algebra, no small advantage in itself. Finally, as was suggested in Section 3-5, many important nonlinear systems can be approximated by linearization more or less adequately, depending on the purposes to which the approximation is to be put. Whether or not any real systems are truly linear is in any case something of a moot point, since all continuous behavior is linear in the small, whereas no physical-chemical system remains linear over a wide enough range of its variables. The question is not whether a system is linear, but rather how linear over the range of interest of the variables.

The equation (3-24) can be solved in a number of ways. For the purposes of this book, it is most convenient to proceed by paralleling where possible the known results of single-variable problems, and begin by assuming a solution form.

Considering as a candidate solution the vector

$$\mathbf{x}(t) = e^{\lambda t}\mathbf{c} \tag{3-26}$$

differentiation gives

$$\dot{\mathbf{x}}(t) = \lambda e^{\lambda t}\mathbf{c} = \lambda \mathbf{x}$$

and substitution back into (3-24) yields

$$\lambda \mathbf{x} = \mathbf{A}\mathbf{x} \tag{3-27a}$$

or, by rearrangement

$$(\lambda \mathbf{I} - \mathbf{A})\mathbf{x} = \mathbf{0} \tag{3-27b}$$

In order that this set of simultaneous linear algebraic equations has any solution other than $\mathbf{x} = \mathbf{0}$, it is necessary that

$$\det(\lambda \mathbf{I} - \mathbf{A}) = 0 \tag{3-28}$$

The values of λ that satisfy the *characteristic equation* (3-28) are the *eigenvalues* of the problem, and with distinct λ's the complete solution to (3-24) is the linear combination of the set of solutions of the form (3-26).

But equation (3-26) is not the only simple assumed form that yields results. The desired vector $\mathbf{x}(t)$ can also be the product of an $n \times n$ square matrix $\mathbf{X}$ and an n-dimensional vector, such as

$$\mathbf{x}(t) = \mathbf{X}(t)\mathbf{x}(0) \tag{3-29}$$

where by identifying the constant vector as the initial condition, it is implied that

$$\mathbf{X}(0) = \mathbf{I} \tag{3-30}$$

Additional properties of the $\mathbf{X}$ matrix can be brought out by differentiating equation (3-29)

$$\dot{\mathbf{x}} = \dot{\mathbf{X}}\mathbf{x}(0) \tag{3-31}$$

and substituting into (3-24):

$$\dot{\mathbf{X}}\mathbf{x}(0) = \mathbf{A}\mathbf{X}\mathbf{x}(0) \tag{3-32}$$

Examination of (3-32) will show that the postulated $\mathbf{X}$ must have the property ordinarily associated with scalar exponentials, that is

$$\dot{\mathbf{X}} = \mathbf{A}\mathbf{X} \tag{3-33}$$

This property can be emphasized by using the symbol $e^{\mathbf{A}t}$ as equivalent to $\mathbf{X}$, this alternative having the added advantage of suggesting the formal definition:

$$\mathbf{X}(t) = e^{\mathbf{A}t} = \mathbf{I} + t\mathbf{A} + \frac{1}{2}t^2\mathbf{A}^2 + \frac{1}{6}t^3\mathbf{A}^3 + \dots \tag{3-34}$$

It is a simple matter to use this definition to demonstrate the validity of equations (3-30) and (3-33), but because it is in the form of an infinite series it is not a convenient form to use for the evaluation of $\mathbf{X}$ to solve a specific problem.

An algebraically more satisfactory algorithm may be developed by Laplace transformation of equation (3-24) to give

$$s\mathscr{L}[\mathbf{x}(t)] - \mathbf{x}(0) = \mathbf{A}\mathscr{L}[\mathbf{x}(t)] \tag{3-35}$$

Rearranging:

$$\mathscr{L}[\mathbf{x}(t)] = (s\mathbf{I} - \mathbf{A})^{-1}\mathbf{x}(0) \tag{3-36}$$

which inverted back to the time domain gives the solution

$$\mathbf{x}(t) = \mathscr{L}^{-1}[(s\mathbf{I} - \mathbf{A})^{-1}]\mathbf{x}(0) \tag{3-37}$$

By direct comparison with equation (3-29), it is evident that

$$\mathbf{X}(t) = \mathscr{L}^{-1}[(s\mathbf{I} - \mathbf{A})^{-1}] \tag{3-38}$$

and that this *principal matrix* solution can be obtained term by term. Note that the inverse signs in equation (3-38) refer to inverse transformation and to matrix inversion, respectively.

EXAMPLE 3-2. Solve for $\mathbf{x}(t)$, given that

$$\dot{x}_1 = x_2, \qquad x_1(0) = -1$$
$$\dot{x}_2 = -5x_1 - 6x_2, \quad x_2(0) = +5$$

Solution: From the problem statement

$$\mathbf{A} = \begin{bmatrix} 0 & 1 \\ -5 & -6 \end{bmatrix}, \quad \mathbf{x}(0) = \begin{bmatrix} -1 \\ +5 \end{bmatrix}$$

and

$$(s\mathbf{I} - \mathbf{A})^{-1} = \begin{bmatrix} s & -1 \\ 5 & (s+6) \end{bmatrix}^{-1} = \frac{1}{s^2 + 6s + 5}\begin{bmatrix} (s+6) & 1 \\ -5 & s \end{bmatrix}$$

Applying equation (3-38):

$$\mathbf{X}(t) = \tfrac{1}{4}\begin{bmatrix} (-e^{-5t} + 5e^{-t}) & (-e^{-5t} + e^{-t}) \\ (5e^{-5t} - 5e^{-t}) & (5e^{-5t} - e^{-t}) \end{bmatrix}$$

and introducing the initial conditions through (3-37):

$$x_1(t) = -e^{-5t}$$
$$x_2(t) = 5e^{-5t}$$

It may be noted in passing that the $\mathbf{X}(t)$ matrix of this problem is exactly the one studied in Example 3-1.

3-7. Linear Second-Order Systems

The formal developments of the previous section reduce to particularly simple computations when $\mathbf{A}$ is a 2×2 matrix. The characteristic equation (3-28) is then

$$\det\begin{bmatrix} (\lambda - a_{11}) & -a_{12} \\ -a_{21} & (\lambda - a_{22}) \end{bmatrix} = \lambda^2 - (a_{11} + a_{22})\lambda + (a_{11}a_{22} - a_{12}a_{21}) = 0 \tag{3-39}$$

which gives for the eigenvalues

$$\lambda = \frac{1}{2}[(a_{11} + a_{22}) \pm \sqrt{(a_{11} + a_{22})^2 - 4(a_{11}a_{22} - a_{12}a_{21})}] \tag{3-40}$$

The graphical interpretation of linear second-order systems is also straightforward and relatively simple. From equations (3-20) one obtains

$$\frac{dx_2}{dx_1} = \frac{a_{21}x_1 + a_{22}x_2}{a_{11}x_1 + a_{12}x_2} = m \tag{3-41}$$

from which the isocline equation is linear:

$$x_2 = \left[\frac{a_{21} - a_{11}m}{a_{12}m - a_{22}}\right]x_1 \tag{3-42}$$

Since $\mathbf{A}$ is known, every choice of m determines a line through the origin, and it does not require much effort to cover enough of the phase-plane to make possible trajectory interpolation.

Among the infinity of isoclines that can be drawn on the phase-plane, there are some that have a special property: the slope of the isocline is identical to the slope of the trajectories that intersect it. When this is the case the isocline in question becomes an asymptote for nearby trajectories, and itself becomes a straight line trajectory from any initial condition along its length. Such an isocline trajectory is called an *eigenvector*.

The slopes of a system's eigenvectors may be determined analytically by using the defining property. Equating the trajectory slope m to that of the isocline equation (3-42):

$$m = \frac{a_{21} - a_{11}m}{a_{12}m - a_{22}} \tag{3-43}$$

and solving for m:

$$\begin{aligned} m &= \frac{(a_{22} - a_{11}) \pm \sqrt{(a_{11} - a_{22})^2 + 4a_{12}a_{21}}}{2a_{12}} \\ &= \frac{(a_{22} - a_{11}) \pm \sqrt{(a_{11} + a_{22})^2 - 4(a_{11}a_{22} - a_{12}a_{21})}}{2a_{12}} \end{aligned} \tag{3-44}$$

But from equation (3-40) this can also be written in terms of the eigenvalues:

$$m = \frac{(a_{22} - a_{11}) + [2\lambda - (a_{11} + a_{22})]}{2a_{12}} = \frac{\lambda - a_{11}}{a_{12}} \tag{3-45}$$

Knowledge of the eigenvalues can in this way aid the sketching of the phase-plane diagram. Equation (3-45) shows also that second-order systems with two real eigenvalues have two eigenvectors, that identical eigenvalues produce only a single eigenvector, and that complex eigenvalue systems have no real eigenvectors.

A more general form of equation (3-45) may be found by using the fact that along the eigenvector

$$m = \frac{x_2}{x_1} \tag{3-46}$$

As a result:

$$\lambda x_1 = a_{11}x_1 + a_{12}x_2 \tag{3-47}$$

which is the first row of the vector-matrix equation (3-27a). This more general statement can be used to define eigenvectors for linear systems of any order.

The various relationships are perhaps best presented through an illustration.

EXAMPLE 3-3. Sketch the phase-plane portrait of the CSTR linearized about its intermediate steady state at $T_s = 400$, $C_s = 0.5$ g-mol/liter, using the parameters chosen by Aris and Amundson and given in Example 2-3:

Solution: Substituting the given numbers into equations (3-23):

$$a_{11} = 4.25$$
$$a_{12} = 1$$
$$a_{21} = -6.25$$
$$a_{22} = -2$$

The eigenvalues are found from equation (3-40):

$$\lambda_1 = 3.00$$
$$\lambda_2 = -0.75$$

and these in turn determine the eigenvector slopes from equation (3-45):

$$m_1 = -1.25 \tag{3-48a}$$
$$m_2 = -5 \tag{3-48b}$$

This information is sufficient for a quite reasonable sketch, as shown in Figure 3-10. The two eigenvectors are drawn first from knowledge of their respective slopes.

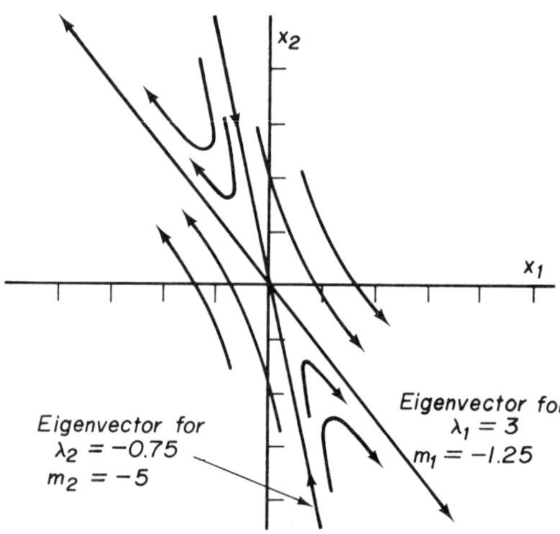

Figure 3-10. Sketch of phase-plane for linearized CSTR of Example 3-3.

Each has a direction associated with it by means of the eigenvalue used in its computation: the vector of slope -5 is directed toward the origin because it is coupled to a negative eigenvalue which contributes to stability. On the contrary, the other eigenvector is an outward bound trajectory since it is linked to a positive (unstable) eigenvalue. The local behavior is obviously of the saddle type and may be sketched by interpolation between the eigenvector asymptotes. For many purposes this degree of accuracy will suffice; however, if further detail is needed it may be readily obtained by filling in one or more other isoclines, the slope of which would be computed from equation (3-42). A comparison between Figure 3-10 and the result of the nonlinear analysis shown as Figure 3-4 confirms the validity of the linearization for local behavior, and emphasizes also the inadequacy of linearization for any but local information about a steady state.

3-8. Comparison of Linear and Nonlinear Systems

One of the more attractive features of the vector-matrix notation is its relative versatility, being applicable to systems of any order whether linear or nonlinear. In view of this, and of the successful extension of the single-variable approach to multi-variable linear problems, one might wonder whether the use of the $e^{\mathbf{A}t}$ function would not be fruitful in studying a model of the form

$$\dot{\mathbf{x}} = \mathbf{A}\mathbf{x} + \mathbf{g} \tag{3-49}$$

such as has already arisen for example as equation (3-6) for the CSTR. Premultiplying each term in (3-49) by $e^{-\mathbf{A}t}$, the inverse of $\mathbf{X}$, introduces a matrix integration factor, as follows:

$$e^{-\mathbf{A}t}\dot{\mathbf{x}} = e^{-\mathbf{A}t}\mathbf{A}\mathbf{x} + e^{-\mathbf{A}t}\mathbf{g} \tag{3-50}$$

$$\frac{d}{dt}(e^{-\mathbf{A}t}\mathbf{x}) = e^{-\mathbf{A}t}\mathbf{g} \tag{3-51}$$

Integrating both sides from 0 to t:

$$e^{-\mathbf{A}t}\mathbf{x} - \mathbf{I}\mathbf{x}(0) = \int_0^t e^{-\mathbf{A}t}\mathbf{g}\,dt$$

or

$$\mathbf{x} = \mathbf{X}(t)\mathbf{x}(0) + \mathbf{X}\int_0^t \mathbf{X}^{-1}\mathbf{g}\,d\sigma \tag{3-52}$$

This result will be most useful in the development of parts of the next chapter, but for the moment, it can hardly be called a closed-form solution, since the integral in equation (3-52) can only be evaluated for a pitifully small number of special cases of $\mathbf{g}$, and then only when $\mathbf{g} = \mathbf{g}(t)$. In the more usual case $\mathbf{g} = \mathbf{g}(\mathbf{x})$, and the integral is merely an alternative notation to equation (3-49). There is therefore no simple counterpart in nonlinear analysis to the eigenvalue-eigenvector description of linear systems, except in as much as one can make use of the linearization approximation. The more numerical character of the phase-plane accordingly becomes that much more important to the designer. There is on the other hand a richness of possibilities in nonlinear cases, which cannot occur in a linear system. These include limit cycles and multiple steady states, and related to the latter, adjacent regions in the phase-plane from which trajectories have widely differing paths and ultimate goals. Any hope of describing such subtlety of behavior must of necessity be lost when a nonlinear system is linearized.

That ambiguous interpretations have sometimes been attached to the term stability has already been pointed out. Regarding linear systems, however, no such problems arise, because it is so simple a matter to categorize the strictly limited possibilities in the form that solutions can take. With $\mathbf{A}$ constant the terms in $(s\mathbf{I} - \mathbf{A})$ are linear in s, and as a consequence the terms in $(s\mathbf{I} - \mathbf{A})^{-1}$ have polynomial numerators and denominators, the highest power of the latter being the order of the $\mathbf{A}$ matrix. But polynomials in s can always be inverted into exponential time functions having terms such as equation (3-26). The well-known special cases for complex or repeated eigenvalues introduce multiplicative sinusoids or power terms, but the growth or decay of the exponential is still the controlling feature. A linear system is stable if the real parts of all its eigenvalues are negative. Such a definition is inadequate for nonlinear systems where the possible forms of solution are myriad; nevertheless, the idea of linearization can serve as a link between the two, if it is applied with due regard to its limitations. This will be in large part the business of Chapter 4.

REFERENCES

Aris, R. and N. R. Amundson, *Chem. Eng. Sci.*, **7**, 121 (1958).

Beckenbach, E. and R. Bellman, *An Introduction to Inequalities*, New Haven, Conn.: Random House (1961).

Bellman, R., *Stability Theory of Differential Equations*, New York: McGraw-Hill, Inc. (1953).

Gurel, O. and L. Lapidus, *Chem. Eng. Prog. Symp. Ser., No. 55*, **61**, 78 (1965).

Hardy, G. H., J. E. Littlewood, and G. Polya, *Inequalities*, 2nd ed., Cambridge University Press (1952).

Luyben, W. L., *AIChE Journal*, **12**, 662 (1966).

Perlmutter, D. D., *Introduction to Chemical Process Control*, New York: Wiley (1965).

Struble, R. A., *Nonlinear Differential Equations*, New York: McGraw-Hill, Inc. (1962).

Truxal, J. G., *Automatic Feedback Control System Synthesis*, New York: McGraw-Hill, Inc. (1955).

Local Dynamic Stability

4

4-1. Intuitive Argument

What is meant by the term *stability* as applied to a nonlinear system? If one appeals to phase-plane portraits such as were illustrated in Figures 3-1 to 3-6, a steady state may be called stable if the trajectories in its neighborhood all approach it, and unstable if any one trajectory is directed away. Then the nodes and foci shown in these figures correspond to stable steady states, the saddles to unstable ones. The possibilities for movement in the opposite sense also exist, so that one may speak of unstable nodes or foci (but not a stable saddle).

But note that although such a local stability is a property associated with a point (a steady state), its definition involves the trajectories in a neighborhood. These are in general the results of some prior numerical computations, and as a matter of practicality one might well ask how small or large the region of study has to be before the definition of stability is to be considered satisfied. This is not merely an academic fine point, since unless one traces the trajectory far enough toward the steady state there is no guarantee that it will not after a time turn around and begin to move away. Indeed, this is precisely what we have observed for the trajectories that form a saddle.

The resolution to this puzzle lies in the recognition that the required property must be demonstrable for a region: $||\mathbf{x}|| \leq \epsilon$. Such reasoning leads to the formal and more rigorous definitions of Section 4-2.

4-2. Liapunov Stability

A steady state is stable by Liapunov's definition, if for any arbitrary choice of an ϵ-neighborhood, it is always possible to find a δ-region, such that trajectories starting initially within δ will never exceed the bounds of ϵ. From another viewpoint this definition can be thought of as a game for two players, the challenger picking a region: $||\mathbf{x}|| \leq \epsilon$, the defender answering with a region: $||\mathbf{x}|| \leq \delta$. The defender wins (establishes stability) if he is able to find a $\delta(\epsilon)$ for each ϵ, such that

$$||\mathbf{x}(0)|| \leq \delta \tag{4-1a}$$

implies

$$||\mathbf{x}(t)|| \leq \epsilon \tag{4-1b}$$

that is, any initial condition within the δ-region leads to a trajectory confined by the ϵ-contour. If the challenger is able to find even just a single ϵ for which there is no δ, he will win the game: the steady state will be called unstable.

As a test of this definition, consider the node of Figure 3-4. The steady state in question must be judged locally stable, because the defender will always be able to place his δ-contour entirely within any ϵ-contour. Since the shape of the δ-contour can also be chosen so that no trajectories cross it in the outward direction, the defender will be able to argue that if no trajectories leave the δ-region they can never exceed the larger ϵ-region. Formally, if

$$||\mathbf{x}(t)|| \leq \delta$$

and $||\mathbf{x}|| \leq \delta$ is a sub-region of $||\mathbf{x}|| \leq \epsilon$, then

$$||\mathbf{x}(0)|| \leq \delta$$

$$||\mathbf{x}(t)|| \leq \epsilon$$

proving the point.

As a second exercise consider the saddle of Figure 3-4. The challenger can establish local *instability* by choosing an ϵ-region near the steady state, since any trajectory sufficiently close will eventually cross the ϵ bound. It should be noted in passing that if the defender were allowed to choose ϵ he could select one for which a δ could be found, and it is for this reason that the definition demands an arbitrary ϵ chosen by the challenger. It is always the small ϵ-regions that are troublesome for the defender, since if he answers successfully such a challenge, he can always use the same response to the presentation of a larger ϵ-region.

4-3. Tracking Contours

Because the intent of Section 4-2 was to clarify a definition, it was assumed throughout that the entire phase-plane portrait was available. But, of course, if that were in fact the case, much of the incentive for this study would be lost. The engineer's interest is quite the reverse: he seeks if possible to establish stability or instability before the system differential equations are solved; that is, he wants a *direct* method that operates on the differential equation model in the absence of an explicit solution.

Suppose for example that a system of interest is modeled by simultaneous equations of the form of (3-20), but with the coefficients numerically specified:

$$\dot{x}_1 = -x_1 + x_2 \tag{4-2a}$$

$$\dot{x}_2 = -x_1 - 2x_2 \tag{4-2b}$$

and that it is proposed to track a trajectory movement in the (x_1, x_2) plane by its position with respect to a set of concentric circles about the steady state:

$$v(\mathbf{x}) = v(x_1, x_2) = x_1^2 + x_2^2 \tag{4-3}$$

If as time progresses the trajectory intersects circles of successively smaller and smaller v, hence smaller radius, the steady state must be stable. This is equivalent algebraically to asking for $(dv/dt) < 0$, and the details of the problem come down to an examination of the sign of (dv/dt).

By differentiation of equation (4-3):

$$(dv/dt) = \dot{v} = 2x_1\dot{x}_1 + 2x_2\dot{x}_2 \tag{4-4}$$

but since the sign of $\dot{v}$ is needed *along a trajectory path*, it is permissible to substitute for $\dot{x}_1$ and $\dot{x}_2$ their equivalents from equations (4-2) which specifically govern the trajectories. These substitutions produce

$$\begin{aligned} \dot{v} &= 2x_1(-x_1 + x_2) + 2x_2(-x_1 - 2x_2) \\ &= -2(x_1^2 + 2x_2^2) \end{aligned} \tag{4-5}$$

and since this $\dot{v}$ will be negative for any choice of (x_1, x_2), except zero, the steady state at the origin is stable.

Ordinarily, the numbers of a problem do not reduce $\dot{v}$ quite so directly to a negative-definite form, but require some further regrouping. If the foregoing manipulations were for example applied to

$$\dot{x}_1 = -x_1 + x_2 \tag{4-6a}$$

$$\dot{x}_2 = x_1 - 2x_2 \tag{4-6b}$$

the result would be

$$\dot{v} = -2(x_1^2 - 2x_1x_2 + 2x_2^2) \tag{4-7}$$

which is not obviously sign definite. The sought for negative definiteness can be demonstrated by inspection, however, after (4-7) is rearranged by completing the square:

$$\dot{v} = -2[(x_1 - x_2)^2 + x_2^2] \tag{4-8}$$

As a final variation on this theme consider the system modeled by

$$\dot{x}_1 = -x_1 - 2x_2 \tag{4-9a}$$

$$\dot{x}_2 = 10x_1 - x_2 \tag{4-9b}$$

In this case it will not help to complete the square, since

$$\dot{v} = -2(x_1^2 - 16x_1x_2 + x_2^2) \tag{4-10}$$

is in fact not negative-definite (e.g., $\dot{v} = +28$ at $x_1 = x_2 = 1$).

Is the steady state of this system then unstable? Not at all. The analysis has proved only that the system trajectories can move to circles of larger radius as well as to smaller, depending on their position in the phase-plane. The same observation can be made graphically on Figure 4-1a, if we allow ourselves the luxury of looking to the solution of equations (4-9) for con-

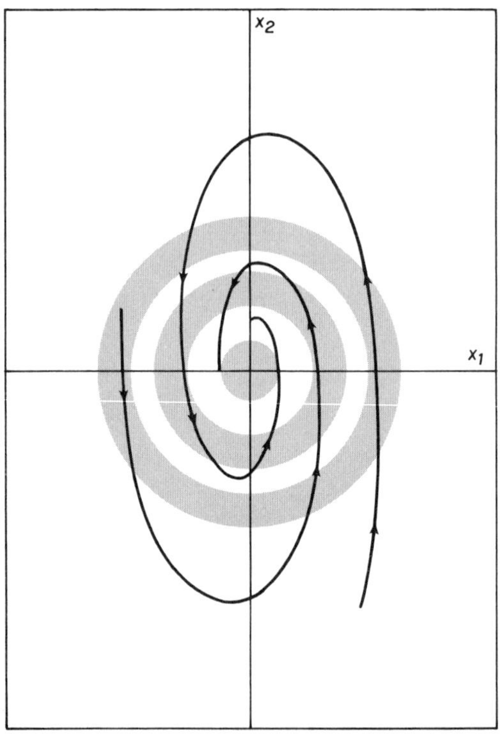

Figure 4-1a. Trajectories of equations (4-9) with superimposed circular v-contours.

firmation. Only if $\dot{v}$ were shown to be positive definite could it justifiably be concluded that the steady state is locally unstable. Lacking a sign definite assurance on $\dot{v}$, the stability or instability questions are left open. In other words, the conditions are sufficient to prove stability (or instability), but not necessary. An inability to find the desirable conditions may be as much a reflection on the analyst as on the system, an uncomfortable conclusion, but not one that leaves the self-confident engineer without hope.

Perhaps the choice of a circle for the tracking contour was too restrictive. If instead of this v, a series of concentric ellipses had been chosen, it might be expected from a glance at Figure 4-1b that the sign definiteness of the corresponding $\dot{v}$ could be demonstrated. That this is indeed the case is a consequence of the development in the next two sections.

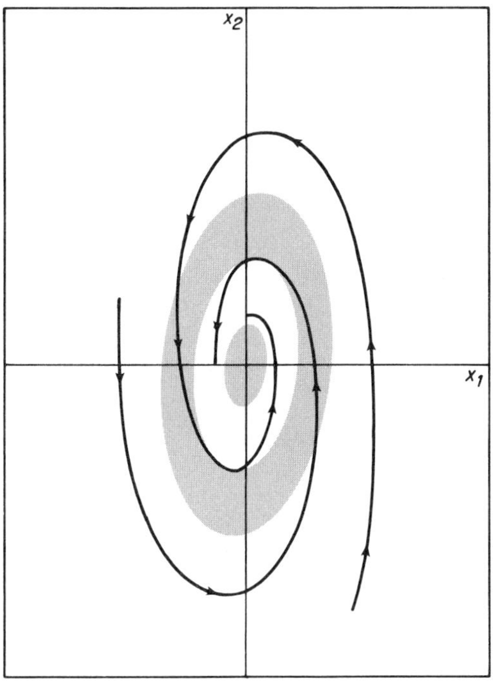

Figure 4-1b. Trajectories of equations (4-9) with superimposed elliptical v-contours.

4-4. Liapunov Functions

The recognition that useful v-functions need not be circular can be developed into a quite general theory. In honor of the maestro (1892) a function $v(\mathbf{x})$ is called a Liapunov function for a given system

$$\dot{\mathbf{x}} = \mathbf{f}(\mathbf{x}), \quad \mathbf{f}(\mathbf{0}) = \mathbf{0}$$

if it provides the sufficient conditions just described; specifically, a series of concentric contours for the family $v = K_i$, decreasing to zero at the steady state, and a time derivative that is negative-definite. Analytically, these requirements can be summarized by:

$$v(\mathbf{x}) > 0, \qquad \mathbf{x} \neq \mathbf{0} \tag{4-11a}$$

$$v(\mathbf{x}) = 0, \qquad \mathbf{x} = \mathbf{0} \tag{4-11b}$$

$$(dv/dt) < 0, \qquad \mathbf{x} \neq \mathbf{0} \tag{4-11c}$$

$$\lim_{\|\mathbf{x}\| \to \infty} v(\mathbf{x}) = \infty \tag{4-11d}$$

The last item of this set guarantees that the contours will be closed, since otherwise v would approach a finite limit as $\|\mathbf{x}\|$ increased in the direction of the open contours. It may be omitted from the list of conditions, if it is substituted by its verbal equivalent. In dealing with circles or ellipses, this point is of course self-evident, requiring no argument for justification.

The condition (4-11c) is frequently relaxed to the semidefinite:

$$(dv/dt) \leq 0, \qquad \mathbf{x} \neq 0 \tag{4-11e}$$

This extension is due to Kalman and Bertram (1960), who observed that the circumstance that $\dot{v}$ might be zero at some $\mathbf{x}$ is not a serious concern so long as the locus of $\dot{v} = 0$ does not overlap a trajectory path. If such a coincidence should occur, the assurance of movement in the direction of decreasing v would be lost, but this is an unlikely event and can in any case be checked by testing whether $\dot{v}(\mathbf{x})$ is a solution to $\dot{\mathbf{x}} = \mathbf{f}(\mathbf{x})$.

To prove formally that the existence of a Liapunov function implies the stability of the steady state, it is convenient to refer to Figure 4-2, constructed as follows:

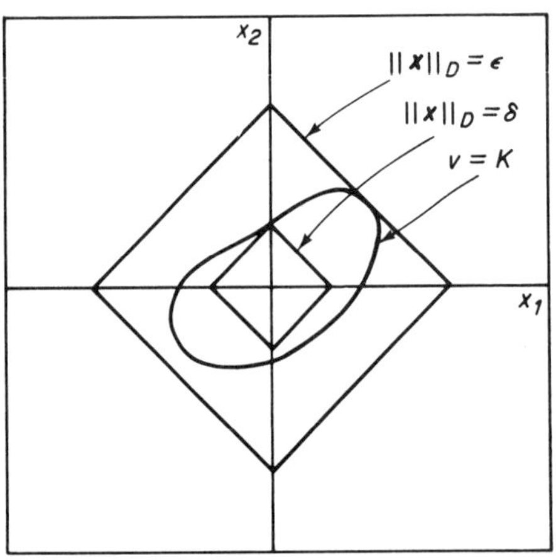

Figure 4-2. Diagram relating Liapunov functions to the definition of stability.

1. Choose any ϵ-contour, $||\mathbf{x}|| = \epsilon$. (Its shape will depend on the norm chosen. The diamond norm is used in the illustration.)
2. Inscribe the largest possible v-contour. (Its shape will depend on the $v(\mathbf{x})$ chosen.)
3. Inscribe a δ-contour, $||\mathbf{x}|| = \delta$ within the v-contour.

Because the conditions (4-11) exclude the possibility of any trajectory crossing the v-contour in an outward direction, no trajectory starting within δ can exceed $v = K$. But then, since this contour is wholly within ϵ, the trajectory can never cross the ϵ boundary, and local Liapunov stability is established from its definition. It may be noted that this line of reasoning uses an entirely arbitrary ϵ, and as such fits quite neatly into the framework of the previously elaborated stability game for two. In more formally mathematical terminology this idea is expressed [Krasovskii (1968)] by requiring that $v(\mathbf{x})$ must *possess an infinitely small upper bound*; that is to say, the argument must be valid for any ϵ, no matter now small.

4-5. Linear Systems

The Liapunov proof of local stability is a non-constructive proof; that is to say, it does not teach us how to find the v-functions that are so essential. It must be admitted that this can be a serious limitation. On the other hand, there are systematic methods by which to approach specific classes of problems, some of these quite broad in their applicability.

As a first step in this direction consider the linear system

$$\dot{\mathbf{x}} = \mathbf{A}\mathbf{x} \tag{4-12}$$

The brief study of several such systems in Section 4-3 suggested that a family of concentric ellipses might be a suitable candidate for a Liapunov function. In the vector-matrix notation the equation for such contours in $\mathbf{x}$-space may be written as

$$v = \mathbf{x}^T\mathbf{P}\mathbf{x} \tag{4-13}$$

where, however, the components of the constant $\mathbf{P}$ matrix must be restricted so as to make v positive-definite. In the two-dimensional (x_1, x_2) plane, for example:

$$\begin{aligned} \mathbf{x}^T\mathbf{P}\mathbf{x} &= [x_1 x_2]\begin{bmatrix} p_{11} & p_{12} \\ p_{21} & p_{22} \end{bmatrix}\begin{bmatrix} x_1 \\ x_2 \end{bmatrix} \\ &= p_{11}x_1^2 + (p_{12} + p_{21})x_1x_2 + p_{22}x_2^2 \end{aligned} \tag{4-14}$$

from which it may be noted that $\mathbf{P}$ may be chosen as symmetric without loss of generality, and (again, by completing the square) with $p_{12} = p_{21}$ the conditions

$$p_{11} > 0 \tag{4-15a}$$

$$p_{11}p_{22} - p_{12}^2 > 0 \tag{4-15b}$$

are necessary and sufficient to make v positive-definite. These requirements are the *Sylvester conditions* for a two-dimensional system and can be generalized for $\mathbf{x}$ of any dimension [Sylvester (1852) and Ogata (1967)]. Any $\mathbf{P}$ that achieves this sign restriction on the quadratic form (4-13) is also referred to as being positive-definite.

When $\mathbf{P}$ is so restricted, all the conditions (4-11) are immediately satisfied with the exception of (4-11c). This last test can also be met, however, by differentiating (4-13) to obtain

$$\dot{v} = \mathbf{x}^T\mathbf{P}\dot{\mathbf{x}} + \dot{\mathbf{x}}^T\mathbf{P}\mathbf{x} \tag{4-16}$$

which by substitution of equation (4-12) becomes

$$\begin{aligned} \dot{v} &= \mathbf{x}^T\mathbf{P}\mathbf{A}\mathbf{x} + \mathbf{x}^T\mathbf{A}^T\mathbf{P}\mathbf{x} \\ &= \mathbf{x}^T(\mathbf{P}\mathbf{A} + \mathbf{A}^T\mathbf{P})\mathbf{x} \end{aligned} \tag{4-17}$$

The essential simplicity of this quadratic form is made even more obvious by letting

$$\mathbf{P}\mathbf{A} + \mathbf{A}^T\mathbf{P} = -\mathbf{Q} \tag{4-18}$$

for now the negative-definite character of

$$\dot{v} = -\mathbf{x}^T\mathbf{Q}\mathbf{x} \tag{4-19}$$

is seen to depend on the association with $\mathbf{Q}$ of precisely the same kind of restrictions that must be applied to $\mathbf{P}$. In brief, the existence of two positive-definite matrices related by equation (4-18) is all that is needed to establish equation (4-13) as a Liapunov function and prove stability. It is always possible to find such a $(\mathbf{P}, \mathbf{Q})$ combination for a stable linear system.

EXAMPLE 4-1. Find a Liapunov function for the system modeled by equations (4-9).

Solution: Use equation (4-18) to find P, given that

$$\mathbf{A} = \begin{bmatrix} -1 & -2 \\ 10 & -1 \end{bmatrix}$$

and assuming as a first trial that $\mathbf{Q} = \mathbf{I}$, the identity matrix. Then:

$$\begin{bmatrix} -1 & +10 \\ -2 & -1 \end{bmatrix}\begin{bmatrix} p_{11} & p_{12} \\ p_{12} & p_{22} \end{bmatrix} + \begin{bmatrix} p_{11} & p_{12} \\ p_{12} & p_{22} \end{bmatrix}\begin{bmatrix} -1 & -2 \\ 10 & -1 \end{bmatrix} = \begin{bmatrix} -1 & 0 \\ 0 & -1 \end{bmatrix}$$

which is the same as the simultaneous set:

$$-p_{11} + 10p_{12} - p_{11} + 10p_{12} = -1$$

$$-p_{12} + 10p_{22} - 2p_{11} - p_{12} = 0$$

$$-2p_{11} - p_{12} - p_{12} + 10p_{22} = 0$$
$$-2p_{12} - p_{22} - 2p_{12} - p_{22} = -1$$

One of these equations (the second or third) is redundant; the other three yield the solution

$$\mathbf{P} = \frac{1}{42}\begin{bmatrix} 61 & 4 \\ 4 & 13 \end{bmatrix}$$

and the Liapunov function

$$v = \mathbf{x}^T\mathbf{P}\mathbf{x} = \frac{1}{42}(61x_1^2 + 8x_1x_2 + 13x_2^2) \tag{4-20}$$

the contours of which are those plotted in Figure 4-1b. As an alternative attempt to meet condition (4-18) one might consider letting $\mathbf{P} = \mathbf{I}$ and solving for $\mathbf{Q}$. Such an analysis leads to the circles

$$v = \mathbf{x}^T\mathbf{I}\mathbf{x} = x_1^2 + x_2^2$$

and from equation (4-18):

$$\mathbf{Q} = -(\mathbf{A}^T + \mathbf{A}) = \begin{bmatrix} 2 & -8 \\ -8 & -1 \end{bmatrix}$$

This $\mathbf{Q}$ does not satisfy condition (4-15b), confirming that the circles of Figure 4-la are inadequate as Liapunov contours.

This comparison demonstrates a relative advantage in choosing $\mathbf{Q}$ rather than $\mathbf{P}$. There is yet another reason for this preference which though not demonstrable in this particular example, has been rigorously proved by Parks (1963): if for a proper choice of $\mathbf{Q}$ the matrix $\mathbf{P}$ turns out not to be positive-definite, the system steady state is *unstable*. That is to say, the paired conditions on $\mathbf{P}$ and $\mathbf{Q}$ that we seek are both necessary and sufficient to establish stability, and as a result a single trial computation is all that is needed to settle the stability question. That the same cannot be said of the procedure that starts with the choice of $\mathbf{P}$ is clear from this example.

4-6. The Fundamental Linearization Theorem

In Chapter 3 it was reasoned that a nonlinear system of the form

$$\dot{\mathbf{x}} = \mathbf{f}(\mathbf{x}) \tag{4-21}$$

could be approximated in the region of the steady state $\mathbf{x} = \mathbf{0}$ by the much more easily manageable linear model

$$\dot{\mathbf{x}} = \mathbf{A}\mathbf{x} \tag{4-22}$$

By the nature of the approximation used, it would be expected that its quality, that is, its adequacy in predicting dynamic behavior, would improve as one approached the steady state, and that in the limit the local stability of equation (4-21) could be determined from analysis of equation (4-22). The formal proof of this conjecture is called the *Fundamental theorem.*

Consider as a starting point that the nonlinear equation (4-21) has been expanded by a Taylor series or other means into a linear term and a nonlinear remainder

$$\dot{\mathbf{x}} = \mathbf{f}(\mathbf{x}) = \mathbf{A}\mathbf{x} + \mathbf{g}(\mathbf{x}) \tag{4-23}$$

If the linear term is to approximate the entire nonlinear function, it seems reasonable to expect that $\mathbf{g}(\mathbf{x})$ will be small as the steady state is approached. Since $\mathbf{g}$ is a vector, its required "smallness" can be described by a norm equation

$$\lim_{\|\mathbf{x}\|\to 0} \frac{\|\mathbf{g}(\mathbf{x})\|}{\|\mathbf{x}\|} = 0 \tag{4-24}$$

that is, the function $\|\mathbf{g}(\mathbf{x})\|$ must decrease to zero more rapidly than does $\|\mathbf{x}\|$, producing a graph such as Figure 4-3. The same idea can also be

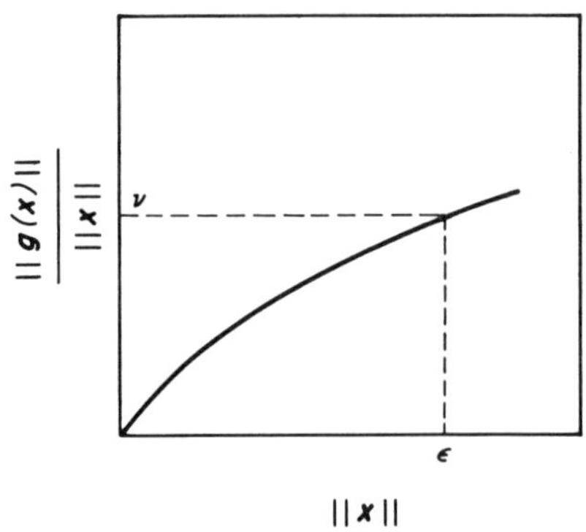

Figure 4-3. Graphical illustration of norm constraint needed to justify linearization.

restated in terms of an inequality bound, since for any $\|\mathbf{x}\| < \epsilon$, a positive ν can always be found (see Figure 4-3) such that

$$\frac{\|\mathbf{g}(\mathbf{x})\|}{\|\mathbf{x}\|} \leq \nu \tag{4-25}$$

or

$$\|\mathbf{g}\| \leq \nu \|\mathbf{x}\| \leq \nu\epsilon \tag{4-26}$$

From equation (3-52) it may be observed that

$$\mathbf{x} = \mathbf{X}\mathbf{x}(0) + \int_0^t \mathbf{X}(t-\sigma)\mathbf{g}(\sigma)d\sigma \tag{4-27}$$

Taking norms on each side:

$$\|\mathbf{x}\| = \|\mathbf{X}\mathbf{x}(0) + \int_0^t \mathbf{X}(t-\sigma)\mathbf{g}(\sigma)d\sigma\|$$

Applying in succession, the triangle inequality (3-9d),

$$||\mathbf{x}|| \leq ||\mathbf{X}\mathbf{x}(0)|| + ||\int_0^t \mathbf{X}(t-\sigma)\mathbf{g}(\sigma)d\sigma||$$

the integral inequality (3-9e),

$$||\mathbf{x}|| \leq ||\mathbf{X}\mathbf{x}(0)|| + \int_0^t ||\mathbf{X}(t-\sigma)\mathbf{g}(\sigma)||\, d\sigma$$

and the Schwartz inequality (3-13f):

$$||\mathbf{x}|| \leq ||\mathbf{X}||\,||\mathbf{x}(0)|| + \int_0^t ||\mathbf{X}(t-\sigma)||\,||\mathbf{g}(\sigma)||\, d\sigma \tag{4-28}$$

The matrix **X** is obtained from the linear analysis of (4-22); this implies (as was illustrated in Example 3-1) that provided (4-22) is stable, it is always possible to find positive c and μ to bound the norm of **X** by

$$||\mathbf{X}|| \leq ce^{-\mu t} \tag{4-29}$$

Using this fact inequality (4-28) becomes

$$||\mathbf{x}|| \leq ce^{-\mu t}||\mathbf{x}(0)|| + \int_0^t ce^{-\mu(t-\sigma)}\,||\mathbf{g}(\sigma)||\, d\sigma \tag{4-30}$$

and including the bound of (4-26):

$$||\mathbf{x}|| \leq ce^{-\mu t}||\mathbf{x}(0)|| + \int_0^t c\nu\epsilon e^{-\mu(t-\sigma)}\, d\sigma \tag{4-31}$$

(This last substitution under the integral is permissible because a norm is always positive.) Now by formal integration

$$\begin{aligned} ||\mathbf{x}|| &\leq ce^{-\mu t}||\mathbf{x}(0)|| + \frac{c\nu\epsilon}{\mu}(1 - e^{-\mu t}) \\ &\leq \frac{c\nu\epsilon}{\mu} + ce^{-\mu t}\left[||\mathbf{x}(0)|| - \frac{\nu\epsilon}{\mu}\right] \end{aligned} \tag{4-32}$$

The proof of stability may now be stated clearly in terms of the Liapunov definition and the challenger-defender viewpoint. In response to the challenger's ϵ, the defender makes the following move. He chooses a μ and c to satisfy (4-29), sets

$$\nu = \mu/c \tag{4-33a}$$

and

$$||\mathbf{x}(0)|| = \delta = \epsilon/c \tag{4-33b}$$

Then he may substitute these choices into (4-32) to obtain

$$||\mathbf{x}(t)|| \leq \epsilon \tag{4-34}$$

which establishes stability by reference to the defining conditions (4-1).

The foregoing does assume that the ϵ-region lies entirely within the ν-region, but this is in no way a restriction, since as pointed out earlier, the challenger's strength always lies in the choice of small ϵ, not a large one.

If the challenger were to pick an ϵ too large to handle by the prior argument, the defender would simply work on a proof for a smaller region and then argue that trajectories trapped within that smaller region must also be confined within the larger.

The Fundamental theorem does break down, however, if the solution to the linearized system involves zero eigenvalues. This marginal case can not properly be considered unstable; nevertheless, it allows (4-29) to be satisfied only by $\mu = 0$ and as a consequence excludes any finite μ for equation (4-33a). The entire nonlinear system can in this circumstance be stable or unstable, but it cannot be studied by linearization. To summarize, the nonlinear system (4-23) has a locally stable steady state, if the solution to the approximation equation (4-22) is asymptotically stable, and if, in addition, condition (4-24) applies.

It is necessary for this analysis to be able to decompose the nonlinear $\mathbf{f}(\mathbf{x})$ into a linear term and a nonlinear remainder, but since this decomposition is not unique, the most obvious choice may not be the best. In examining for example the reactor system modeled by equations (3-6), a direct comparison with (4-23) might lead one to identify:

$$\mathbf{A} = \begin{bmatrix} \left(-\frac{q}{V} - \frac{U}{VC_p}\right) & 0 \\ 0 & -\frac{q}{V} \end{bmatrix}; \ \mathbf{g}(\mathbf{x}) = \begin{bmatrix} +r \\ -r \end{bmatrix}$$

This would be an unfortunate choice, for although it is very simple to show that the eigenvalues of the linear problem defined by this $\mathbf{A}$ are real and negative, it is not easy to establish condition (4-24) for this $\mathbf{g}(\mathbf{x})$. For the Euclidian norm this latter part would call for establishing conditions under which

$$\lim_{\substack{x_1 \to 0 \\ x_2 \to 0}} \frac{\sqrt{r^2 + (-r)^2}}{\sqrt{x_1^2 + x_2^2}} = \lim_{\substack{x_1 \to 0 \\ x_2 \to 0}} \sqrt{2}\, \frac{r(x_1, x_2)}{\sqrt{x_1^2 + x_2^2}} = 0$$

not at all a simple demonstration for kinetic forms resembling equation (1-6). By meeting the linear requirement in the most opportunistic manner, a most unpleasant secondary condition was created. This difficulty can be avoided if the nonlinearity condition is satisfied first. In shifting the primary attention in this way, the final burden will fall on the linear analysis where it can much more readily be handled.

The needed change in orientation is greatly aided by assuming that the nonlinear $\mathbf{f}(\mathbf{x})$ can be expanded in a multi-variable Taylor series

$$\mathbf{f}(\mathbf{x}) = \mathbf{J}_s \mathbf{x} + \sum_{k=2}^{\infty} \frac{1}{k!} \left(\frac{\partial^k \mathbf{f}}{\partial \mathbf{x}^k}\right)_s \mathbf{x}^k \tag{4-35}$$

where

$$\mathbf{J} = \begin{bmatrix} \frac{\partial f_1}{\partial x_1} & \frac{\partial f_1}{\partial x_2} & \cdots & \frac{\partial f_1}{\partial x_n} \\ \frac{\partial f_2}{\partial x_1} & \frac{\partial f_2}{\partial x_2} & \cdots & \frac{\partial f_2}{\partial x_n} \\ \cdots \\ \cdots \\ \frac{\partial f_n}{\partial x_1} & \frac{\partial f_n}{\partial x_2} & \cdots & \frac{\partial f_n}{\partial x_n} \end{bmatrix} = \frac{\partial \mathbf{f}}{\partial \mathbf{x}} \tag{4-36}$$

is the so-called *Jacobian* matrix of first partial derivatives, the subscript s indicates evaluation at the steady state, and the summation is the collection of all the second- and higher-order terms in the infinite series. The requirement of a Taylor-series expansion is ordinarily not a serious restriction, since most nonlinear functions of interest in reactor engineering have continuous derivatives of all orders.

Because the terms in the vector summation all include squares and higher powers in $\mathbf{x}$, they will all approach zero faster than $\mathbf{x}$, automatically satisfying the needed condition (4-24). With this observation in mind, letting

$$\mathbf{J}_s = \mathbf{A} \tag{4-37a}$$

$$\sum_{k=2}^{\infty} \frac{1}{k!}\left(\frac{\partial^k \mathbf{f}}{\partial \mathbf{x}^k}\right)_s \mathbf{x}^k = \mathbf{g}(\mathbf{x}) \tag{4-37b}$$

reduces the nonlinear problem to an "equivalent" linear one.

EXAMPLE 4-2. Is the following system locally stable about its steady state at $\mathbf{x} = \mathbf{0}$?

$$\dot{x}_1 = -x_1^3 + x_2^2$$

$$\dot{x}_2 = -x_2(1 + x_1)$$

Solution: The Jacobian matrix is

$$\mathbf{J} = \begin{bmatrix} -3x_1^2 & 2x_2 \\ -x_2 & -(1 + x_1) \end{bmatrix}$$

Then

$$\mathbf{A} = \mathbf{J}_s = \begin{bmatrix} 0 & 0 \\ 0 & -1 \end{bmatrix}$$

When as is here the case, the $\mathbf{A}$ matrix is diagonal, the solutions to equation (3-28) are simply the diagonal entries. Thus, $\lambda = 0, -1$. The linear system is therefore stable, *but not asymptotically stable*, and the Fundamental theorem is not applicable. Local stability can be established, nevertheless, by the circular Liapunov function (4-3), from which

$$\dot{v} = -2(x_1^4 + x_2^2)$$

thus satisfying the conditions (4-11).

4-7. Local Stability of the CSTR

Having finally justified with some rigor the process of linearization, and having clarified what nonlinear stability implies, it is worthwhile to develop local stability criteria for the nonlinear CSTR, which means in effect, stability criteria for the linearized CSTR.

The starting point for a second-order system is equation (3-40), from which it is clear that both eigenvalues will have negative real parts if and only if

$$a_{11} + a_{22} < 0 \tag{4-38a}$$

$$a_{11}a_{22} - a_{12}a_{21} > 0 \tag{4-38b}$$

These inequalities can be rewritten as stability criteria for the linearized CSTR by making use of equations (3-19) in terms of the reactor parameters. Substitution gives

$$-2\frac{q}{V} - \frac{U}{C_pV} + \frac{\partial r}{\partial x_1} - \frac{\partial r}{\partial x_2} < 0 \tag{4-39a}$$

$$\left[-\frac{q}{V} - \frac{U}{C_pV} + \frac{\partial r}{\partial x_1}\right]\left[-\frac{q}{V} - \frac{\partial r}{\partial x_2}\right] - \left[\frac{\partial r}{\partial x_2}\right]\left[-\frac{\partial r}{\partial x_1}\right] > 0 \tag{4-39b}$$

which may be rearranged to a pair with identical left sides:

$$\left[\frac{\partial r}{\partial x_1} - \frac{\partial r}{\partial x_2}\right] - \left[\frac{q}{V} + \frac{U}{C_pV}\right] < \frac{q}{V} \tag{4-40a}$$

$$\left[\frac{\partial r}{\partial x_1} - \frac{\partial r}{\partial x_2}\right] - \left[\frac{q}{V} + \frac{U}{C_pV}\right] < \frac{U}{qC_p}\frac{\partial r}{\partial x_2} \tag{4-40b}$$

For kinetics obeying rate equation (1-6a) substitution of equations (3-23) into (4-40) produces:

$$\left[\frac{\Delta HQ}{C_pT_s^2} - \frac{n}{C_s}\right](C_0 - C_s) - \left[1 + \frac{U}{qC_p}\right] < 1 \tag{4-41a}$$

$$\left[\frac{\Delta HQ}{C_pT_s^2} - \frac{n}{C_s}\right](C_0 - C_s) - \left[1 + \frac{U}{qC_p}\right] < \frac{nU}{qC_pC_s}(C_0 - C_s) \tag{4-41b}$$

It is a simple matter to determine whether or not a particular steady state is stable by applying these test inequalities. When this is done for a set of parameter combinations, the plane of the operating curves can be separated into regions of different stability character. To illustrate this with a specific system, recall the parameters used for Figure 2-7. With those values and utilizing equation (2-90), it is convenient to express the first term of inequalities (4-41) as

$$
\begin{aligned}
\frac{\Delta HQ}{C_p T_s^2} &= \frac{1}{C_0}\frac{\Delta HC_0}{C_p Q}\left(\frac{Q}{T_s}\right)^2 \\
&= \frac{1}{C_0}\frac{\Delta HC_0}{C_p Q}\left[\ln\left(\frac{k_0 V}{q}\frac{C_s}{(C_0 - C_s)}\right)\right]^2 \\
&= \frac{0.0133}{C_0}\left[\ln\left(\frac{C_s}{C_0 - C_s} \times 10^{11}\right)\right]^2
\end{aligned}
$$

By scanning the various (C/C_0) and (U/qC_p) combinations, the prior figure may be reproduced as Figure 4-4, but now modified to include the *region*

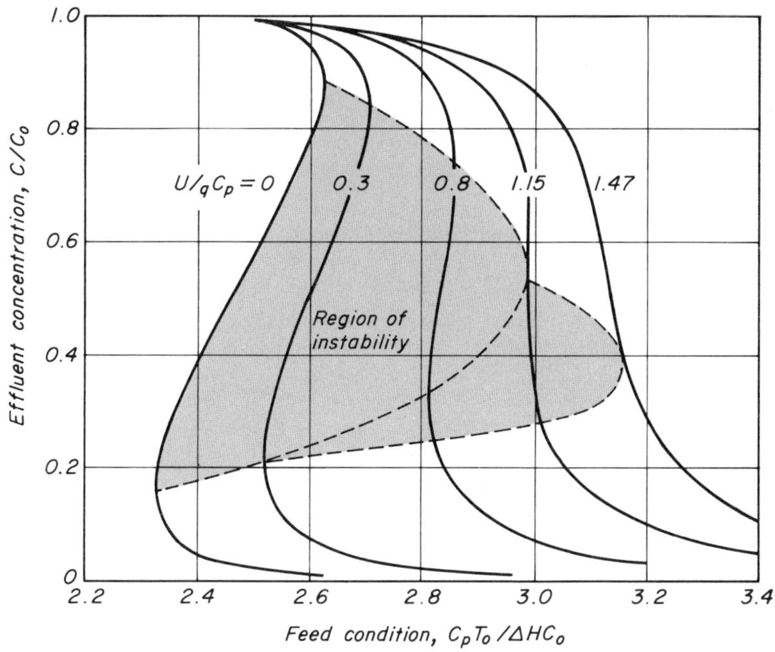

Figure 4-4. Steady-state operating curves for the temperature-dependent CSTR with region of instability.

of instability. The two dashed boundaries of this region arise from the inequality conditions that must both be satisfied for stability. It is especially interesting to note that there is an appreciable range of feed conditions where for $(U/qC_p) > 1.15$ the steady state is *unique and unstable*. This is the range of limit cycle behavior described in Section 3-2. Sustained oscillations of this kind were for example experimentally obtained by Bush (1969) in his extensive study of the gas-phase chlorination of methyl chloride.

EXAMPLE 4-3. Check for stability (a) each of the three steady states of the Aris and Amundson system given in Example 2-3:

Steady state 1: $T_s = 354°\text{K}$, $C_s = 0.964$ g-mol/liter
Steady state 2: $T_s = 400°\text{K}$, $C_s = 0.500$
Steady state 3: $T_s = 441°\text{K}$, $C_s = 0.0885$

(b) the single steady state of Berger's data (also given in Example 2-3):

$$T_s = 550°\text{R}, \quad C_s = 0.165 \text{ g-mol/ft}^3$$

and (c) the single steady state at

$$T_s = 300°\text{K}, \quad C_s = 0.5 \text{ g-mol/liter}$$

computed by Kerr (1966) for the temperature-dependent CSTR with parameters:

$$\Delta H = 79{,}200 \text{ cal/g-mol} \qquad C_0 = 1 \text{ g-mol/liter}$$
$$Q = 12{,}500°\text{K} \qquad U = 30 \text{ cal/sec°K}$$
$$C_p = 1{,}000 \text{ cal/liter°K} \qquad q = 0.03 \text{ liter/sec}$$
$$n = 2$$

Solution: Numerical evaluation of the terms in equalities (4-41) may be summarized in tabular form as follows:

Steady state	LHS of (4-41)	RHS of (4-41a)	RHS of (4-41b)	Conclusion
Part (a) 1	−1.46	1	0.037	Stable
Part (a) 2	+3.25	1	1.0	Unstable
Part (a) 3	−2.91	1	10.3	Stable
Part (b)	−0.99	1	0.032	Stable
Part (c)	+1.50	1	2.0	Unstable

The conclusions are of course in agreement with the findings in the original references. Kerr's unstable steady state is an unstable focus which develops into a limit cycle, the Aris and Amundson instability is a saddle point as observed earlier. It should be noted that in general checks are needed on both inequalities before a conclusion is drawn, but that the calculation can be stopped if the LHS value is negative, since the RHS values are always positive.

An already familiar physical interpretation may be made of inequality (4-39b) if instead of (4-40b) it is rearranged to:

$$qC_p + U > \left[\frac{VC_p \dfrac{\partial r}{\partial x_1}}{1 + \dfrac{V}{q}\dfrac{\partial r}{\partial x_2}}\right] \tag{4-42}$$

The chain rule of calculus may be written as

$$\frac{dr}{dx_1} = \frac{\partial r}{\partial x_2}\frac{dx_2}{dx_1} + \frac{\partial r}{\partial x_1} \tag{4-43}$$

But from equation (3-6b) at steady state

$$x_2 = -\frac{V}{q}r \tag{4-44}$$

and by differentiation

$$\frac{dx_2}{dx_1} = -\frac{V}{q}\frac{dr}{dx_1} \tag{4-45}$$

Combining (4-43) with (4-45) and rearranging gives

$$\frac{dr}{dx_1} = \left[\frac{\dfrac{dr}{dx_1}}{1 + \dfrac{V}{q}\dfrac{\partial r}{\partial x_2}}\right] \tag{4-46}$$

which may be substituted into (4-42) to obtain

$$qC_p + U > VC_p\frac{dr}{dx_1} = \Delta HV\frac{dR}{dT} \tag{4-47}$$

This inequality is identical with (2-47) established earlier as sufficient for uniqueness, and interpreted in terms of the sensitivities to temperature of heat generation and heat removal. The uniqueness condition was more restrictive in that it had to apply at all temperatures of possible interest, whereas the stability condition need only apply at the steady state. As a result a CSTR system can have a unique steady state that is unstable (if, for example, (4-47) is true at all temperatures but (4-40a) is violated at the steady state), or it can have a stable steady state that is not a unique steady state (if (4-47) is satisfied at the steady state but not at other temperatures.) The physical heat balance argument is in fact older than the stability analysis presented here and goes back at least to van Heerden (1953).

4-8. Other Direct Methods

The Liapunov Second Method is called *direct* because it treats the stability problem as a question in its own right without first solving a differential equation for explicit time dependence. As we have seen, the technique is applicable to linear or nonlinear systems. For the former the option also exists of examining the signs of the real parts of the eigenvalues, but since this seems to require a complete solution of the characteristic equation (3-28), it is not a welcome prospect for higher-order systems. In point of fact this complete solution is not necessary because it is possible to find the signs of interest without solving for the λ's numerically. The technique is due to Schwartz (1956), Kalman and Bertram (1960), and Parks (1963) and connects the Liapunov arguments with the famous Routh-Hurwitz conditions.

Basic to Schwartz's development is first of all the well-known observation that any matrix $\mathbf{B}$ will have the same eigenvalues as $\mathbf{A}$ if they are related by

$$\mathbf{B} = \mathbf{TAT}^{-1} \tag{4-48}$$

a so-called *similarity transformation*. The evidence for this statement is that

$$\begin{aligned}\det(\lambda\mathbf{I} - \mathbf{B}) &= \det(\lambda\mathbf{I} - \mathbf{TAT}^{-1}) \\ &= \det(\mathbf{T}\lambda\mathbf{IT}^{-1} - \mathbf{TAT}^{-1}) \\ &= \det\mathbf{T}\det(\lambda\mathbf{I} - \mathbf{A})\det\mathbf{T}^{-1} \\ &= \det(\lambda\mathbf{I} - \mathbf{A})\end{aligned} \tag{4-49}$$

Since $\mathbf{A}$ and $\mathbf{B}$ have identical characteristic equations, they must have identical λ's, which is in effect to say that no difference exists between the stability character of the systems

$$\dot{\mathbf{x}} = \mathbf{Ax} \tag{4-50a}$$

and

$$\dot{\mathbf{x}} = \mathbf{Bx} \tag{4-50b}$$

More to the point, Chen and Chu (1966) have presented rules (repeated with examples in Appendix A) by which a given $\mathbf{A}$ can be transformed into a $\mathbf{B}$ of the Schwartz form:

$$\mathbf{B} = \begin{bmatrix} 0 & 1 & 0 & 0 & \cdots & 0 & 0 & 0 \\ -b_n & 0 & 1 & 0 & \cdots & 0 & 0 & 0 \\ 0 & -b_{n-1} & 0 & 1 & \cdots & 0 & 0 & 0 \\ \cdots & & & & & & & \\ \cdots & & & & & & & \\ 0 & 0 & 0 & 0 & \cdots & -b_3 & 0 & 1 \\ 0 & 0 & 0 & 0 & \cdots & 0 & -b_2 & -b_1 \end{bmatrix} \tag{4-51}$$

This form is of particular interest because if they are all positive the various b's in this array can be used to form the positive-definite Liapunov function $v = \mathbf{x}^T\mathbf{Px}$, where

$$\mathbf{P} = \begin{bmatrix} b_1b_2b_3\cdots b_n & 0 & \cdots & 0 & 0 & 0 \\ 0 & b_1b_2b_3\cdots b_{n-1} & \cdots & & & \\ \cdots & & & & & \\ \cdots & & & & & \\ 0 & 0 & \cdots & b_1b_2b_3 & 0 & 0 \\ 0 & 0 & \cdots & 0 & b_1b_2 & 0 \\ 0 & 0 & \cdots & 0 & 0 & b_1 \end{bmatrix} \tag{4-52}$$

As was shown in Section 4-5:

$$\dot{v} = \mathbf{x}^T(\mathbf{PB} + \mathbf{B}^T\mathbf{P})\mathbf{x} \tag{4-53}$$

but as a result of the many zero entries in **B** this reduces to

$$\dot{v} = -2b_1^2 x_n^2 \tag{4-54}$$

Finally, because an $x_n = 0$ is not a solution, $\dot{v}$ must be negative definite and the system stable. The only requirement was the necessary and sufficient condition that all the b's > 0.

Another direct method, simpler to apply because it does not require the transformation of **A** into any special form, is based on the fact that all the eigenvalues of **A** are located in the complex plane within or on the circles with

$$\text{centers at } a_{ii}$$

and

$$\text{radii equal to } \sum_{\substack{j \neq i \\ i=1}}^{n} |a_{ij}|$$

Bodewig (1959) credits this theorem to Gerschgorin (1931); MacFarlane (1964) showed how it may be applied to stability analysis in the form of a fairly conservative sufficient condition.

All the eigenvalues of **A** will be located in the left half-plane (hence the system will be stable) if the diagonal entries of **A** are all negative, and

$$|a_{ii}| > \sum_{\substack{j \neq i \\ i=1}}^{n} |a_{ij}| \tag{4-55}$$

for all $i = 1, 2 \ldots n$. The same theorem applies to $\mathbf{A}^T$ since the transpose matrix has the same eigenvalues as **A**. In spite of its conservatism this approach is so simple that it deserves some attention in a linear problem. In particular it should be considered when the off-diagonal elements of a matrix are small in magnitude in comparison to the negative diagonal terms, or if the system matrix has many zero elements. It may also be useful in conjunction with numerical methods in which sequential steps increase the magnitude of diagonal entries at the expense of the off-diagonal elements [Carnahan, Luther, and Wilkes (1968)].

EXAMPLE 4-4. Is the system $\dot{\mathbf{x}} = \mathbf{A}\mathbf{x}$ stable where

$$\mathbf{A} = \begin{bmatrix} 6 & -1 & -5 \\ 0 & 3 & 1 \\ 6 & 1 & 6 \end{bmatrix}$$

Solution: Following the rules of Appendix A, it is shown in Example A-1 that

$$\mathbf{B} = \begin{bmatrix} 0 & 1 & 0 \\ -12.4 & 0 & 1 \\ 0 & -88.6 & 15 \end{bmatrix}$$

from which by inspection: $b_1 = -15$, $b_2 = 88.6$, $b_3 = 12.4$. Since these b's are not all positive, the system is unstable (in fact, it has three unstable λ's.) The same conclusion could also have been reached by applying Gerschgorin's theorem to the columns of matrix **A**, since the three eigenvalues of **A** must be located in the complex plane within or on the circles centered at 6, 3, 6 with radii 6, 2, 6, respectively. The possibility of a zero eigenvalue may be eliminated by the observation that det $\mathbf{A} \neq 0$.

EXAMPLE 4-5. Apply Gerschgorin's theorem to estimate the control gain K_p needed to establish stability for Luyben's model at the steady state: $T_s = 600°R$, $C_s = 0.246$ lb/lb.

Solution: Equations (3-19) describe Luyben's model, if the first coefficient is modified by the substitution of equation (2-108a). Then application of inequality (4-55) calls for:

$$|a_{11}| > |a_{21}|$$

$$|a_{22}| > |a_{12}|$$

which becomes via equations (3-19)

$$\frac{q}{V} + \frac{\Delta H_v K_p}{C_p V} > 2\frac{\partial r}{\partial x_1}$$

$$\frac{q}{V} > 0$$

The second of these is obviously always true, and the use of (3-22a) produces the single sufficient condition

$$K_p > \frac{qC_p}{\Delta H_v}\left[\frac{2\Delta HQ(C_0 - C_s)}{C_p T_s^2} - 1\right]$$

The same condition also assures that a_{11} and a_{22} are both negative. Numerical substitution of the values provided with Example 2-8 gives

$$K_p > 87.3 \frac{(\text{lb/hr})}{°\text{F}}$$

REFERENCES

BODEWIG, E., *Matrix Calculus*, 2nd Ed., North Holland, Amsterdam (1959) p. 67.

BUSH, S. F., *Proc. Roy. Soc.*, **A309**, 1 (1969).

CARNAHAN, B., H. A. LUTHER, and J. C. WILKES, *Applied Numerical Methods*, New York: Wiley (1964).

CHEN, C. F. and H. CHU, *IEEE Trans. Automatic Control*, **AC-11**, No. 2, 303 (April, 1966).

GERSCHGORIN, S., *Bull. Acad. Sci., USSR, Leningrad, Class Math., Ser.* 7, 749 (1931); discussed by O. Taussky, *Amer. Math. Monthly*, **54**, 672 (1949).

KALMAN, R. E. and J. E. BERTRAM, *ASME Trans. Ser. D, J. Basic Eng.*, **82**, 371, 394 (1960).

KERR, R. B., M. S. Thesis, University of Manchester, England (Oct. 1966).

KRASOVSKII, N. N., *Stability of Motion*, Stanford University Press (1963).

LIAPUNOV, A. M., "Probleme General de la Stabilité du Mouvement" (1892); reprinted in *Ann. Math. Study*, No. 17, Princeton University Press (1949).

MACFARLANE, A. G. J., *Engineering Systems Analysis*, London: Harrap and Co. (1964) p. 189.

OGATA, K., *State Space Analysis of Control Systems*, Englewood Cliffs, New Jersey: Prentice-Hall, Inc. (1967) p. 277.

PARKS, P. C., *Proc. Joint Automatic Control Conference*, p. 471–476 (1963).

SCHWARTZ, H. R., *Z. Angew. Math. Phys.*, **7**, 473 (1956).

SYLVESTER, J. J., *Phil. Mag. London.* 4 (1852).

VAN HEERDEN, C., *Ind. Eng. Chem.* **45**, 1242 (1953).

Regions of Stability 5

5-1. Finite Disturbances

Besides formalizing an otherwise intuitive concept, the Liapunov definition of stability has the great virtue that it leads to the direct method of analysis, avoiding the usually ardous task of differential equation integrations. For all that, it must be admitted that from the point of view of practical engineering application, the proof that a steady state is locally stable is only of limited use in itself. This is precisely because the information is local, and a guarantee of such behavior says nothing about the behavior of a trajectory while on its way to a steady state.

A typical problem might arise as follows: a CSTR, designed to operate in a steady-state condition, is expected to be subject to impulsive disturbances which would instantaneously carry it to a new state (C_i, T_i). Anticipating such an event, it is desired to know whether and how the system will recover after the disturbance has ceased acting. In effect the question is concerned with the trajectory leaving initial condition (C_i, T_i).

Reference to Figure 3-4 shows that for a system with multiple steady states even relatively small disturbances from steady state A can give rise to trajectories that go to state C. If the system is known to have a unique steady state which is asymptotically stable, it is more likely that the trajectory will eventually return to the desired steady state, but Figure 3-5 demonstrates that even for this case other behavior is possible. Another counter example may be found in the well-known van der Pol equation which has an unstable limit cycle [see e.g. Struble (1962)]. The same situation can arise from movement between the steady-state points that correspond to different values of a parameter. If points A and B are the steady states in the phase-plane that arise from flow rates q_A and q_B, respectively, a step change:

$$q = \begin{cases} q_{A'}, & t < 0 \\ q_{B'}, & t \geq 0 \end{cases}$$

will drive the system from state A which will be its initial condition at $t = 0$, but the trajectory from A will *not* necessarily be asymptotically stable to B.

In any event and whatever the number of steady states, the numerical expectations are for practical design at least as important as the simple yes-or-no local result. Will the trajectory from the given initial condition return to the desired steady state without exceeding permissible limits? A case with a single steady state which is locally stable, but is practically unexceptable, is illustrated pictorially in Figure 5-1. It is evident that steady-state analysis does

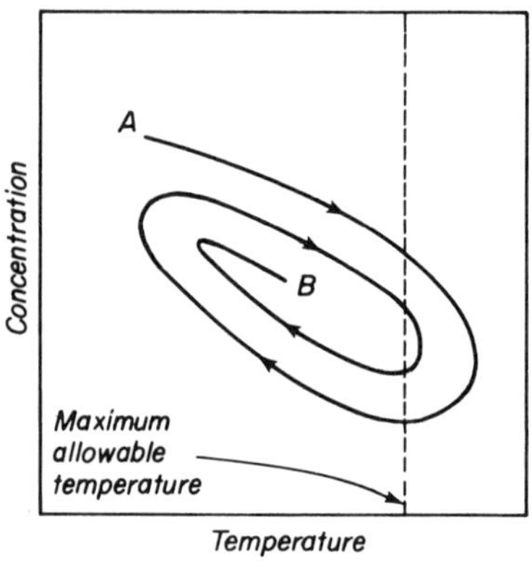

Figure 5-1. Hypothetical system which is Liapunov stable but not practically stable.

not by itself provide the dynamic information needed for design, and the local stability considerations need badly to be extended in some sense.

Reviewing the developments in Section 4-4 that were used to prove local stability, it may be noted that the argument rested on an interpretation of Figure 4-2, in particular on the fact that no trajectory could cross a v-contour in the outward direction. But if that is the case, the same logic establishes the area enclosed by the v-contour as a *region of stability*, that is, a region from which no trajectories can escape. Further, since the same Liapunov function also establishes local asymptotic stability, the region will be a *region of asymptotic stability* (RAS). The distinction between a region of stability and an RAS can only be recognized in the unusual circumstance that a trajectory exists along which $\dot{v} = 0$. If condition (4-11e) is used instead of (4-11c) to include this possibility, the resulting set of conditions would be sufficient to establish a region of stability, but not asymptotic stability. A region of asymptotic stability is also sometimes referred to as a *region of attraction*. When such a region extends over the entire range of possible trajectories, the system is said to be *globally* stable.

Since a successful Liapunov function produces a family of concentric v-contours, it is a simple matter to choose an RAS as large or as small as one wishes. Contrary to what might at first thought seem obvious, neither the largest RAS nor the smallest is necessarily the best. To see that this is so it is helpful to consider two extreme cases. If the RAS is chosen so large that it virtually encompasses the entire phase-plane, it establishes that all trajectories ultimately converge on the steady state, and little more. If, on the other hand, a very small RAS is chosen the conclusion is again hardly better than that implied by local stability, since the feature of the RAS which reflects on finite disturbances has not been exploited to any extent. Typically then the engineer will be most interested in one or more intermediate v-contours, those that bound trajectories of interest because they correspond in numerical terms to the magnitudes of expected disturbances, allowable responses, or both.

5-2. Regions Constrained by the $\dot{v}$ Condition

Of the four conditions (4-11) which confirm that a chosen v is a Liapunov function, the truth of the third (4-11c) is usually the most difficult to demonstrate. For some problems $\dot{v}(\mathbf{x})$ will indeed be negative-definite over some range of $\mathbf{x}$, but not for all $\mathbf{x}$. This would be entirely satisfactory for the local stability proof, provided that the steady state is within the acceptable region defined by $\dot{v}(\mathbf{x}) < 0$ and not on its boundary. Such a condition on $\dot{v}$ can serve as well to fix an RAS, but whether or not such a region is adequately large for engineering purposes will depend solely on the physical reality and numbers of the problem.

EXAMPLE 5-1. Find the largest possible RAS for the system

$$\dot{x}_1 = -x_1 + x_2 \tag{5-1a}$$

$$\dot{x}_2 = -x_1^3 - x_2 \tag{5-1b}$$

using the circular v-function.

Solution: Note as a preliminary that this system of equations has only a single steady state at $x_1 = x_2 = 0$, since the other roots of

$$0 = -x_1 + x_2$$

$$0 = -x_1^3 - x_2$$

are imaginary. If as directed

$$v = x_1^2 + x_2^2$$
$$\dot{v} = 2x_1\dot{x}_1 + 2x_2\dot{x}_2$$

substitution of equations (5-1) yields

$$\begin{aligned}\dot{v} &= 2x_1(-x_1 + x_2) + 2x_2(-x_1^3 - x_2) \\ &= -2[x_1^2 + x_2^2 + x_1x_2(x_1^2 - 1)]\end{aligned} \tag{5-2}$$

To find the region of (x_1, x_2) values that will guarantee a negative $\dot{v}$ it is convenient to first find the locus of $\dot{v} = 0$ for equation (5-2). This line which has been drawn in Figure 5-2 divides the plane into regions of positive and negative $\dot{v}$, respectively. A check of any convenient point in, say, the first quadrant is now sufficient to show which signs ought to be associated with the regions separated by the $\dot{v} = 0$ line. At $x_1 = x_2 = 1$, for example, $\dot{v} = -4$ indicating that this point lies in the desired region about the steady state, and that all points in the darkened region of Figure 5-2 must be excluded as corresponding to positive $\dot{v}$.

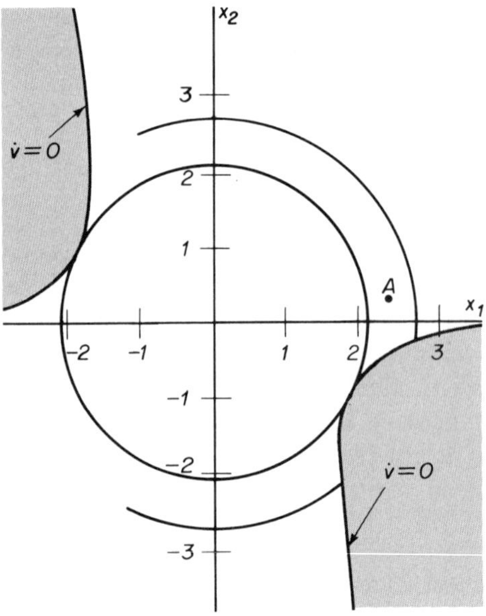

Figure 5-2. Stability region for a constrained $\dot{v}$.

Any v-circle that lies entirely within the $\dot{v} < 0$ region will enclose an RAS, and the largest RAS will therefore be defined by the circle tangent to the $\dot{v} = 0$ curve. This tangency occurs at $x_1 = \pm 1.86$, $x_2 = \mp 0.95$ for a circle of radius $=$ 2.09. It is important to note that any region outside of this tangent circle cannot be guaranteed to be part of an RAS, in spite of the fact that trajectories in such a region may move in the direction of decreasing v. The essential ingredient that is missing is the certainty that no trajectory will wander into the forbidden $\dot{v} > 0$ region. The reader should convince himself that a trajectory starting at point A in Figure 5-2, for example, could move toward smaller v-circles and yet enter the $\dot{v} > 0$ region. If this should occur the trajectory would begin to move away from the origin, and there would be no way to bound its subsequent path.

This is not to say however that the system is necessarily unstable following such disturbances, but merely that since there is no way to be sure in this analysis that such a trajectory could not arise from system (5-1), the region outside of the tangent circle is of indeterminate stability character, so far as this choice of v-function is concerned. Would another choice of v-function have produced a larger RAS? The answer is yes, as for example, if

$$\begin{aligned} v &= x_1^4 + 2x_2^2 \\ \dot{v} &= 4x_1^3\dot{x}_1 + 4x_2\dot{x}_2 \\ &= 4x_1^3(-x_1 + x_2) + 4x_2(-x_1^3 - x_2) \\ &= -4(x_1^4 + x_2^2) \end{aligned} \tag{5-3}$$

The fact that v is positive-definite and $\dot{v}$ negative-definite for *all* $\mathbf{x}$ proves that the system is in fact globally stable, and that there need be no limit placed on the size of permissible RAS contours expressed in terms of this v. There is of course no implication of contradiction in this discovery, but rather it highlights the fact that any result of a Liapunov analysis may be very sensitive to the particular v-function chosen. In other words, this mode of analysis produces sufficient but not necessary conditions for stability. Specifically, in the problem at hand, any trajectory starting within the tangent circle of Figure 5-2 must be asymptotically stable to the steady state at the origin, but as the consideration of the second v-function demonstrated, these are not necessarily the only stable trajectories.

EXAMPLE 5-2. Can a reasonable-size RAS be established for the reactor of equations (3-6) by using the circular v-function?

Solution: By defining the combined constants

$$a = (q/V) + (U/C_pV) \tag{5-4a}$$

$$b = q/V \tag{5-4b}$$

Equations (3-6) become

$$\frac{dx_1}{dt} = r - ax_1 \tag{5-5a}$$

$$\frac{dx_2}{dt} = -r - bx_2 \tag{5-5b}$$

For the given $v(x_1, x_2)$ differentiation and substitution of equations (5-5) lead to

$$\begin{aligned} \dot{v} &= 2x_1\dot{x}_1 + 2x_2\dot{x}_2 \\ &= 2x_1(r - ax_1) + 2x_2(-r - bx_2) \\ &= -2[ax_1^2 + bx_2^2 + r(x_2 - x_1)] \end{aligned} \tag{5-6}$$

As in the preceding example, it is instructive to use the $\dot{v} = 0$ locus to divide the phase-plane into regions of positive and negative $\dot{v}$. This necessitates, however, that specific choices be made for the until now general reaction rate $r(x_1, x_2)$ and system parameters. Assuming kinetics and Arrhenius temperature dependence of the form of equation (1-6b), the numerical values of Berger and Perlmutter given in Example 2-3 correspond to a system with a single steady state at $C_s = 0.165$ lb-mol/ft^3, $T_s = 550°$R. Setting equation (5-6) equal to zero produces a quadratic in x_2 for every choice of temperature and leads to the curve of Figure 5-3. A Region of

Asymptotic Stability therefore exists about the origin within any circle that does not enter the darkened $\dot{v} > 0$ region. The largest such circle is shown in Figure 5-3, indicating that momentary increases in temperature of up to about 20°F are tolerable from a stability viewpoint. Whether or not this deviation is to be considered entirely acceptable is a matter for the engineer's professional judgment and is likely to include other factors than those under discussion here.

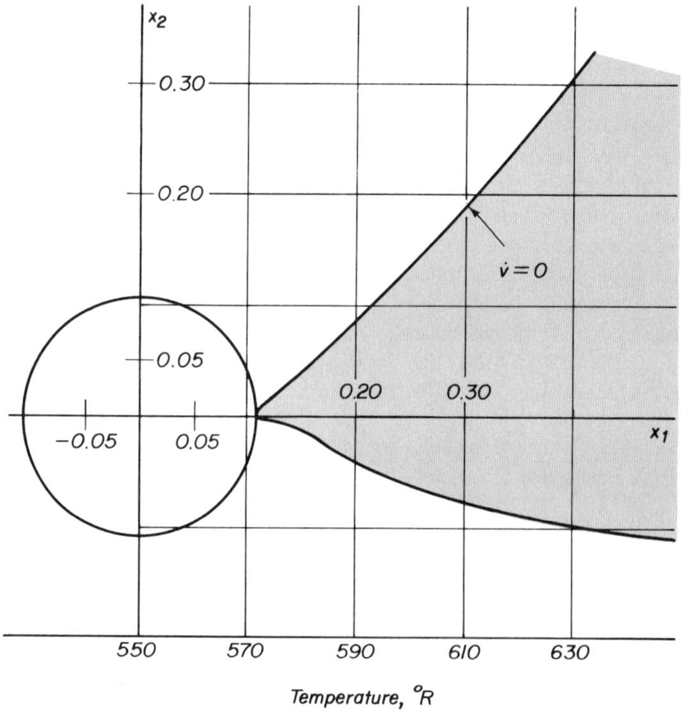

Figure 5-3. Circular RAS for CSTR example.

Should the analyst decide that the area enclosed by the tangent circle is in fact too small for his purposes, he can appreciably widen both the scope of his investigation and the RAS by replacing the circular v by the more general ellipse, the quadratic form:

$$v = \mathbf{x}^T\mathbf{P}\mathbf{x} \tag{5-7}$$

As an illustration, results are presented in Figure 5-4 for the case of

$$\mathbf{P} = \begin{bmatrix} 1 & \frac{1}{2} \\ \frac{1}{2} & 1 \end{bmatrix} \tag{5-8}$$

for which

$$v = x_1^2 + x_1x_2 + x_2^2 \tag{5-9a}$$

and

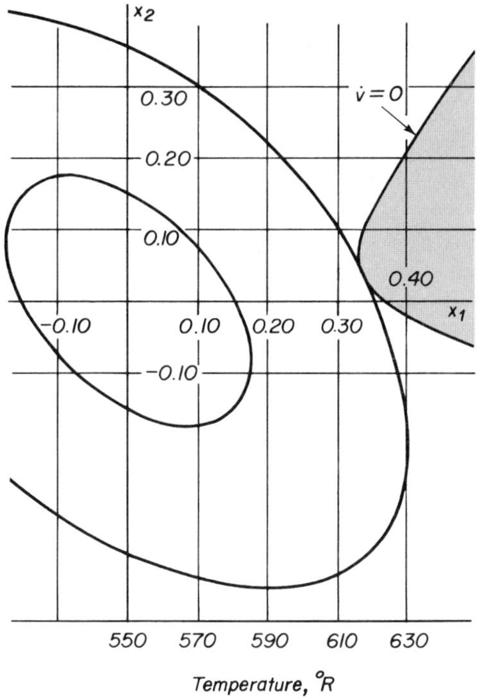

Figure 5-4. Elliptic RAS for CSTR example.

$$\dot{v} = -2\left[ax_1^2 + bx_2^2 + \left(\frac{a-b}{2}\right)x_1x_2 + \frac{r}{2}(x_2 - x_1)\right] \tag{5-9b}$$

It is apparent by comparison that this modification has had a major effect on the result, serving to approximately triple the range of disturbances for which asymptotic stability has been proved.

As the reader will probably expect, every choice of constants for the entries in **P** will not be as fortunate as the foregoing. For any **P** that has $+1$ as the equal off-diagonal elements, the $\dot{v} = 0$ locus will pass through the origin. In such a circumstance, it is impossible to find any v-ellipse that does not include a region for which $\dot{v} > 0$, but as emphasized above the inadequacy of any proposed v-function should not lead one to infer that the system is unstable. In point of fact this reactor model has only a single steady state which trajectories approach asymptotically from every initial condition in the (x_1, x_2) plane. Further study of this kind may be found in the work of Warden, Aris, and Amundson (1964).

Before leaving the subject of $\dot{v} = 0$ loci, a caveat. Because the origin is usually an isolated solution of the $\dot{v} = 0$ equation, it may appear to superficial algebraic examination that the $\dot{v} = 0$ locus passes through the origin even when in fact it does not do so. If one were to apply this incorrect reasoning, he would have prematurely rejected the quite satisfactory v-functions of the last example.

5-3. Krasovskii's Theorem

In dealing with local stability questions as in Chapter 4, the systematic procedure for finding a Liapunov function which culminated in equation (4-18) could be thought of as a luxury. For after all, the matter could almost always be resolved by linearization, and the v-function development seems rather heavy machinery to bring to bear for the sole purpose of handling the rare case of a zero eigenvalue.

However, if the name of the game is to be *regions* of stability, the Fundamental linearization theorem is so far entirely irrelevant, and methods for finding satisfactory Liapunov functions take on new importance. To this point we have recourse to a systematic approach only for linear systems, but of course virtually all real reactors are nonlinear. If our efforts are not to be overly dependent on guesswork, a technique is needed for attacking the system:

$$\dot{\mathbf{x}} = \mathbf{f}(\mathbf{x}) \tag{5-10}$$

The details, which parallel in large part the logic of Section 4-5, are from Krasovskii (1954).

As an alterative to the elliptical v-function in $\mathbf{x}$-space that was used in Example 5-2, consider as a candidate the ellipse in $\mathbf{f}$-space

$$\dot{v} = \mathbf{f}^T\mathbf{P}\mathbf{f} \tag{5-11}$$

Because $\mathbf{f}$ is a function of $\mathbf{x}$, there is a unique mapping between any ellipse in $\mathbf{f}$ and a closed (but non-elliptical) contour in $\mathbf{x}$, as illustrated in Figure 5-5. Evidently the conditions (4-11a, b, and d) are satisfied by this v-function, but as usual some work is needed on condition (4-11c). By differentiation

$$\dot{v} = \mathbf{f}^T\mathbf{P}\dot{\mathbf{f}} + \dot{\mathbf{f}}^T\mathbf{P}\mathbf{f} \tag{5-12}$$

But

$$\dot{\mathbf{f}} = \frac{d\mathbf{f}}{dt} = \mathbf{J}\dot{\mathbf{x}} = \mathbf{J}\mathbf{f} \tag{5-13}$$

where $\mathbf{J}$ is defined as in equation (4-36). Thus

$$\begin{aligned}\dot{v} &= \mathbf{f}^T\mathbf{P}\mathbf{J}\mathbf{f} + \mathbf{f}^T\mathbf{J}^T\mathbf{P}\mathbf{f} \\ &= \mathbf{f}^T(\mathbf{P}\mathbf{J} + \mathbf{J}^T\mathbf{P})\mathbf{f}\end{aligned} \tag{5-14}$$

a result so similar in form to (4-17) that it immediately calls forth the identification of

$$\mathbf{P}\mathbf{J} + \mathbf{J}^T\mathbf{P} = -\mathbf{Q}(\mathbf{x}) \tag{5-15}$$

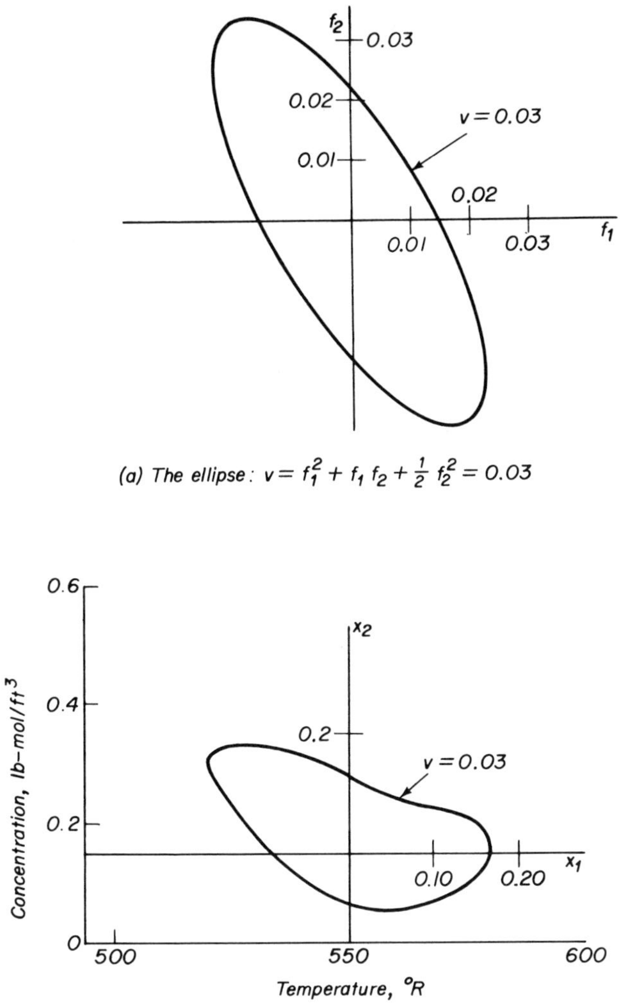

(a) The ellipse: $v = f_1^2 + f_1 f_2 + \frac{1}{2} f_2^2 = 0.03$

(b) The mapping in x-space (using the numerical values of Example 5-2)

Figure 5-5. Comparison of contours in f-and x-spaces.

Note, however, that this **Q** is a function of **x** (because **J** is) whereas the **Q** of equation (4-18) was a constant. This difference notwithstanding

$$\dot{v} = -\mathbf{f}^T\mathbf{Q}(\mathbf{x})\mathbf{f} \tag{5-16}$$

is still negative-definite for a positive-definite $\mathbf{Q}(\mathbf{x})$, but now the possibility exists that the sign-definite features may be confined to certain regions of the phase-space. Indeed this will be the case in the example that follows.

EXAMPLE 5-3. Reexamine the reactor problem of Example 5-2, but now use the Liapunov function, $v = \mathbf{f}^T\mathbf{P}\mathbf{f}$, suggested by Krasovskii.

Solution: For simplicity, consider initially the v-function which is a circle in $\mathbf{f}$-space; that is, let $\mathbf{P} = \mathbf{I}$ the identity matrix. The sought for RAS will then depend on the existence of a region in (x_1, x_2) in which according to equations (5-15) and (5-16):

$$\mathbf{Q}(\mathbf{x}) = -(\mathbf{J} + \mathbf{J}^T) > 0 \tag{5-17}$$

From equations (5-5):

$$\mathbf{J} = \begin{bmatrix} \left(-a + \dfrac{\partial r}{\partial x_1}\right) & \dfrac{\partial r}{\partial x_2} \\ -\dfrac{\partial r}{\partial x_1} & \left(-b - \dfrac{\partial r}{\partial x_2}\right) \end{bmatrix}$$

Then

$$\mathbf{Q}(\mathbf{x}) = \begin{bmatrix} 2\left(a - \dfrac{\partial r}{\partial x_1}\right) & \left(\dfrac{\partial r}{\partial x_1} - \dfrac{\partial r}{\partial x_2}\right) \\ \left(\dfrac{\partial r}{\partial x_1} - \dfrac{\partial r}{\partial x_2}\right) & 2\left(b + \dfrac{\partial r}{\partial x_2}\right) \end{bmatrix} \tag{5-18}$$

Applying to this case the Sylvester conditions given as equations (4-15) yields the inequalities

$$a - \left(\frac{\partial r}{\partial x_1}\right) > 0 \tag{5-19a}$$

$$4\left(a - \frac{\partial r}{\partial x_1}\right)\left(b + \frac{\partial r}{\partial x_2}\right) - \left(\frac{\partial r}{\partial x_1} - \frac{\partial r}{\partial x_2}\right)^2 > 0 \tag{5-19b}$$

which establish the desired positive sign of $\mathbf{Q}(\mathbf{x})$. Once more it is desirable to choose specific kinetics and numbers to make clearer the significance of these restrictions. Adopting the same reaction equation as was used in Example 5-2, it may be recalled from equations (3-21) that

$$\frac{\partial r}{\partial x_1} = \frac{\partial r}{\partial \eta} = \frac{\Delta H}{C_p}\frac{\partial r}{\partial T} = \frac{\Delta H}{C_p}\frac{Q}{T^2}R \tag{5-20a}$$

$$\frac{\partial r}{\partial x_2} = \frac{\partial r}{\partial C} = \frac{nR}{C} \tag{5-20b}$$

which by substitution into equations (5-19) produce the algebraic limitations:

$$\frac{\Delta H}{C_p}\frac{Q}{T^2}R < a \tag{5-21a}$$

$$R^2\left(\frac{\Delta H}{C_p}\frac{Q}{T^2} - \frac{n}{C}\right) < 4\left(a - \frac{\Delta H}{C_p}\frac{Q}{T^2}R\right)\left(b + \frac{nR}{C}\right) \tag{5-21b}$$

These inequalities can be interpreted as regions in the phase-plane, as in Figure 5-6,

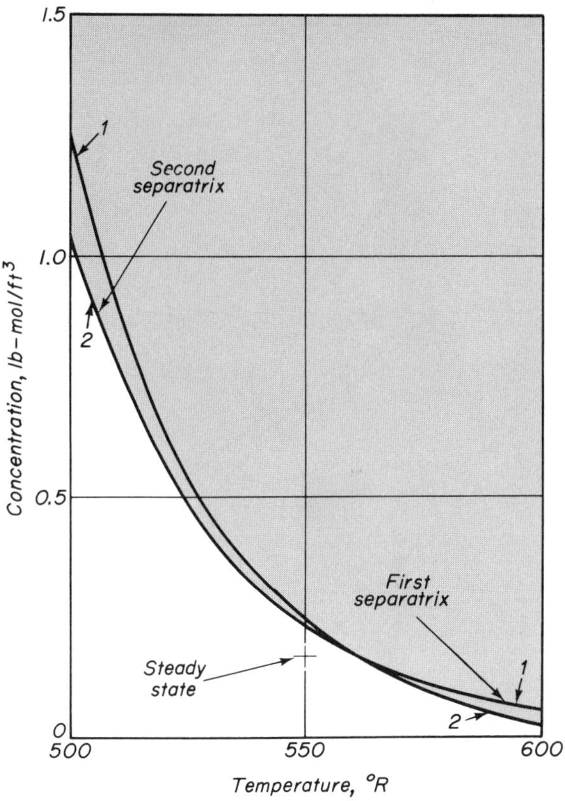

Figure 5-6. The region that satisfies Sylvester's conditions in Example 5-3.

bounded respectively by the (C, T) loci obtained when the inequalities are replaced by equations.

The undarkened region to the left and below both lines satisfies both inequalities. Consequently it is only in this region of the (x_1, x_2) plane that we have proved $\dot{v} < 0$. It only remains to superimpose on this diagram the family of concentric v-contours:

$$v = f_1^2 + f_2^2 = (r - ax_1)^2 + (-r - bx_2)^2 \tag{5-22}$$

As before the largest RAS will be formed within the contour that is tangent to the $\dot{v}$ restrictions. The result of such a computation by Berger (1964) is given in Figure 5-7, where as expected, the closed contours which arise from circles in (f_1, f_2) are not circles in (x_1, x_2).

If improvement over this result is desirable, it may be sought in the manner of Example 5-2, that is, by letting **P** be other than the identity matrix and studying the regions arising from ellipses in **f**. Such an investigation for this system was also carried out in detail by Berger (1965) who systematically varied the two parameters in:

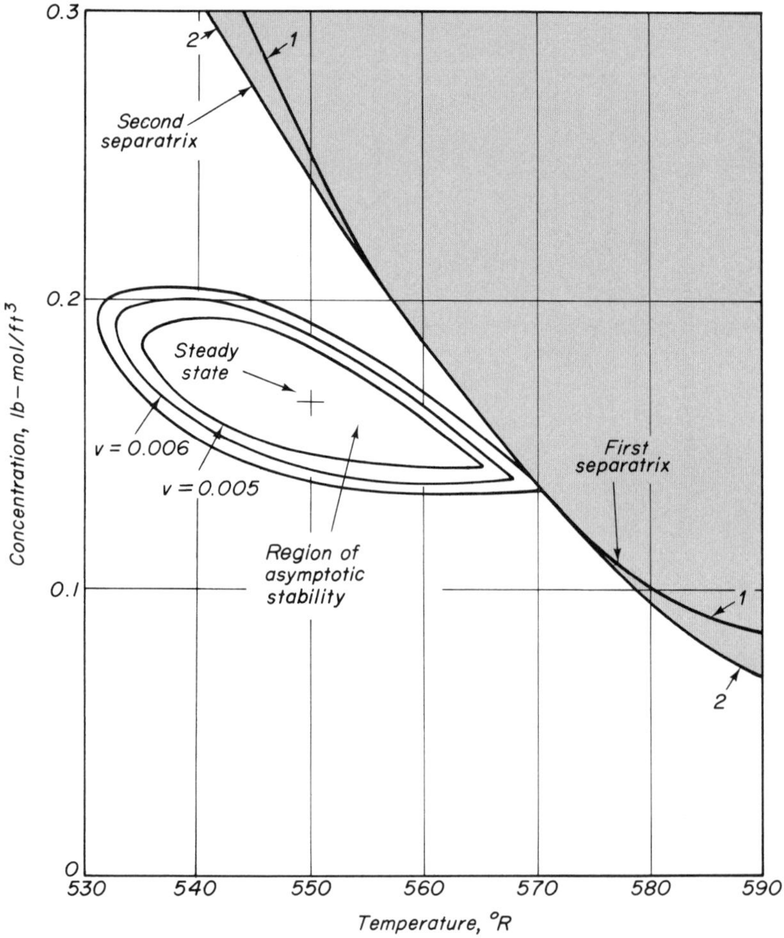

Figure 5-7. Region of asymptotic stability for the CSTR of Example 5-3.

$$\mathbf{P} = \begin{bmatrix} 1 & p_{12} \\ p_{12} & p_{22} \end{bmatrix} \tag{5-23}$$

Typical results are shown in Figure 5-8 in terms of the largest RAS obtainable for each choice of **P**.

Yet another approach to this problem was taken by Paradis (1966) and by Leucke and McGuire (1965, 1967), who independently recognized the conservatism inherent in the use of the Sylvester conditions, and elected to develop the $\dot{v} = 0$ locus directly by solving the algebraic statement obtained by setting equation (5-14) equal to zero. Samples of their results are reproduced in Figures 5-9 and 5-10.

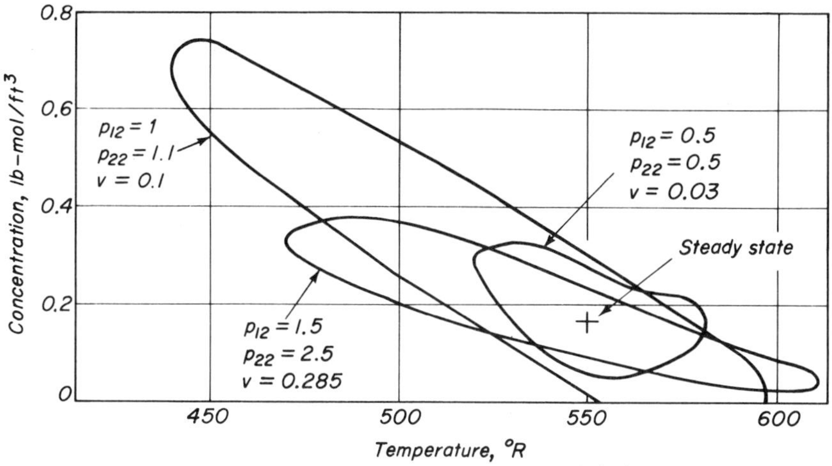

Figure 5-8. Regions of asymptotic stability for different Liapunov function parameters.

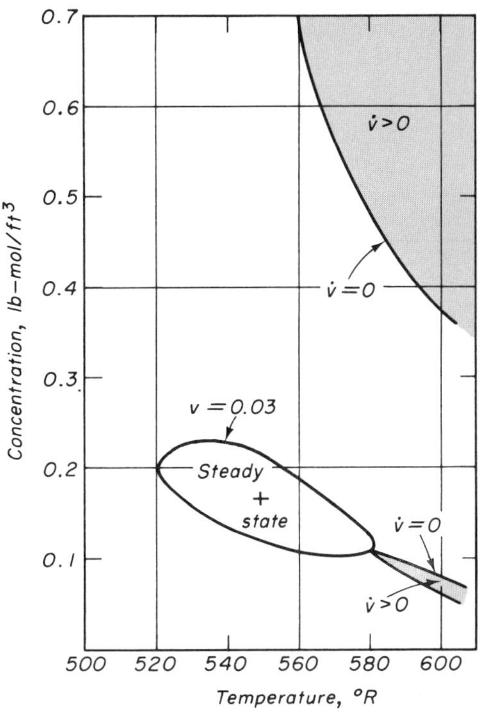

Figure 5-9. Largest RAS for $v = f_1^2 + f_2^2$. (After Paradis, 1966 and Leucke and McGuire, 1965.)

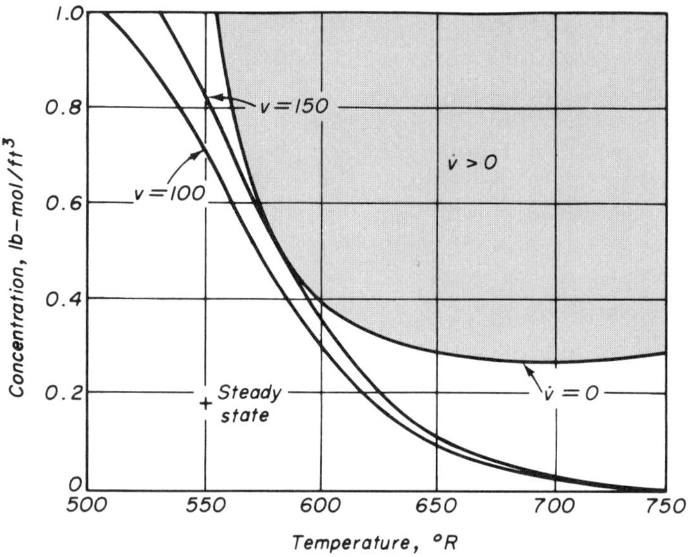

Figure 5-10a. Liapunov contours for $v = \mathbf{f}^T\mathbf{P}\mathbf{f}$ with $p_{12} = 1$, $p_{22} = 3$. (After Paradis, 1966.)

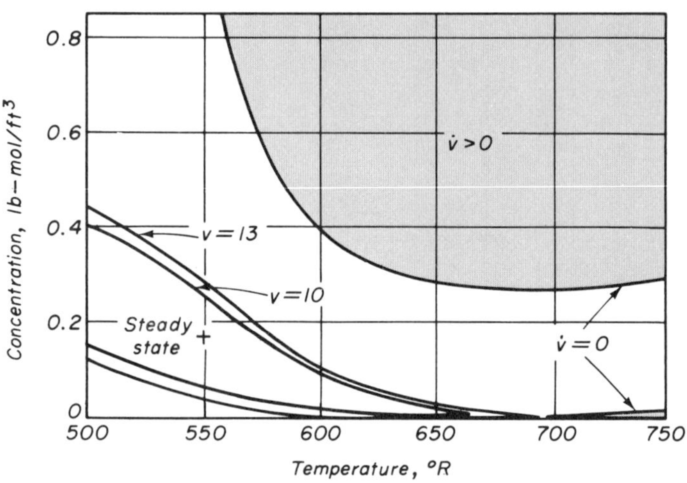

Figure 5-10b. Liapunov contours for $v = \mathbf{f}^T\mathbf{P}\mathbf{f}$ with $p_{12} = 2$, $p_{22} = 9$. (After Paradis, 1966.)

5-4. The Largest RAS

In spite of the fact that the largest RAS is not necessarily the best for engineering design purposes, there are good reasons for finding the largest region for a given choice of v. Above all, such a result provides an upper limit and could indicate (if the RAS is unreasonably small) that further work on the v in question will be unproductive. Secondly, since the family of v-contours is nested about a common center, it is always a simple matter to find a smaller RAS concentric with the largest one. Finally, the largest RAS can be thought of as an estimate of the entire region of initial conditions from which the system is asymptotically stable. If the latter is known from an independent analysis—probably detailed numerical integration of the system equations—the relative size of the largest RAS is in a sense a measure of the degree of conservatism associated with the chosen Liapunov function.

The interest in a largest RAS naturally suggests the applicability of optimization theory. The key to this connection lies in formalizing the procedure used in the last few examples, that of finding the v-contour that is tangent to the $\dot{v} = 0$ locus. This can be done in either of two ways: (1) by minimizing v subject to the constraint that $\dot{v} = 0$, or (2) by maximizing $\dot{v}$ subject to the constraint that $v = K$ at a sequence of K values, increasing K from low values until the maximum $\dot{v} = 0$. Either approach can be framed into the formalism of the classical Lagrange multiplier treatment, as is demonstrated in the following.

EXAMPLE 5-4. Reconsider the problem of Example 5-1 as an optimization.

Solution: In accordance with the first approach outlined above

$$\text{Minimize:} \quad v = x_1^2 + x_2^2 \tag{5-24a}$$

$$\text{Subject to:} \quad \dot{v} = -[x_1^2 + x_2^2 + x_1 x_2(x_1^2 - 1)] = 0 \tag{5-24b}$$

The Lagrangian for this problem is:

$$L = v + \lambda \dot{v} \tag{5-25}$$

and by requiring that

$$\frac{\partial L}{\partial x_1} = \frac{\partial L}{\partial x_2} = 0 \tag{5-26}$$

one arrives at the necessary condition

$$\left(\frac{x_2}{x_1}\right)^2 = \frac{x_1^2 - 1}{3x_1^2 - 1} \tag{5-27}$$

As a last step, equations (5-24b) and (5-27) are solved simultaneously (numerically) to identify the (x_1, x_2) coordinates of the point of tangency shown in Figure 5-2.

Following the alternative approach noted above

$$\text{Maximize: } \dot{v} = -[x_1^2 + x_2^2 + x_1x_2(x_1^2 - 1)] \tag{5-28a}$$

$$\text{Subject to: } v = x_1^2 + x_2^2 = K \tag{5-28b}$$

$$\text{And: K chosen to give maximum } \dot{v} = 0 \tag{5-28c}$$

The Lagrangian for this case is

$$L = \dot{v} + \lambda(v - K) \tag{5-29}$$

but by coincidence for this problem, the necessary conditions (5-26) lead to the identical result given as equation (5-27). Now, however, this relation must be solved simultaneously with equation (5-28b) in terms of the parameter K. The result is this time analytic:

$$x_1^2 = \frac{1}{8}[2 + 3K \pm \sqrt{4(1 - K) + 9K^2}] \tag{5-30}$$

but K must still be chosen by numerical trials to give a value of x_1 which will in turn yield x_2 by equation (5-28b), and then $\dot{v} = 0$ by substitution into (5-28a). The reader can verify that $K = 4.36$ will give the expected answer.

The Lagrangian approaches just presented are formally elegant in that they permit apparently closed-form solutions for necessary optimization conditions. The details of solution are, however, far from trivial and require numerical iterations for even the simple illustration used above. More complex systems would accordingly lead to simultaneous nonlinear equations of yet higher-order, suggesting that alternative formulations of the problem might have appreciable advantages. One such alternative was exploited by Berger and Lapidus (1968), who recast equations (5-28) in the form of an unconstrained minimization that includes a penalty function:

$$\text{Minimize: } F = -\dot{v} + c(v - K) \tag{5-31}$$

For a sufficiently large c, the minimum F can only occur for v close to K, for otherwise the second term in (5-31) would make a significant contribution. In accordance with equation (5-28c) a numerical search must still be made for the K which gives $\dot{v} = F = 0$. Electing to examine the same system as was studied in Examples 5-2 and 5-3, Berger and Lapidus duplicated the results shown in Figure 5-10 for the Liapunov function

$$v = f_1^2 + f_2^2 \tag{5-32}$$

in excellent agreement with the independent computations cited above.

Because the formulation of equation (5-21) is free of graphical limitations, it lends itself to consideration of higher-order systems. By taking advantage of this feature, Berger and Lapidus were able to study a system of three CSTR's in series and a sixteen-stage approximation to a TRAM. The number of state variables were six and thirty-two, respectively, since concentration and temperature were considered in each stage. Equation (5-32) served as the

Liapunov function in both cases, and the results reflect the increasing conservatism that follows increase in dimensionality. For the numbers chosen in the six state-variable problem, the maximum RAS based on equation (5-32) corresponded to a first-stage temperature disturbance of 11.2°F, whereas numerical integration yielded a value of 79.2°F. The parallel numbers for the sixteen-stage model were 0.18°F and 30.2°F. Similar inadequately small regions were developed by Sabo and Dranoff (1970) via an alternative search method. They applied a Krasovskii-form Liapunov function to a CSTR model that required three state variables to describe a system of non-isothermal consecutive reactions. Evidently, this mode of analysis is workable for systems of any order, but loses its usefullness as dimensionality grows. The result of the higher-order analysis is practically no better than a finding of local stability.

5-5. Regions of Practical Stability

It was noted in Section 5-1 with reference to Figure 5-1 that even a finding of unique local stability does not insure acceptable trajectories. The same may be said, too, for an RAS, since the mere existence of a stability region still leaves open the all-important questions of numerical design. Indeed a system with an RAS might be useless in a practical sense if its trajectories exceed a design specification, whereas even a locally unstable system might be entirely satisfactory if actual trajectories remain bounded within a region of practical interest. The latter might well be the case, for example, if the system goes into a limit cycle of acceptably small magnitudes. What is really needed then is yet another definition of stability, one which will have a strong resemblance to the engineering outlook. For this purpose, LaSalle and Lefschetz (1961) have defined *practical stability*, a term which is neither weaker nor stronger, but different from Liapunov stability.

To make clear the essential differences between these two stability definitions, it is most instructive to return to the orientation of Section 4-2, where the decision on whether or not a steady state was to be classed as Liapunov stable rested on the outcome of a game for two players. The challenger, it may be recalled, attempted to pick a state-space ϵ-region for which the defender could match no δ-region. His opponent could win (and thus establish Liapunov stability) by countering each and every ϵ with an appropriate $\delta(\epsilon)$.

The essential characteristics of practical stability will emerge, if we recast this game in a somewhat friendlier atmosphere, for there is after all no reason that the two players must remain antagonists. It is usually the case that the ϵ-region can be fixed by the engineer who writes design specifications, the δ-region estimated by his colleague whose experience includes familiarity with likely disturbances. Indeed, the two skills will, as often as not, be found in the same individual.

The two designers (it is no longer meaningful to call them challenger and defender) can establish that a system exhibits practical stability by co-operatively finding a (δ, ϵ)-pair of regions with the same property as in the Liapunov definition, provided, however, that the chosen regions are of dimensions that correspond to useful design specifications. More formally, a system meets the definition of practical stability, if *any* practical-sized (δ, ϵ) pair can be found, such that:

$$||\mathbf{x}(0)|| \leq \delta \tag{5-33a}$$

implies

$$||\mathbf{x}(t)|| \leq \epsilon \tag{5-33b}$$

Moreover, if it is advantageous, it is entirely acceptable to have $\delta = \epsilon$.

It should be noted that the conditions for practical stability are much less stringent than those required for the Liapunov definition, since they call only for *some* (δ, ϵ) pair whereas the latter requires a δ for *all* possible choices of ϵ. Moreover, it is of significance for the focus of this chapter that *regions* of stability are an inherent part of the practical definition, a feature that reenforces our identification of this definition with the more realistic needs of engineering design. This is not to argue that a demonstration of Liapunov stability is not useful. Often a system will be both Liapunov stable and practically stable, the evidence for one being valuable also in establishing the other. The essential points are that one kind of stability cannot be derived from the other, and that the test that matters is the practical.

EXAMPLE 5-5. Is the system studied in Examples 5-2 and 5-3 practically stable?

Solution: Several regions of asymptotic stability have already been demonstrated for this problem. Any of these RAS has within it a family of concentric contours. If the corresponding Liapunov function is thought of as a norm, any pair of these concentric contours can be identified with the definition of equations (5-33). This leaves only the question of design specifications of practical size. From either Figures 5-3, 5-4, or 5-7 to 5-10, for example, the system is practically stable if no disturbances exceed the boundry of an RAS, and any trajectory within such a region is acceptable.

5-6. Tracking Functions

As we have seen, the geometric arguments associated with Liapunov analysis are based in essence on the idea that contours can be found to serve as bounds on possible trajectories. The task of finding a Liapunov function is made more difficult by the need for closed contours, and LaSalle and Lefschetz (1961) have suggested that in some cases it is preferable to form a desirable closed contour by piecing together several functions, each applicable in only part of the phase-plane.

Paradis (1966) has carried this idea to its ultimate simplicity by demonstrating the utility of straight line segments that can be assembled to form closed polygonal contours. In effect, a single Liapunov function is replaced by several *tracking functions* which though less powerful individually, yield collectively the desired stability information. Considering a two-dimensional system:

$$\dot{x}_1 = f_1(x_1, x_2) \tag{5-34a}$$

$$\dot{x}_2 = f_2(x_1, x_2) \tag{5-34b}$$

the straight line tracking function

$$E_1 = m_1 x_1 + x_2 \tag{5-35}$$

has a time-derivative (along a trajectory):

$$\dot{E}_1 = \frac{dE_1}{dt} = m_1 \dot{x}_1 + \dot{x}_2 = m_1 f_1 + f_2 \tag{5-36}$$

As before in the examination of Liapunov functions, it is useful for this tracking function to identify the phase-plane region where the time-derivative is negative; such a region is bounded by the locus

$$\dot{E}_1 = m_1 f_1(x_1, x_2) + f_2(x_1, x_2) = 0 \tag{5-37}$$

which is simply a line through the origin for a particular choice of the constant m_1. The elementary geometry of this discussion is illustrated in Figure 5-11 where a hypothetical $\dot{E}_1 = 0$ line is shown with the corresponding family of E_1 straight line "contours." If the darkened part of the plane is assumed to be the side of $\dot{E}_1 = 0$ where the time-derivative is negative, it is clear that a trajectory from point A will move in a generally downward direction to lines of smaller E_1, and similarly except for sign reversal, movement from point B will be upward to more positive E_1. In fact the lines *aa* and *bb* can be considered bounds that no trajectory originating in the strip between them can cross, but it should be observed that as yet we have no guarantee of stability, since trajectories are free to move right and left without any limit.

To remedy this last deficiency it is only necessary to consider another tracking function of different slope:

$$E_2 = m_2 x_1 + x_2 \tag{5-38}$$

and repeat the prior development. The composite Figure 5-12 shows how the two tracking functions can complement each other to divide the plane into four regions, each associated with particular bounds on trajectory movement. The arrow-angle symbol in each region indicates the spectrum of possible

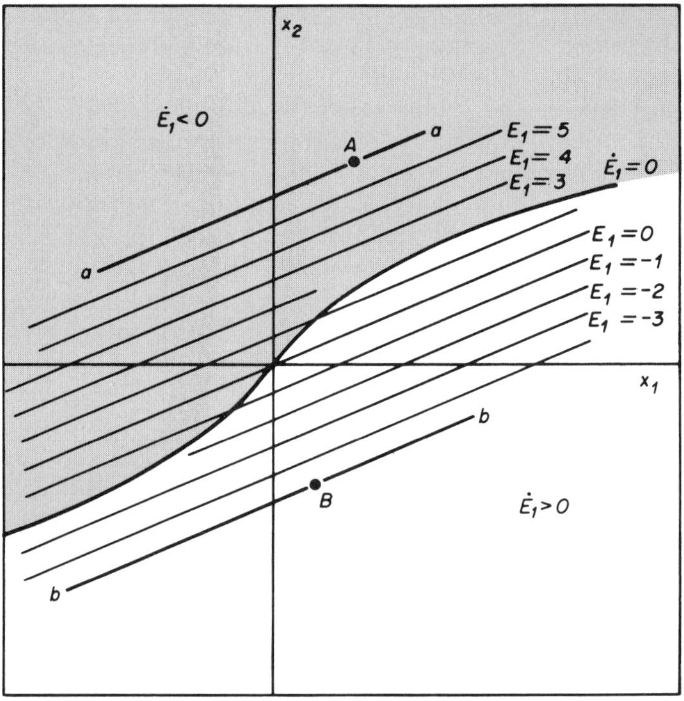

Figure 5-11. A typical tracking function on the phase-plane.

directions of movement that would remain after the constraints of these hypothetical *E*-functions are applied. It would remain now only to use this information graphically to put together a closed region. In practice this construction would most conveniently be superimposed on Figure 5-12, but for clarity it is here presented as Figure 5-13.

This latter figure was constructed as follows. Starting at an arbitrary point *A* in region IV, lines were drawn in the directions of the two tracking functions (as shown by the arrow-angle symbols) terminating in *B* and *D* which are also arbitrary except that they must lie a convenient distance into the respective adjacent regions, I and III. Point *C* is determined by the completion of the parallelogram *ABCD*, but it must be in the as yet unentered region II. The point of this discussion is that because the direction of movement is confined to each region, no trajectory originating within the parallelogram can ever leave it. If this discussion were based on a known physical problem instead of the hypothetical case, we could conclude that the area *ABCD* is a *region of practical stability* (RPS) provided its dimensions were practical in terms of the particular x_1 and x_2 variables.

The considerable freedom associated wih the placement of points *ABCD*

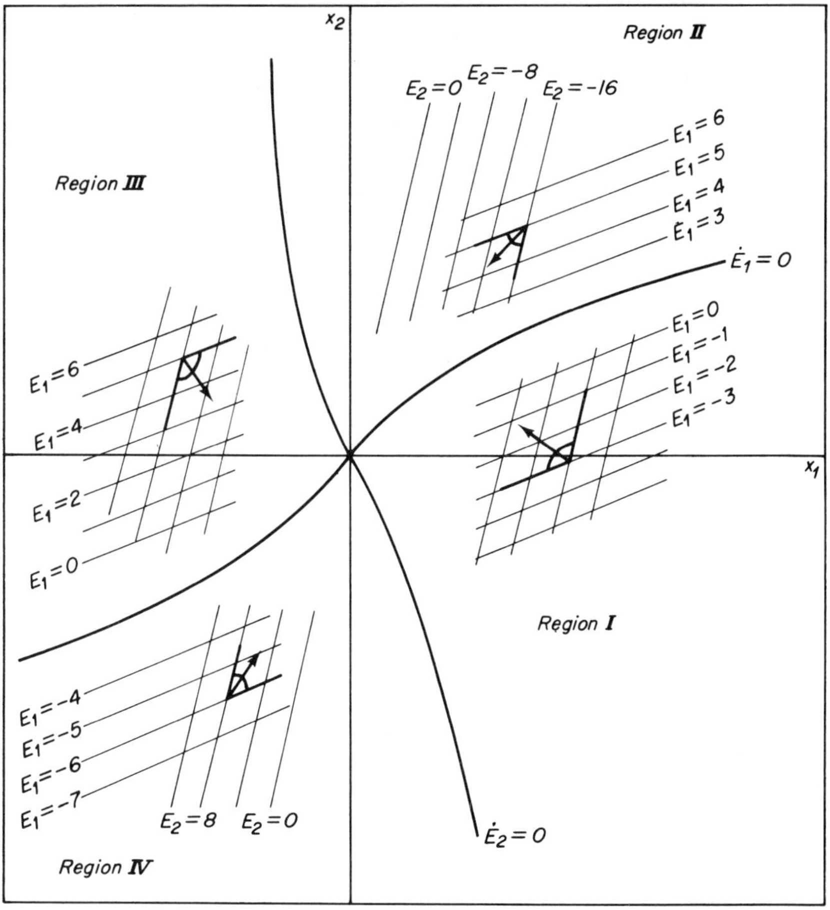

Figure 5-12. Two superimposed tracking functions.

is an advantage to the designer, since he can alter the size and shape of the parallelogram within wide limits. Furthermore, he can change the shape of the RPS if he wishes as for example in the inner lines of Figure 5-13. In this latter case lines *EF*, *FG*, and *GH* terminate on the $\dot{E} = 0$ loci, and all the lines are parallel to the prior construction. The boundary *AEFGHIA* encloses a region from which trajectories cannot escape for the same reasons as were detailed above. An application to a reactor problem follows.

EXAMPLE 5-6. Use the tracking function approach to establish a RPS for the system of Example 5-2.

Solution: The simplest tracking functions are the vertical and horizontal sets of lines

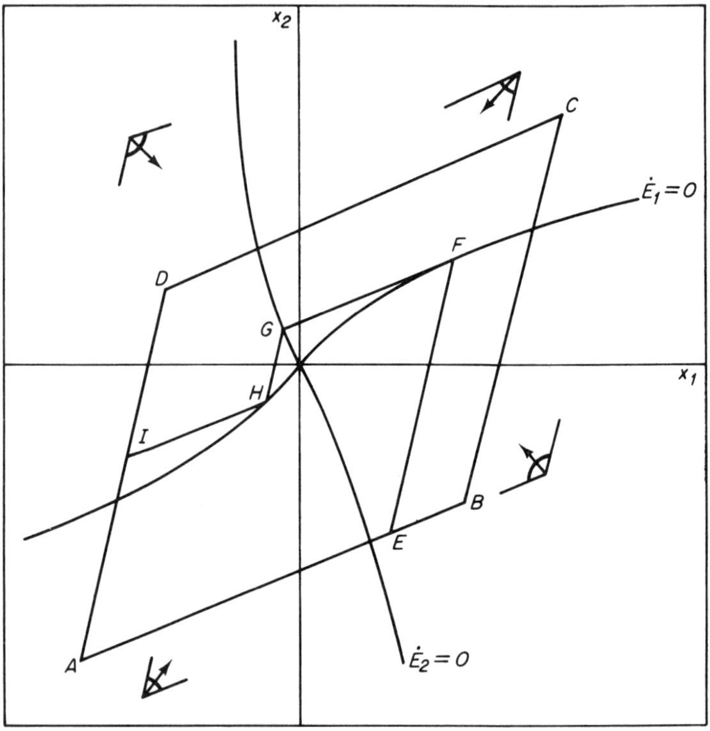

Figure 5-13. Regions of stability for hypothetical example.

$$E_1 = x_1$$

$$E_2 = x_2$$

with corresponding derivatives

$$\dot{E}_1 = \dot{x}_1 = r - ax_1$$

$$\dot{E}_2 = \dot{x}_2 = -r - bx_2$$

Using again the same numbers as before, the loci for $\dot{E}_1 = 0$ and $\dot{E}_2 = 0$ are computed and graphed in Figure 5-14. Once this is done it is a simple matter to establish the regions in the plane corresponding to positive and negative $\dot{E}$ by computing the values at a single convenient point, and the arrow-angle symbols may be put on the graph as handy signposts.

In accordance with the procedure outlined above, the rectangular region $ABCD$ is drawn by locating A, B, and C on the $\dot{E} = 0$ loci and using the parallels to fix D. As before this region is a bound on trajectories interior, but is it of practical size? Specifically, is there reason for concern with deviations from steady state greater than about 10°F in temperature or 0.01 lb-mol/ft³ in concentration? Surely this is a matter of judgment, but for the sake of argument, let us say that the region found is too small to be of practical interest. In seeking a larger closed region, it is evident

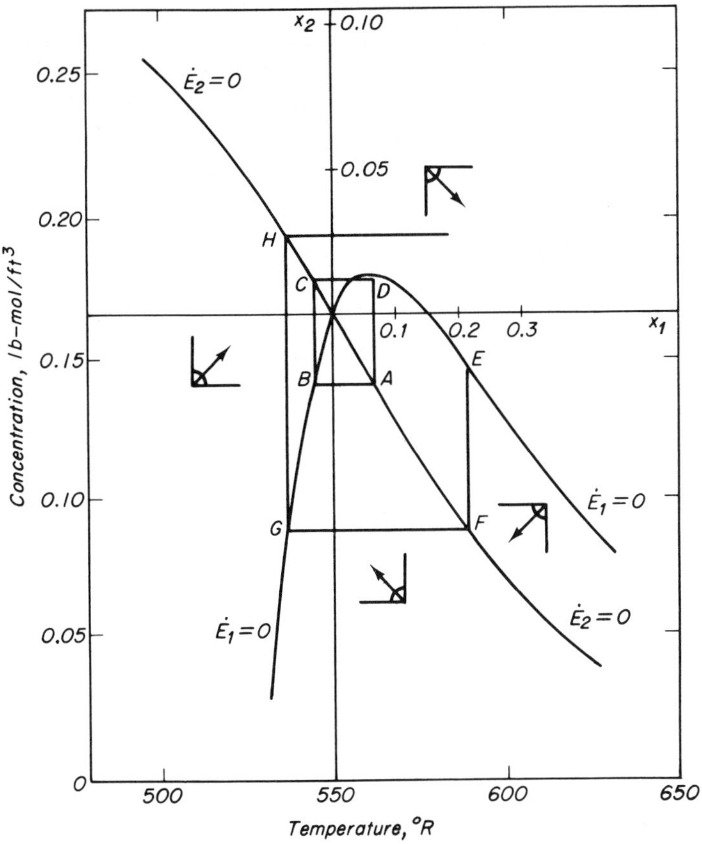

Figure 5-14. Tracking function analysis of a CSTR.

that this system is significantly different from the prior hypothetical example. The shape of the $\dot{E}_1 = 0$ locus is such in Figure 5-14 that attempts at appreciable enlargement of the region lead to lines such as *EFGH* which do not close upon themselves.

The basic technique is adaptable, however, as can be seen by consideration of the additional tracking function

$$E_3 = x_1 + x_2$$

which has derivative

$$\dot{E}_3 = \dot{x}_1 + \dot{x}_2 = -ax_1 - bx_2$$

It may be noted in passing that (coincidentally) the locus of $\dot{E}_3 = 0$ is a straight line. This has no special significance but does make its graphing virtually effortless. The result is shown as Figure 5-15, where attention is now focused on the darkened region between the $\dot{E}_3 = 0$ and $\dot{E}_1 = 0$ loci. Because of the addition of the third tracking function, trajectories in this region are now restricted not to cross the lines of -1 slope. As a result it is possible to limit the prior open construction and produce

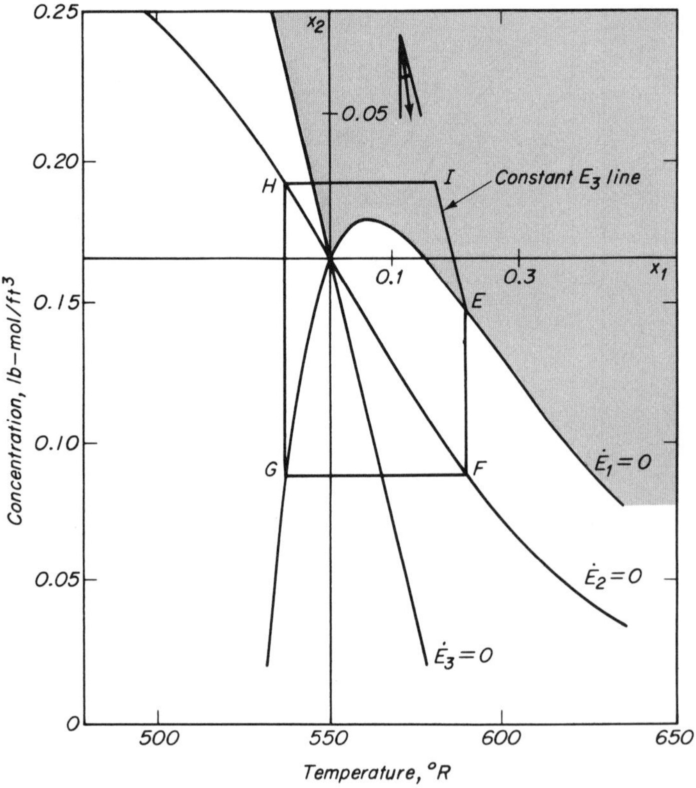

Figure 5-15. Region of practical stability for Example 5-6.

the closed larger region *EFGHIE*, and others of comparable magnitude. Other advantages of the tracking function E_3 have also been presented by Sabo and Dranoff (1966).

For ease of visualization and computation the tracking function technique is the most attractive of those developed for finding regions of practical stability. Still further convenience is available through the computer program worked out by Hewit and Storey (1966, 1967); however, it should be emphasized that the tracking function approach is limited to systems of two dimensions where the planar arguments are valid. As has been demonstrated, this restriction is not present in dealing with Liapunov functions. A generalization of these ideas to systems of any dimension has been presented by Denn (1970), but computational algorithms are not yet developed for such cases.

5-7. Regions via the Fundamental Theorem

The proof of the Fundamental linearization theorem that was offered in Section 4-6 was intended there to show the great breadth of conditions under which the greatly simplified equation (4-22) can be used in place of (4-21) in studying local stability. This is the context in which this powerful theorem is usually presented, however, as recognized by Gura (1965) the same arguments can be interpreted geometrically to establish regions of stability for nonlinear systems of the form of equation (4-23).

Gura's application of the development in Section 4-6 builds primarily on equation (4-26) combined with the choice of ν suggested by (4-33a):

$$||\mathbf{g}|| \leq \frac{\mu\epsilon}{c} \tag{5-39}$$

In a two-dimensional problem, for example, this inequality represents an area of the (x_1, x_2) phase-plane. This is schematically indicated by the darkened region in Figure 5-16, where assuming a Euclidian norm, a family of $||\mathbf{x}|| \leq \epsilon$ circles is also superimposed.

Any region within one of these circles, which is simultaneously within the darkened area, will satisfy both inequalities (4-25) and (4-34), and as such, qualify as an ϵ-region of stability. The corresponding δ-region is a concentric circle as dictated by equation (4-33b). Moreover, by evaluating equation (4-31) at $t = 0$ (where the integral is zero):

$$||\mathbf{x}(0)|| \leq c\,||\mathbf{x}(0)||$$

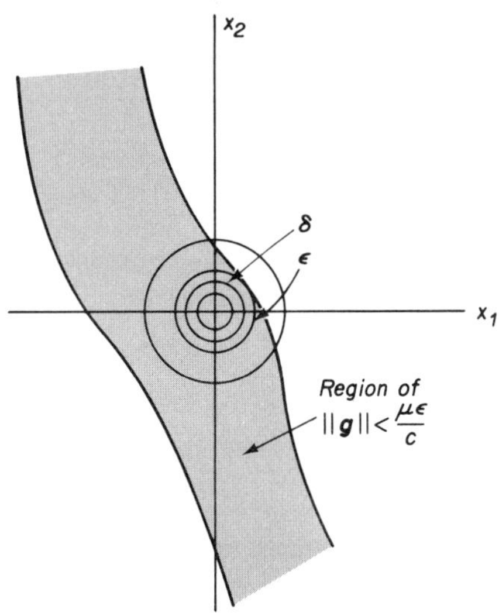

Figure 5-16. Region of stability via the Fundamental theorem.

from which it may be concluded immediately that $c \geq 1$ and as a consequence that $\delta \leq \epsilon$; that is, the δ-circle is necessarily equal to or smaller than the ϵ-region. In effect these constructions meet all the requirements of the derivation of the Fundamental theorem and provide the assurance that all trajectories starting in δ will be bounded as indicated in equation (4-32).

Numerically, one may reasonably ask for some guidance in choosing values of μ, c, and ϵ for a stability study. To find the largest region, equation (5-39) shows that the (μ/c) ratio should be maximized, as should also the value of ϵ, consistent however with the observation that the ϵ-circle of Figure 5-16 must remain within the darkened region. Graphically, this calls simply for the tangent circle of the family. The computation of a largest (μ/c) ratio has already been illustrated in Example 3-1.

Several illustrations of this approach to chemical reactor problems were worked out by Gura. One of these follows.

EXAMPLE 5-7. Find a region of practical stability for the reactor system modeled by equations (1-18) with

$$R = k_1 C_A^n - k_2 C_B^m$$

and parameter values:

$$\begin{aligned} (V/q) &= 0.4 \text{ hr} & k_1 &= 1(\text{lb-mol/ft}^3)^{1/2}/\text{hr} \\ C_{B_0} &= 0 & k_2 &= 0.2(\text{lb-mol/ft}^3)^{1/2}/\text{hr} \\ C_{A_0} &= 0.270 \text{ lb-mol/ft}^3 & m &= 0.5 \\ \gamma &= 1 & n &= 0.5 \end{aligned}$$

Solution: For the numbers given this model has a single steady state at $C_A = 0.146$, $C_B = 0.124$ lb-mol/ft^3. At this point

$$\mathbf{A} = \mathbf{J}_s = \begin{bmatrix} -\dfrac{q}{V} - \left(\dfrac{\partial R}{\partial C_A}\right)_s & -\left(\dfrac{\partial R}{\partial C_B}\right)_s \\ \left(\dfrac{\partial R}{\partial C_A}\right)_s & -\dfrac{q}{V} + \left(\dfrac{\partial R}{\partial C_B}\right)_s \end{bmatrix} = \begin{bmatrix} -3.80 & +0.284 \\ 1.30 & -2.78 \end{bmatrix}$$

which shows by inspection that the steady state is locally stable (Gerschgorin's theorem). From equation (3-38)

$$\mathbf{X}(t) = \begin{bmatrix} (0.176e^{-2.50t} + 0.825e^{-4.09t}) & (0.179e^{-2.50t} - 0.179e^{-4.09t}) \\ (0.818e^{-2.50t} - 0.818e^{-4.09t}) & (0.818e^{-2.50t} + 0.182e^{-4.09t}) \end{bmatrix} \tag{5-40}$$

If the diamond norm is adopted:

$$||\mathbf{x}(t)||_D = 1.99 \exp(-2.5t) + 0.010 \exp(-4.09t)$$

which by the procedure detailed in Example 3-1 gives the largest (μ/c) ratio as

$$\text{Max}\left[\frac{\mu}{c}\right] = 1.25 \qquad \text{at } c = 2 \tag{5-41}$$

The **g** vector for this problem consists of two components where

$$g_1 = -g_2 = R_s - R + nk_1 C_{As}^{n-1}(C_A - C_{As}) - mk_2 C_{Bs}^{m-1}(C_B - C_{Bs})$$

For the diamond norm and the values given:

$$\begin{aligned} \|\mathbf{g}\| &= |g_1| + |g_2| = 2|g_1| \\ &= |-2\sqrt{C_A} + 0.4\sqrt{C_B} + 2.62C_A - 0.568C_B + 0.312| \end{aligned} \tag{5-42}$$

Substitution of equations (5-41) and (5-42) into (5-39) provides the algebraic relation that ordinarily bounds the diamond-norm contour. In this particular problem, however, the physical constraints prohibit consideration of negative concentrations, as do also the requirement of real solutions to equation (5-42). As a result, the largest ϵ-region of practical stability as shown in Figure 5-17 is limited by the C_A-axis rather than by contact with the $\|\mathbf{g}\| = (\epsilon\mu/c)$ line. The inner line in the same figure is the δ-bound on initial conditions found from equation (4-33b):

$$\delta = \frac{\epsilon}{c} = \frac{0.124}{2} = 0.062$$

A series of other results for various norms and various orders of reaction are also reported in the reference cited.

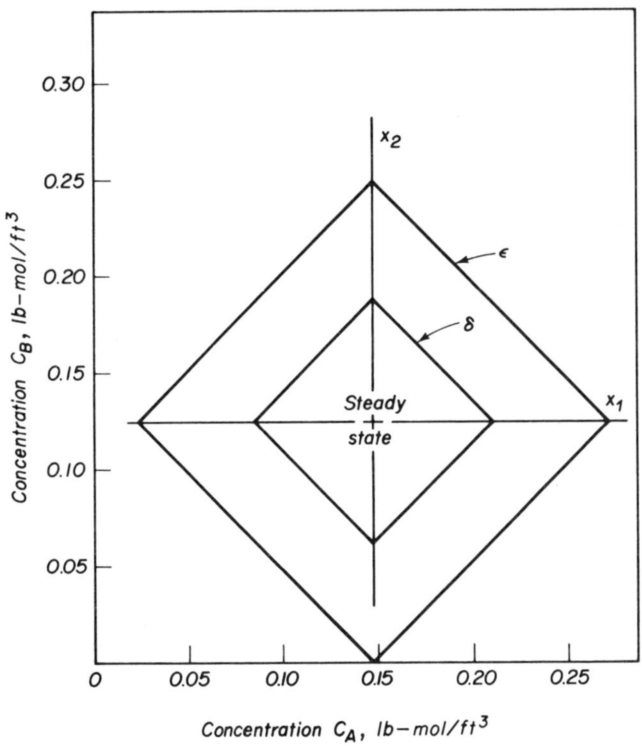

Figure 5-17. Region of stability for isothermal CSTR of Example 5-7.

5-8. System Quality

For some purposes it is convenient to think of a Liapunov function $v(\mathbf{x})$ as similar to a vector norm. The connection is most apparent if the comparison is made between the circular

$$v(\mathbf{x}) = x_1^2 + x_2^2$$

and the Euclidian norm

$$||\mathbf{x}||_E = \sqrt{x_1^2 + x_2^2}$$

but the connection applies as well to the more complex algebraic choices. The essential element of similarity is that each of these functions is a measure of distance from the (x_1, x_2) origin. One application of this concept was presented in Example 5-5. Another advantage of this perspective is that one can formulate bounds on system behavior and regions in a sense different from those of the phase-plane, more closely allied to equation (3-16).

Since as is required by the defining equations (4-11) a Liapunov function is monotonically decreasing with time, it is always possible to find an exponential decay curve which will form an upper bound on the scaler $v(\mathbf{x})$:

$$v(\mathbf{x}) \leq v(0)e^{-\mu t} \tag{5-43}$$

where $\mu > 0$ and the symbol

$$v(0) = v[\mathbf{x}(0)]$$

In fact, as is illustrated in Figure 5-18 the decay bound is not uniquely defined, and there are an endless number of such curves possible; but to insure that the decaying exponential never crosses the v-curve, the conservative restriction is used that the exponential must at all times be shallower than the v-curve:

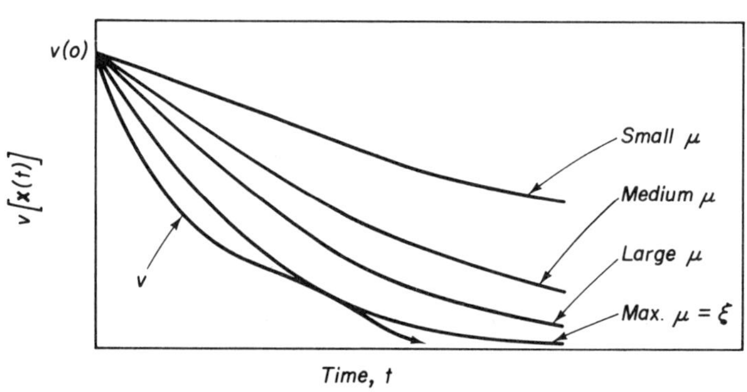

Figure 5-18. Various bounds on a hypothetical v-function.

$$\frac{dv}{dt} \leq -\mu v(0)e^{-\mu t} \tag{5-44}$$

Combining (5-43) and (5-44):

$$-\frac{dv}{dt} \geq \mu v \tag{5-45}$$

To make the bound as least conservative as possible it is desirable to choose from the infinite number that bound which approaches the v-curve most closely. This is accomplished by increase of μ up to some maximum. Above this maximum value the decay curve might intersect the v-curve and lose its significance as an upper bound. Using ξ to designate the maximum value of μ, equation (5-45) shows that

$$\xi = \text{Max } \mu = \text{Min } [-(\dot{v}/v)] \tag{5-46}$$

where the minimization refers to the possible range of t or its equivalence in terms of state variables $\mathbf{x}$. The parameter ξ is variously described as an estimate of a transient [Kalman and Bertram (1960)], as a reciprocal convergence time [Grayson (1963)], or as a measure of *system quality* [Letov (1961)].

EXAMPLE 5-8. Find ξ for the reactor system modeled by equations (1-18) for each of the Liapunov functions:

$$\text{(a)} \quad v = x_A^2 + x_B^2$$
$$\text{(b)} \quad v = 4x_A^2 + x_B^2$$

Assume first-order kinetics, forward and backward, $(q/V) = 1$, $k_1 = 2$, $k_2 = 1$, and $\gamma = 2$.

Solution: The system will have a steady state (C_{As}, C_{Bs}) determined by the solution of

$$0 = q(C_{0A} - C_{As}) - VR(C_{As}, C_{Bs})$$
$$0 = q(C_{0B} - C_{Bs}) + \gamma\, VR$$

With the numerical values given:

$$R = k_1 C_A - k_2 C_B = 2C_A - C_B$$
$$C_{As} = \tfrac{1}{5}(3C_{0A} + C_{0B})$$
$$C_{Bs} = \tfrac{1}{5}(4C_{0A} + 3C_{0B})$$

but these values are not explicitly needed for what is sought. Instead define deviation variables

$$x_A = C_A - C_{As}$$
$$x_B = C_B - C_{Bs}$$
$$r = R - R(C_{As}, C_{Bs}) = k_1 x_A - k_2 x_B$$

and rewrite the system equations (1-18) in the form

$$\frac{dx_A}{dt} = -\frac{q}{V}x_A - r$$

$$\frac{dx_B}{dt} = -\frac{q}{V}x_B + \gamma r$$

which has a steady state at the origin $x_{As} = x_{Bs} = 0$.

Part (a): Adopting as a Liapunov function

$$v = x_A^2 + x_B^2$$

one finds that

$$\dot{v} = \frac{2q}{V}\left[-(x_A^2 + x_B^2) - \frac{rV}{q}(x_A - \gamma x_B)\right]$$

and

$$\frac{\dot{v}}{v} = \frac{2q}{V}\left[-1 - \frac{rV(x_A - \gamma x_B)}{q(x_A^2 + x_B^2)}\right]$$

or, after substituting for r in terms of x_A and x_B and using the numerical values given

$$\frac{\dot{v}}{v} = -6 + \frac{10x_Ax_B}{x_A^2 + x_B^2}$$

It remains only to find ξ by applying the definition:

$$\xi = \underset{x_A, x_B}{\text{Min}}\left[\frac{-\dot{v}}{v}\right] = \underset{x_A, x_B}{\text{Min}}\left[6 - \frac{10x_Ax_B}{x_A^2 + x_B^2}\right]$$

In this example, the desired minimum is easily found by setting partial derivatives equal to zero. In the process one also obtains a maximum value from the same necessary condition:

$$\text{Min}\left[\frac{-\dot{v}}{v}\right] = \xi = 1 \text{ at } x_A = x_B$$

$$\text{Max}\left[\frac{-\dot{v}}{v}\right] = 11 \text{ at } x_A = -x_B$$

As to system quality, it may be concluded that the v-function will decay with time at least as rapidly as $\exp(-t)$, but at no time more rapidly than $\exp(-11t)$. The detailed trajectory is still not known, but it has been established that the values of v will be confined to the darkened area of Figure 5-19, or alternatively the same information may be shown in terms of the state variables by a diagram such as Figure 5-20. The trajectories are confined to remain in the space between the two horn-shaped surfaces.

Part (b): For the Liapunov function

$$v = 4x_A^2 + x_B^2$$

the ratio

$$\frac{\dot{v}}{v} = -6 + \frac{16x_Ax_B}{4x_A^2 + x_B^2}$$

and

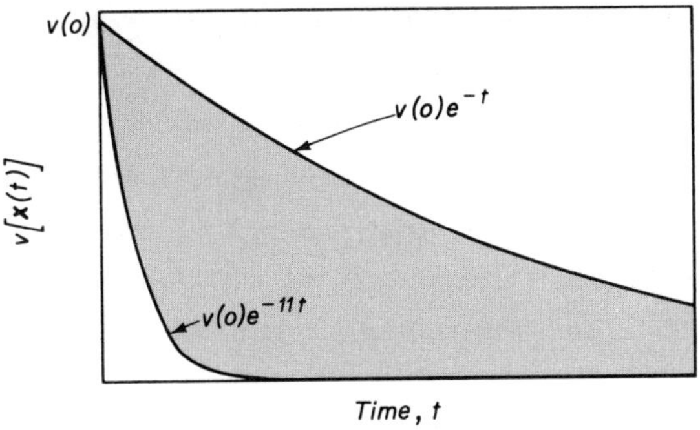

Figure 5-19. Bounds on trajectories of Example 5-8.

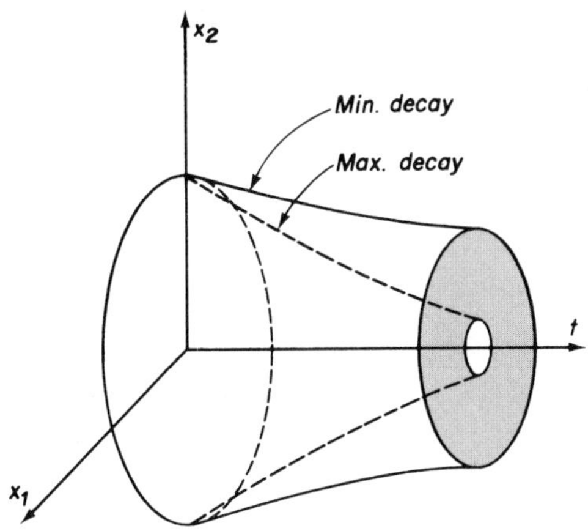

Figure 5-20. Isometric view of trajectory bounds of Example 5-8 (not to scale).

$$\text{Min}\left[\frac{-\dot{v}}{v}\right] = \xi = 2 \text{ at } x_A = x_B/2$$

$$\text{Max}\left[\frac{-\dot{v}}{v}\right] = 10 \text{ at } x_A = -x_B/2$$

Evidently, the value of ξ depends on the v-function as well as the system. In this case the elliptical v-function is a better tracking function than the circular one and decays more rapidly along system trajectories.

REFERENCES

BERGER, J. S. and D. D. PERLMUTTER, *AIChE Journal*, **10**, 233 (1964).

BERGER, J. S. and D. D. PERLMUTTER, *Chem. Eng. Sci.*, **20**, 147 (1965).

BERGER, A. J. and L. LAPIDUS, *AIChE Journal*, **15**, 171 (1969).

DENN, M. M. *AIChE Journal*, **16**, 670 (1970).

GRAYSON, L. P., *Proc. Joint Automatic Control Conference, 589* (1963).

GURA, I. A. and D. D. PERLMUTTER, *AIChE Journal*, **11**, 474 (1965).

HEWIT, J. R. and C. STOREY, *Electronics Letters*, **2**, 408 (1966).

HEWIT, J. R. and C. STOREY, *Proc. I.E.E.*, **114**, 1347 (Sept. 1967).

KALMAN, R. E. and J. E. BERTRAM, *ASME J. Basic Engr.*, **82**, Ser. D, No. 2, 371 (1960).

KRASOVSKII, N. N., *Prikl. Mat. Mekh.*, **18**, 735 (1954).

LASALLE, Jr. and S. LEFSCHETZ, *Stability by Liapunov's Direct Method*, New York: Academic Press (1961).

LETOV, A. M., *Stability in Nonlinear Control Systems*, Princeton University Press (1961).

LEUCKE, R. H. and M. L. MCGUIRE, *AIChE Journal*, **11**, 749 (1965).

LEUCKE, R. H. and M. L. MCGUIRE,, *Ind. Eng. Chem. Fund.*, **6**, 432 (1967).

PARADIS, W. O., Ph. D. Thesis, University of Illinois, Urbana (1966).

PARADIS, W. O. and D. D. PERLMUTTER, *AIChE Journal*, **12**, 130 (1966).

SABO, D. S. and J. S. DRANOFF, *AIChE Journal*, **12**, 1223 (1966).

SABO, D. S. and J. S. DRANOFF, *AIChE Journal*, **16**, 211 (1970).

STRUBLE, R. A., *Nonlinear Differential Equations*, New York: McGraw-Hill, Inc. (1962).

WARDEN, R. B., R. ARIS and N. R. AMUNDSON, *Chem. Eng. Sci.*, **19**, 149 (1964).

Part II: Distributed-Parameter Systems Modeled by Partial Differential Equations

Steady States in Distributed Systems

6

6-1. State-Function Space

Perhaps the single most significant distinction that one can draw between lumped and distributed models is that of dimensionality. A set of n ordinary differential equations in terms of n state variables can be viewed geometrically as defining paths in an n-dimensional space. Thus when $n = 2$ the trajectories are expressible as lines in the phase-plane, and though the visualization becomes increasingly more difficult as n increases, it is conceptually not difficult to think of paths in high- but finite-dimensional space.

When one attempts to deal in a similar way with partial differential equations, at least two alternatives are possible. Either each of the dependent variables is thought of as an infinite set of state variables, one for each spacial position; or else the state of a system is redefined as a set of distributed profiles, and the trajectory is taken to be the surface generated as the profile line moves with time in the *state-function space*. The first of these alternatives is conceptually related to the process of discretization that is commonly used in numerical analysis of partial differential equations, but the latter formulation is more adaptable to the purposes of this chapter because it uses a relatively small number of state variables, making it possible to retain some convenient geometric arguments.

It is useful to recall that the steady state of a lumped-parameter system

(LPS) is a point in state space, defined by solution of the set of algebraic equations that arise when all time derivatives are set equal to zero in the ordinary differential equation model of the system. In dealing with the CSTR for example, the steady state of the system modeled by equations (1-1) and (1-3) was defined as the solution of algebraic equations (1-5). Parallel reasoning applied to distributed-parameter systems (DPS) leads to the observation that the steady state must be a function of spacial position, since it is the solution of a set of ordinary differential equations. The TRAM modeled by equations (1-8) may be used for illustration. A steady state is the solution of

$$0 = D\frac{d^2C}{dZ^2} - u\frac{dC}{dZ} - R \tag{6-1a}$$

$$0 = \alpha\frac{d^2\eta}{dZ^2} - u\frac{d\eta}{dZ} + R - \frac{h}{C_p}(\eta - \eta_w) \tag{6-1b}$$

which with the appropriate boundary constraints will result in the set of profiles

$$C_s(Z), \eta_s(Z) \tag{6-2}$$

an answer entirely different in character from the steady-state point(s) that satisfied the algebraic CSTR steady-state equations. In this case the system dimensionality is small enough for the steady state to be represented graphically, as in Figure 6-1, by a line in the (C, η, Z) *state-function space*. The projections of this line on the (η, Z) and (C, Z) planes are identically the profiles (6-2). The pictorial aspect of the state-function space is lost if the number of dependent variables exceeds two, but the concept of profiles as projections is generally applicable, so long as only a single variable is involved. Generalization to greater spacial complexity can be examined by means of the radial gradient model (1-10). Here the steady state is determined by partial differential equations in the two spacial variables of radial and axial distance. The solutions in the form

$$C_s(Z, Y), \qquad \eta_s(Z, Y) \tag{6-3}$$

are no longer profiles, but rather functions over a plane. Their projections on convenient coordinates will produce families of profiles such as are for example shown in Figures 6-2. These particular numerical results were obtained by McGowin (1971b) for the TRRM with parameters:

$$\begin{aligned}
D/\alpha &= 1\\
uL_0^2/DL &= 25\\
UL_0/\alpha C_p &= 25\\
C_pT_0/\Delta HC_0 &= 3.50\\
C_pT_w/\Delta HC_0 &= 3.25\\
C_pQ/\Delta HC_0 &= 60\\
k_0L/u &= 1.0 \times 10^7
\end{aligned}$$

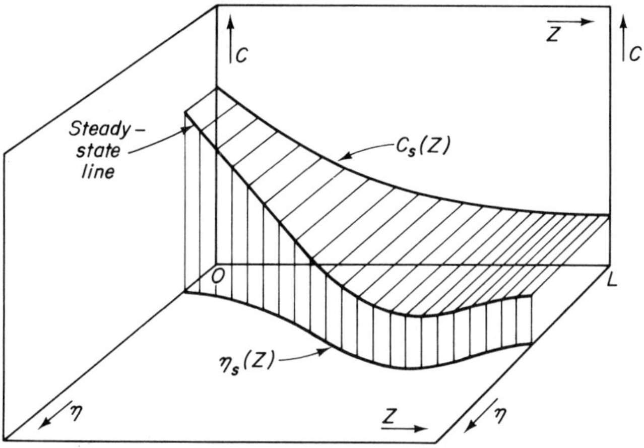

Figure 6-1. A steady-state line in function space and its projections.

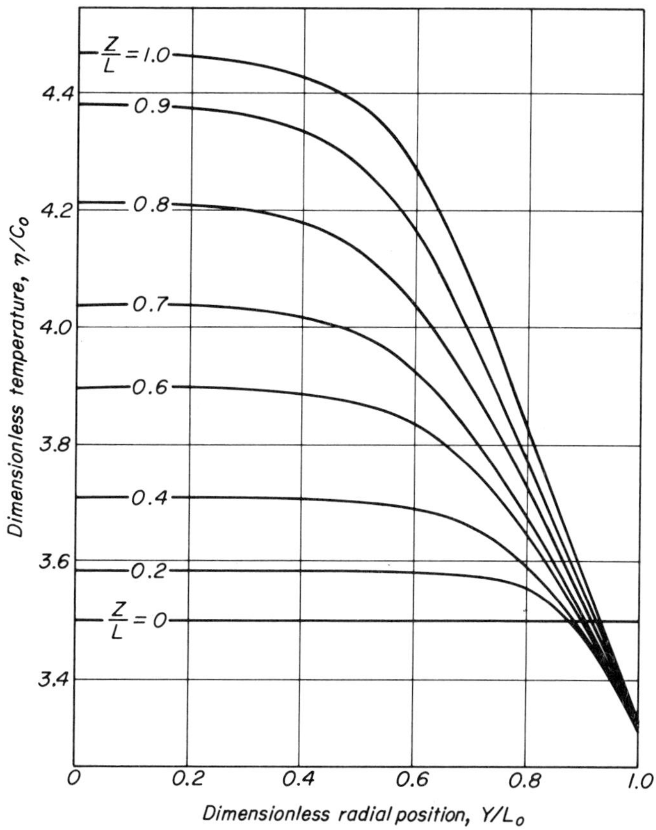

Figure 6-2a. Steady-state temperature profiles for TRRM example.

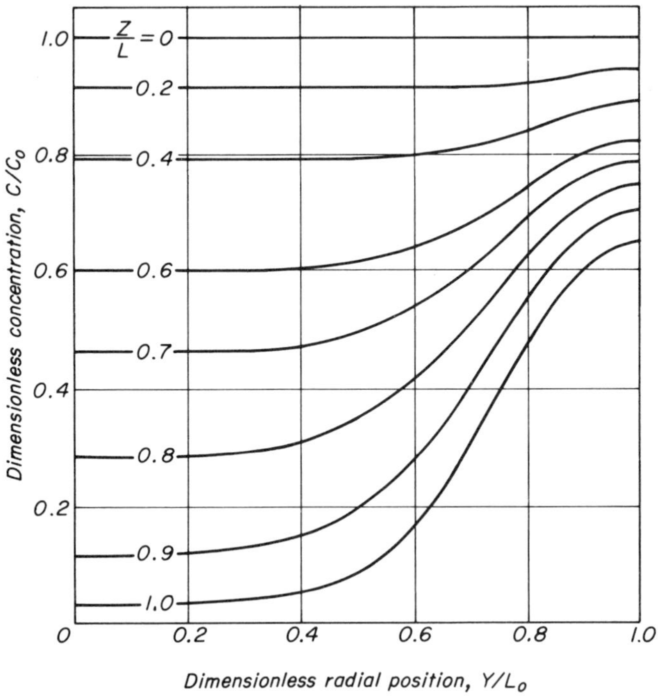

Figure 6-2b. Steady-state concentration profiles for TRRM example.

The concentration and temperature profiles in Figures 6-2 demonstrate the effects of radial mass and heat transfer resistance. A large radial temperature gradient develops downstream from the reactor entrance. This in turn causes a concentration gradient to develop, since the rate of depletion of reactant is higher along the reactor axis than along the wall.

In view of the complexities in analysis and visualization that arise from increase in system dimensionality, whether in dependent or independent variables, there is ample reason to look for simplification of either logical or mathematical origin. Such is the subject of Section 6-2.

6-2. Uncoupling the Energy and Mass Balances

The reactor models presented to this point, whether for flow reactors or catalyst particles, are complicated primarily by the reaction rate terms which appear in each equation as nonlinear functions in both dependent variables. These terms incorporate in effect the interactions that occur between heat and mass, and it would to a considerable extent lose the essential nature of the problem to ignore them. Nevertheless, there is sometimes a way in a steady-state analysis to absorb the concentration effect into a modified rate term explicitly dependent only on temperature. As might be expected, the result is a considerable simplification of the subsequent work. Following van Heerden (1958) consider the adiabatic TRAM model for a steady state:

$$0 = \alpha\frac{d^2\eta}{dZ^2} - u\frac{d\eta}{dZ} + R \tag{6-4a}$$

$$0 = D\frac{d^2C}{dZ^2} - u\frac{dC}{dZ} - R \tag{6-4b}$$

with boundary conditions:

$$\frac{dC}{dZ} = \frac{u}{D}(C - C_0), \qquad Z = 0 \tag{6-5a}$$

$$\frac{dC}{dZ} = 0, \qquad Z = L \tag{6-5b}$$

$$\frac{d\eta}{dZ} = \frac{u}{\alpha}(\eta - \eta_0), \qquad Z = 0 \tag{6-5c}$$

$$\frac{d\eta}{dZ} = 0, \qquad Z = L \tag{6-5d}$$

Adding equations (6-4) to eliminate the reaction term

$$\alpha\frac{d^2\eta}{dZ^2} + D\frac{d^2C}{dZ^2} = u\left[\frac{d\eta}{dZ} + \frac{dC}{dZ}\right] \tag{6-6}$$

Because the same mechanisms are responsible for both the mass and thermal energy dispersion in a highly turbulent flow, it is reasonable to assume that $(D/\alpha) = 1$. This ratio is usually called the Lewis number. Then the new variable

$$w = (\eta + C) \tag{6-7}$$

will reduce equation (6-6) to the ordinary differential equation

$$\frac{d^2w}{dZ^2} = \frac{u}{D}\frac{dw}{dZ} \tag{6-8}$$

with boundary conditions

$$(dw/dZ) = 0, \qquad Z = L \tag{6-9a}$$

$$(dw/dZ) = \frac{u}{D}(w - w_0), \qquad Z = 0 \tag{6-9b}$$

A general solution to an equation of the form of (6-8) is

$$w(Z) = A + B\exp(uZ/D) \tag{6-10}$$

but for the specific boundary conditions required here, $B = 0$, and $A = w_0$.

Thus the solution is

$$w(Z) = w_0 = C_0 + \eta_0 = \text{constant}$$

that is to say, the changes in concentration and temperature complement one another such that the disappearance of reactant produces an equivalent amount of heat. Perhaps after all this should have been expected in an adiabatic reactor (the advantage of hindsight), but note that this intuition is not necessarily correct if $\alpha \neq D$, or if an unsteady state is considered.

By taking advantage of the very simple relationship just derived

$$C = C_0 + \eta_0 - \eta \tag{6-11}$$

Equation (6-4a) can be written as a function of only one dependent variable

$$\alpha \frac{d^2\eta}{dZ^2} - u\frac{d\eta}{dZ} + R(\eta) = 0 \tag{6-12}$$

which though still nonlinear because of $R(\eta)$ has been simplified in being uncoupled from the explicit concentration dependence of equation (6-4b).

A parallel result was first noted by Prater (1958) for the steady state of the catalyst particle system

$$0 = \alpha \frac{d^2\eta}{dZ^2} + R(C, \eta) \tag{6-13a}$$

$$0 = D\frac{d^2C}{dZ^2} - R(C, \eta) \tag{6-13b}$$

with boundary conditions

$$\frac{dC}{dZ} = \frac{d\eta}{dZ} = 0, \qquad Z = 0 \tag{6-14a}$$

$$C = C_0, \eta = \eta_0, \qquad Z = L \tag{6-14b}$$

Adding equations (6-13) as before gives

$$\alpha \frac{d^2\eta}{dZ^2} + D\frac{d^2C}{dZ^2} = 0 \tag{6-15}$$

This time the substitution of

$$w = \alpha\eta + DC \tag{6-16}$$

produces

$$\frac{d^2w}{dZ^2} = 0 \tag{6-17}$$

and the general solution

$$w(Z) = A + BZ \tag{6-18}$$

Finally, the use of boundary conditions (6-14) show that $A = (DC_0 + \alpha\eta_0)$,

$B = 0$, and thus

$$w(Z) = DC_0 + \alpha\eta_0 = \text{constant} \tag{6-19a}$$

or

$$C = C_0 + \frac{\alpha}{D}\eta_0 - \frac{\alpha}{D}\eta \tag{6-19b}$$

This is virtually the same result as for the tubular reactor when the Lewis number is one, but note that here the linear dependence of temperature on concentration does not require that $D = \alpha$. This is a fortunate circumstance since in this case where there is no turbulent mixing it is more difficult to justify the assumption that mass and heat transfer occur within the particle by the same mechanism.

As before the application of equation (6-19) to equation (6-13a) reduces the problem to a single (uncoupled) equation in a single dependent variable (temperature):

$$0 = \alpha\frac{d^2\eta}{dZ^2} + R(\eta) \tag{6-20}$$

6-3. The PFTR and the Batch Reactor

Applying the definition, the steady state of a PFTR is determined by setting the time derivatives in equations (1-7) equal to zero:

$$0 = -u\frac{dC}{dZ} - R \tag{6-21a}$$

$$0 = -u\frac{d\eta}{dZ} + R - \frac{h}{C_p}(\eta - \eta_w) \tag{6-21b}$$

The solution to this set of ordinary differential equations will be the two profiles: $C_s(Z)$ and $\eta_s(Z)$. Such solutions are shown for example in Figures 6-3, both based on kinetics of the first-order Arrhenius form. The results of Wang (1968) are based on the parameter values:

$$
\begin{aligned}
k_0 &= 1.2 \times 10^{10}\ \text{hr}^{-1}\\
Q &= 9{,}400°\text{R}\\
\Delta H &= 1 \times 10^5\ \text{Btu/mol}\\
u &= 7{,}200\ \text{ft/hr}\\
L &= 5\ \text{ft}\\
h &= 7{,}000\ \text{Btu/ft}^3\ °\text{F hr}\\
C_p &= 50\ \text{Btu/ft}^3\ °\text{F}\\
T_w &= 530°\text{R}\\
T_0 &= 600°\text{R}\\
C_0 &= 0.035\ \text{mol/ft}^3
\end{aligned}
$$

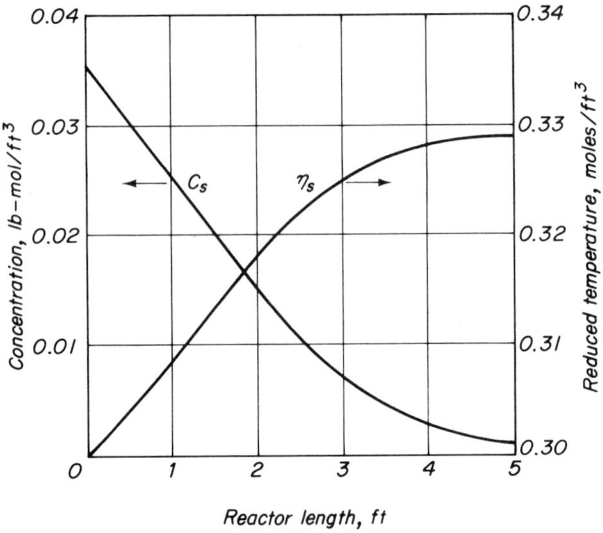

Figure 6-3a. Steady-state profiles of a PFTR (after Wang, 1968).

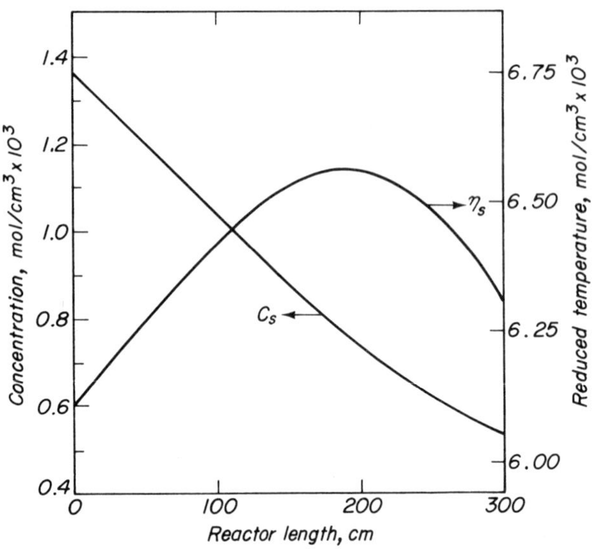

Figure 6-3b. Steady-state profiles of a PFTR (after Reilly and Schmitz, 1966).

The curves obtained by Reilly and Schmitz (1966) show that temperature can also go through a maximum for a relatively higher heat of reaction. The corresponding parameter groups are:

$$\frac{k_0 L}{u} = 1.0 \times 10^{11}$$

$$\frac{C_p Q}{\Delta H C_0} = 75$$

$$\frac{C_p T_w}{\Delta H C_0} = 2.5$$

$$\frac{hL}{uC_p} = 0.79$$

When a more direct comparison with CSTR conversion is desirable, a set of PFTR steady-state operating curves may be prepared. Some results of McGowin (1971a) are reproduced as Figure 6-4, based on the same coordinates as Figure 2-7, but with the heat transfer coefficient modified to account for the tubular geometry. McGowin used the same parameter groups as Reilly and Schmitz.

An alternative approach to the problem highlights physical arguments by viewing the reaction and heat transfer as occurring in a plug sliding along the tube. Defining

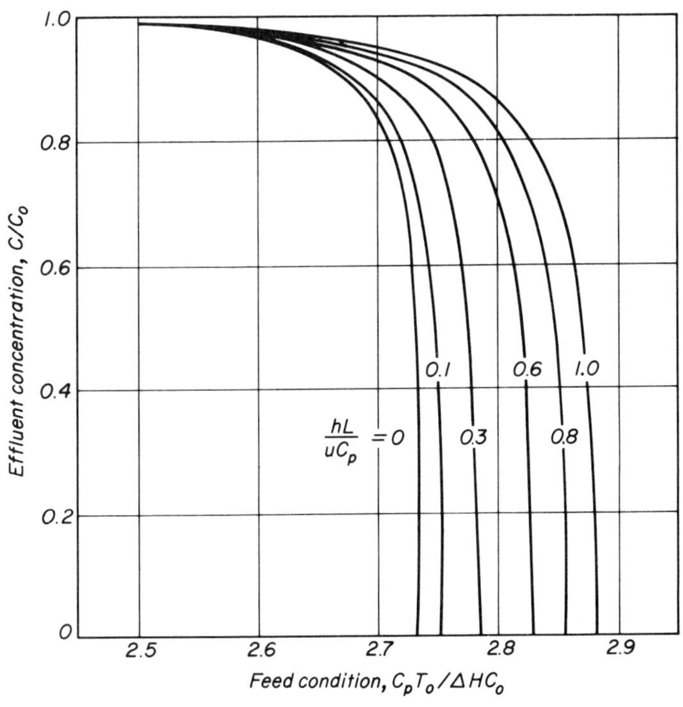

Figure 6-4. Steady-state operating curves for the PFTR.

$$\tau = \frac{Z}{u} \tag{6-22}$$

as the time that the plug has been in the tube at any position Z, equations (6-21) become:

$$\frac{dC}{d\tau} = -R \tag{6-23a}$$

$$\frac{d\eta}{d\tau} = +R - \frac{h}{C_p}(\eta - \eta_w) \tag{6-23b}$$

which except for the small difference in the definition of the heat transfer coefficient bears a striking resemblance to the batch reactor equations (1-4). The near-identity is of course far from coincidental, for the plug of the PFTR has been treated as a small batch reactor moving along the tube axis. What is true for the batch as a function of time may be studied in the steady-state PFTR as a function of distance, and vice versa.

Adding equations (6-23) to eliminate the nonlinear rate term:

$$\frac{d}{d\tau}(C + \eta) = -\frac{h}{C_p}(\eta - \eta_w) \tag{6-24}$$

For the special case of adiabatic reaction, $h = 0$, and

$$(C + \eta) = \text{constant} = C_0 + \eta_0 \tag{6-25}$$

a result not unexpected in view of the previous equation (6-11). A reaction path may be exhibited graphically in the (C, η) plane as in Figure 6-5 where the relationship (6-25) is a straight line of slope $= -1$. Detailed information for a more general nonadiabatic path requires numerical integration, but a qualitative description may be obtained from equation (6-24). Since η_w is the coolant (or wall temperature), $(\eta - \eta_w) > 0$ and

$$\frac{d}{d\tau}(C + \eta) < 0 \tag{6-26}$$

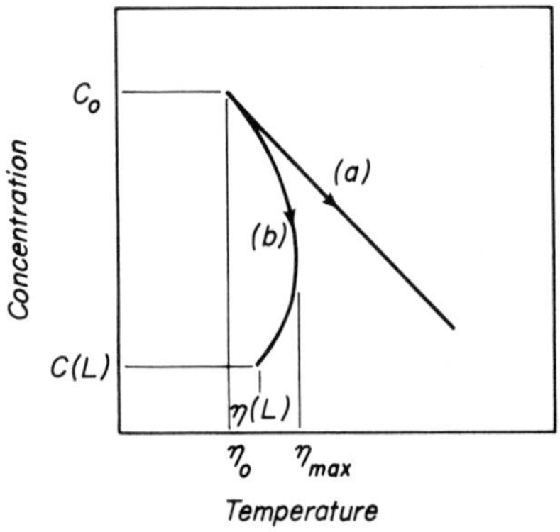

Figure 6-5. Reaction paths: (a) adiabatic reaction; (b) nonadiabatic path.

In other words the reaction path as τ increases must be in the direction of decreasing $(C + \eta)$, as in Figure 6-5. Further discussion in terms of such graphs may be found in Aris (1965) especially as they relate to constraints of chemical equilibrium and maximum rate considerations. For the adiabatic case in particular, Douglas and Eagleton (1962) have shown that concentration and temperature profiles can be computed by using closed-form solutions in terms of tabulated functions.

More immediately to the point here is the matter of multiple solutions. Is it possible for a PFTR to have more than one steady-state profile for given parameters and feed condition? Since the variation of batch contents with time is precisely the variation of tubular reactor contents with distance, the question may be restated in the equivalent (and more intuitively obvious) form: Is it possible for a batch reactor to follow more than one reaction path from a given charge? Common experience says no; the reproducibility of chemical experiments would have to be seriously questioned were this not the answer. A more rigorous proof would quote the theorem of Chapter 3 for sets of first-order differential equations:

$$\frac{d\eta}{d\tau} = f_1(C, \eta); \qquad \eta(0) = \eta_0 \tag{6-27a}$$

$$\frac{dC}{d\tau} = f_2(C, \eta); \qquad C(0) = C_0 \tag{6-27b}$$

to the effect that the solution is always unique, if f_1 and f_2 have continuous first partial derivatives. The functions in equations (6-23) usually do satisfy these rather mild conditions at all C and η of possible interest. When they do not it is appropriate to ask whether the author of the particular kinetic form intended to allow for possible nonunique batch behavior.

6-4. Parametric Sensitivity

To say that a PFTR has a unique steady-state profile implies that one considers a fixed set of system parameters. These are primarily the flow rate u, coolant temperature T_w, and the heat transfer coefficient h, but include also the various thermodynamic and kinetic values that appear in equations (6-21).

The effect of small changes in such parameters on the steady-state profiles can sometimes be surprisingly significant, as demonstrated by Bilous and Amundson (1956) in the results reproduced as Figure 6-6. A mere 2.5°K change in T_w from 335°K to 337.5°K, for example, produces a 67°K increase in maximum temperature. Similar high sensitivities have been reported for changes in heat transfer coefficient. Such *parametric sensitivity* should be avoided in a good design. If this possibility is not anticipated, unexpected

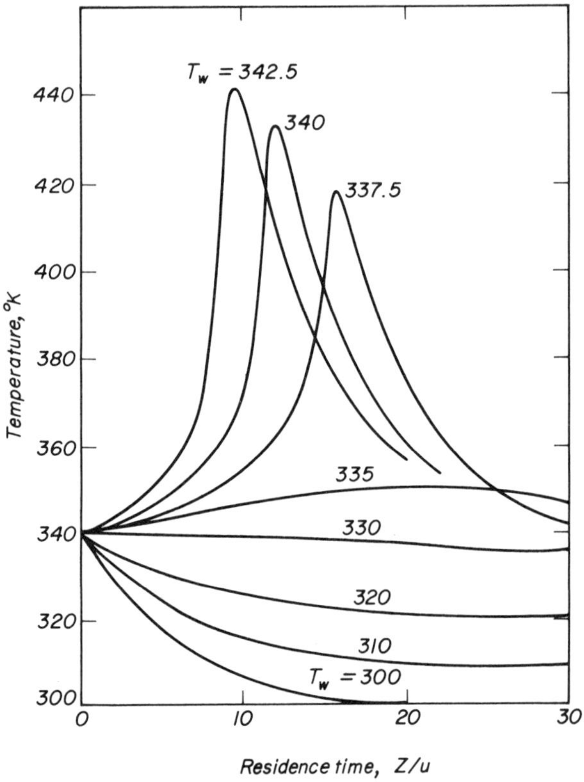

Figure 6-6. Effect of T_w on the temperature profile of a PFTR (after Bilous and Amundson, 1956).

local "hot spots" can develop in a PFTR as following some small parameter fluctuation, the steady-state profile is reestablished with a much higher temperature peak. Equivalently, for a batch reactor for which time dependence takes the place of spacial variation, systems with high parametric sensitivity sometimes give rise to so-called "temperature runaways." On occasion the new temperature peak is so high that explosive side reactions or other catastrophe is possible.

A most direct treatment of this design problem focuses on T_{max}, the maximum of the temperature profile which occurs at the reactor position Z^*. At this location, the heat generated by reaction

$$Q_G = \Delta HR(Z^*) \tag{6-28}$$

is exactly equal to the heat removed by heat transfer

$$Q_R = h(T_{max} - T_w) \tag{6-29}$$

Equating these terms gives

$$R(Z^*) = \frac{h}{\Delta H}(T_{\max} - T_w) = \frac{h}{C_p}(\eta_{\max} - \eta_w) \tag{6-30}$$

which is consistent with a zero derivative in equation (6-21b), as expected. Any choice of the parameters h and T_w that tends to increase Q_R beyond that called for by equation (6-30) will serve to reduce $T_{\max}$ until the balance is reestablished. For a given $T_{\max}$ a safe design may therefore be based on

$$\frac{h}{\Delta H}(T_{\max} - T_w) \geq R(Z^*) \tag{6-31}$$

However, since $R(Z^*)$ is dependent on concentration as well as temperature, one cannot evaluate it without first solving for $C_s(Z)$. To avoid this it is convenient to note that since $C_0 \geq C$ at any position Z (for irreversible reaction):

$$R(C_0, T_{\max}) \geq R(C, T_{\max}) = R(Z^*) \tag{6-32}$$

As a result the equation

$$\frac{h}{\Delta H}(T_{\max} - T_w) = R(C_0, T_{\max}) \tag{6-33}$$

is sufficient to establish (6-31) and provides a conservative basis for design.

Harriot (1961) suggested a useful graphical treatment of equation (6-33). For a given value of C_0 the right side is a function of only $T_{\max}$, usually rising steeply in accordance with the Arrhenius form. The left side is linear in $T_{\max}$, having a slope of $(h/\Delta H)$ and intercept at T_w on the temperature axis. Both functions are shown schematically in Figure 6-7a, on the assumption that the curves intersect at a given $T_{\max}$. Whether they will do so is determined by the designer who will need to choose the slope and intercept parameters for specified values of $T_{\max}$ and C_0. Also shown on the same diagram is a typical

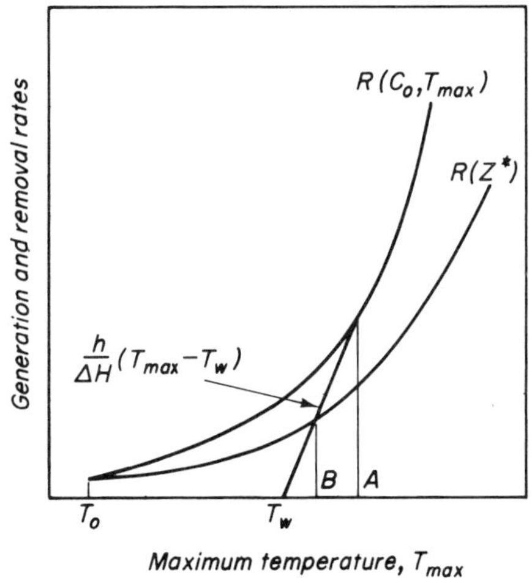

Figure 6-7a. Design diagram for PFTR with large heat transfer coefficient (after Harriot, 1961).

curve for the function $R(Z^*)$. The degree of conservatism inherent in the use of design equation (6-33) may therefore be estimated by noting that with the T_{max} specification at A, this conservative design produces a system whose actual T_{max} would be located at temperature B.

As in many design problems the choices are open within a range of possibilities, and the limitations are at least as significant as the specific numerical values. If the designer works with too steep a straight line he has overdesigned the system by calling for excessive heat transfer. In effect this means unnecessarily small tubes, since the heat transfer coefficient h varies approximately inversely with tube diameter [Hanratty (1954)]. If on the contrary the designer attempts to reduce capital investment and operating cost by taking advantage of a lower coolant temperature, Figure 6-7b shows that

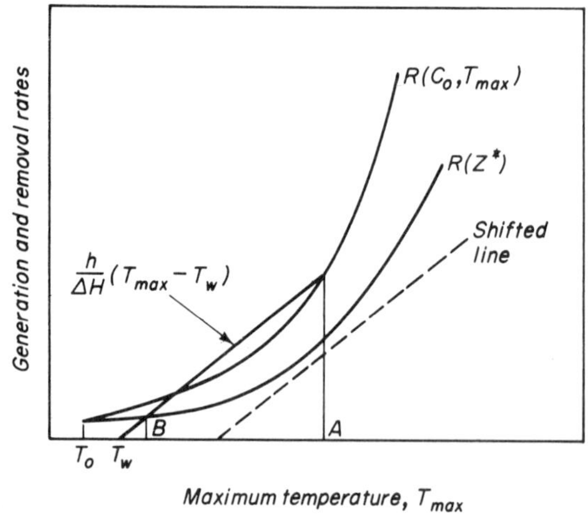

Figure 6-7b. Design diagram for PFTR with low coolant temperature (after Harriot, 1961).

the conservatism of the design has increased to the point that the reacting material at temperature B is not even close to the desired level. Moreover, the designer runs the risk of losing condition (6-31). His relatively flat heat removal line will meet this requirement only so long as the system parameters are within a narrow band. Were a random fluctuation in T_w for example to shift the line to the parallel dashed position shown in Figure 6-7b, T_{max} would rise precipitously. It is clear that a compromise is called for between the extremes of overdesign and oversensitivity. Of the two the latter error is of course the more dangerous.

Several of the remedies commonly used to avoid overheat can be rationalized in the framework of this discussion. Dilution of reactor feed or catalyst has the effect of lowering $R(Z^*)$. The use of staged reaction with intercooling serves the same purpose, since the feed to all stages after the first is at a lower concentration level than C_0. In either case, the designer's problem remains one of properly assessing the heat removal potential of the system.

As a rough but very safe rule-of-thumb one can place the heat removal line to be tangent to $R(C_0, T_{\max})$ at the $T_{\max}$ specification. Then with Arrhenius temperature dependence

$$\frac{h}{\Delta H} = \frac{d}{dT_{\max}} R(C_0, T_{\max}) = \frac{Q}{T_{\max}^2} R(C_0, T_{\max}) \tag{6-34}$$

which may be combined with equation (6-33) to yield

$$(T_{\max} - T_w) = \frac{T_{\max}^2}{Q} \tag{6-35}$$

This design equation was derived for zero-order reactions by Boynton, Nichols and Spurlin (1959), and anticipated earlier by Wilson (1946).

Some of the conservatism in the foregoing discussion can be avoided by using a better estimate of $R(Z^*)$ than is provided by inequality (6-32). The reaction paths of Figure 6-5 suggest that actual concentration levels are always lower than C_{adi} established by adiabatic conversion, a result of the higher adiabatic temperature. Consequently

$$C_0 \geq C_{\text{adi}} \geq C \tag{6-36}$$

at any position in the reactor and

$$R(C_0, T_{\max}) \geq R(C_{\text{adi}}, T_{\max}) \geq R(Z^*) \tag{6-37}$$

This intermediate estimate of $R(Z^*)$ was used by Harriot (1961) to develop his graphical interpretation of equation (6-33). An answer obtained on this basis will be less conservative than the one satisfying (6-35), but at the cost of the additional algebraic detail of finding concentrations from equation (6-25).

With the investment of considerably more computational effort, it is feasible to evaluate $R(Z^*)$ directly and replace the uncertain degree of conservatism by a well-defined safety factor. A systematic attack on this problem was made by Barkelew (1959), who used results of many numerical integrations of equations (6-21) to construct charts that delineate the regions of parameter values for which dangerous sensitivity exists. His results are correlated in a series of design graphs, a composite sample of which is reproduced in terms of the nomenclature of this chapter as Figure 6-8. The three solid lines show the effect of order of reaction when the reactor feed temperature equals the coolant temperature. The darkened band around the first-order line shows the effect of various feed temperatures in the range

$$T_w - \frac{T_w^2}{2Q} \leq T_0 \leq T_w + \frac{T_w^2}{Q} \tag{6-38}$$

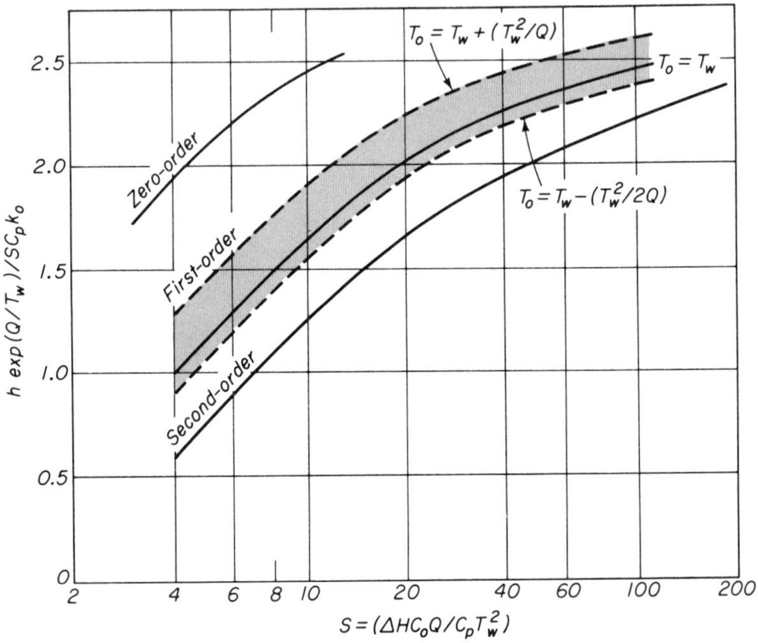

Figure 6-8. Parametric sensitivity correlation (after Barkelew, 1959).

In every case the region below the line is a region of potential danger where Barkelew found the maximum temperature to be highly sensitive to parameter changes.

Above all it should be appreciated that the various design approaches to the PFTR become dangerously undependable as the assumption of negligible radial gradients breaks down, and this in spite of the conservative estimates that are usually used. Consider for instance that the "safe" design equation (6-33) rests on the notion that the same T_{max} that enters the evaluation of reaction rate also affects the driving force for heat loss. It would be closer to the truth when appreciable radial gradients exist to rewrite this equation as

$$\frac{h}{\Delta H}[T(L_0) - T_w] = R(C_0, T_{max}) \tag{6-38}$$

where by the nature of the heat transfer requirements $T(L_0)$ at the tube wall is less than T_{max}. Examining Figure 6-2a as a case in point, it may be noted that

$$\frac{1}{C_0}[\eta_{\max} - \eta_w] = 4.47 - 3.25 = 1.22$$

whereas the temperature driving force at the wall corresponds to about

$$\frac{1}{C_0}[\eta(L_0) - \eta_w] = 3.35 - 3.25 = 0.10$$

Evidently an estimate of heat transfer coefficient based on the former would be too low by over an order of magnitude. In recognition of this effect, more recent computational studies have looked at parametric sensitivities for the more elaborate tubular reactor models—those that account for radial mass and heat transfer. Several parameters were found to have very significant effects in this regard, particularly the wall heat transfer coefficient [Froment (1967)] and the thermal Peclet number in the radial direction [Carberry and White (1969)].

EXAMPLE 6-1. Compare the results of applying the several procedures of this section to the design of a PFTR for the following steady-state specifications:

$$Q = 8{,}400$$
$$k_0 = 3.94 \times 10^{12}\ \text{hr}^{-1}\ \text{(first-order kinetics)}$$
$$\frac{\Delta H}{C_p} = 7{,}300\ \text{lit}\ ^\circ\text{C/g-mol}$$
$$T_{\max} = 418^\circ\text{K}$$
$$C_0 = 0.02\ \text{g-mol/lit}$$
$$T_0 = 340^\circ\text{K}$$

What changes are needed to make the analysis applicable to a batch reactor?

Solution: Part (a): To evaluate Wilson's rule-of-thumb, equation (6-35) is solved for

$$T_w = T_{\max}\left(1 - \frac{T_{\max}}{Q}\right) = 396^\circ\text{K}$$

and equation (6-34) is rearranged to

$$\frac{h}{C_p} = \frac{\Delta H}{C_p}\frac{Q}{T_{\max}^2}k_0C_0\exp\left(-\frac{Q}{T_{\max}}\right) = 8.36\ \text{sec}^{-1}$$

Part (b): The arbitrariness of Wilson's rule is not present if one returns to equation (6-33), but then some other basis is needed for the choice of T_w. If it is assumed that

$$T_w = T_0 = 340^\circ\text{K}$$

equation (6-33) gives

$$\frac{h}{C_p} = \frac{\Delta H}{C_p}\frac{k_0C_0}{(T_{\max} - T_w)}\exp\left(-\frac{Q}{T_{\max}}\right) = 3.90\ \text{sec}^{-1}$$

Part (c): For a less conservative design inequality (6-37) shows that the part (b) computation can be modified by replacing C_0 with C_{adi} from equation (6-25)

$$C_{\text{adi}} = C_0 + \eta_0 - \eta_{\max} = C_0 + \frac{C_p}{\Delta H}(T_0 - T_{\max}) = 0.0093 \text{ g-mol/lit}$$

Then

$$\frac{h}{C_p} = 3.90\frac{C_{\text{adi}}}{C_0} = 1.81 \text{ sec}^{-1}$$

Part (d): Barkelew's correlation uses the dimensionless group

$$S = \frac{\Delta H}{C_p}\frac{C_0 Q}{T_w^2} = 10.6$$

For the case where the feed and coolant temperatures are the same, Figure 6-8 gives

$$\frac{h}{C_p} = 1.6\, Sk_0 \exp\left(-\frac{Q}{T_w}\right) = 0.331 \text{ sec}^{-1}$$

as the critical value below which dangerous sensitivity occurs. The estimate agrees very well with the Bilous and Amundson results cited above, in which excessive sensitivity to T_w was observed for a choice of

$$\frac{h}{C_p} = 0.200 \text{ sec}^{-1}$$

A comparison of the four design bases appears to favor Barkelew's method, since it is the least conservative. Note, however, that with a temperature rise of 78°K there is reason to hold serious reservations regarding the neglected radial gradients. With this in mind, the estimates of parts (a) and (b) are probably more realistic.

The results are also applicable to a batch reactor design if the heat transfer coefficient is replaced by one that is appropriate to the stirred vessel geometry. Comparing equations (6-23) with (1-4) shows that (U/V) should replace h, and it is understood that the maximum rate and maximum temperatures referred to occur at a particular time rather than a particular place. The assumption of uniformity in temperature and concentration seems well-founded in the stirred batch reactor and accordingly one can have greater confidence in the part (d) estimate:

$$(U/VC_p) = 0.331 \text{ sec}^{-1}$$

A word is in order here on the use of the term stability in the context of parametric sensitivity studies. In as much as stability may be understood to mean steadiness, the reactor with great parametric sensitivity is indeed unstable, a usage favored by Wilson (1946), Barkelew (1959), Boynton, Nichols, and Spurlin (1959), and Harriot (1961). More precisely one should perhaps specify that it is the PFTR steady-state profiles that are unstable with respect to changes in the system parameters. Some arbitrary measure is implied in this description, since there are no completely stable profiles in this sense, all steady states being sensitive in some degree to parameter changes. Furthermore, though it is reasonable in itself this use of the name stability is not consistent with the usage in dynamics and control, where it refers to *time* dependence, in response to a change in input or initial condition. Since this interpretation was used in the previous discussion of lumped-parameter systems, and will be appropriate in later chapters, there will be less likelihood of confusion if the term stability is avoided entirely in discussions of parametric sensitivity in the PFTR.

A connection can be established between these concepts by considering the temperature and concentration profiles for one value of some parameter to be the initial conditions of time-dependent changes, and the profiles corresponding to a second value to be a final steady state. From this viewpoint the PFTR and batch reactors are both stable in the usual sense of dynamics, even when their parametric sensitivity is high, because the initial deviations always decay to the new steady state. A fuller exploration of this subject belongs in the next chapters on stability in distributed-parameter systems.

6-5. Steady-State Multiplicity

Since the term steady state is nothing more than the condition that all time derivatives of the state variables are zero, the questions of stability and multiple solutions remain open until the system is more narrowly defined. For the PFTR it was easy to demonstrate that only unique profiles are possible, but this simple argument breaks down when the changes in the reactor are more complicated than uni-directional and downstream. Whenever heat or mass can move upstream there exists the possibility of multiple steady states (van Heerden, 1958). The upstream movement can be the result of physical recycle or feedback control, but can arise even more simply from diffusional or dispersive backmixing effects, as demonstrated in the experiments of Wicke and Vortmeyer (1959), Wicke (1961), Padberg and Wicke (1967), and Wicke, Padberg, and Arens (1968).

Even when affects such as backmixing allow the possibility of multiple steady-state solutions, their occurrence is not in general guaranteed, but will depend on the parameter values specific to the system under study. A numerical study of equation (6-12) by Raymond and Amundson (1964) produced either unique or multiple solutions according to the parameter values chosen. It was assumed that the reaction rate was of the first-order Arrhenius form, which when combined with the linear relation (6-11) gives

$$R(\eta) = k_0(C_0 + \eta_0 - \eta)\exp\left(-\frac{C_pQ}{\Delta H\eta}\right) \tag{6-39a}$$

That such findings of multiplicity are not limited to adiabatic reactors was demonstrated by McGowin (1971a), who numerically integrated the coupled TRAM equations (6-1) subject to boundary conditions (1-9). Selected results of his study are shown in Figure 6-9a and 6-9b in the form of families of steady-state operating curves. The first of these emphasizes the effect of heat transfer at constant axial dispersion at the level of

$$\frac{uL}{D} = \frac{uL}{\alpha} = 10$$

The second of these figures uses axial dispersion as the parameter for constant heat transfer to the extent of

$$\frac{hL}{uC_p} = 0.10$$

In all cases the kinetic parameters are the same as those reported with Figure 6-4.

Even a cursory inspection of Figures 6-9 leads to the conclusion that multiple steady states exist over a wide range of the parameters, and that ignition and extinction phenomena can occur when feed conditions fluctuate in a critical range. A more careful comparison of Figures 6-9a, 6-4, and 2-7 indicates that multiplicity in the TRAM occurs over a much wider range of the heat transfer coefficient than for either the CSTR or the PFTR. A further

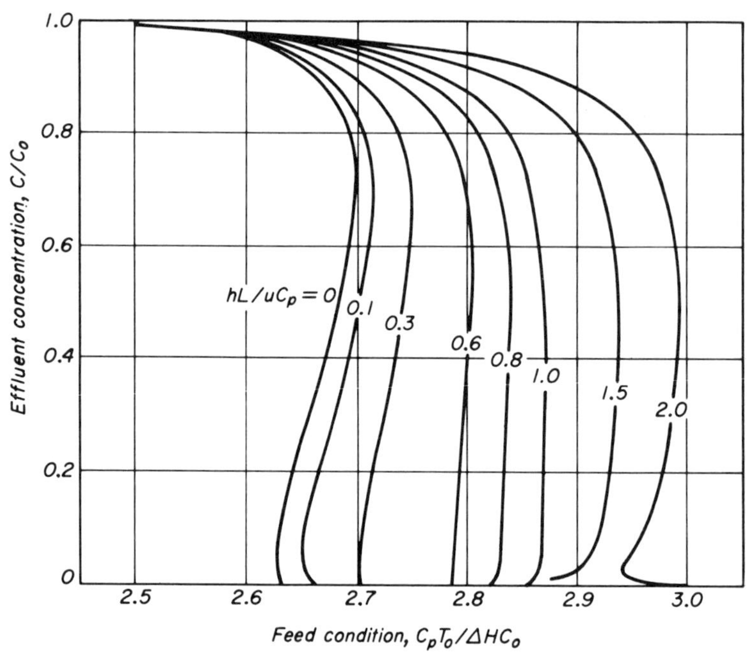

Figure 6-9a. Steady-state operating curves for the TRAM; $\alpha = D$, $(uL/D) = 10$.

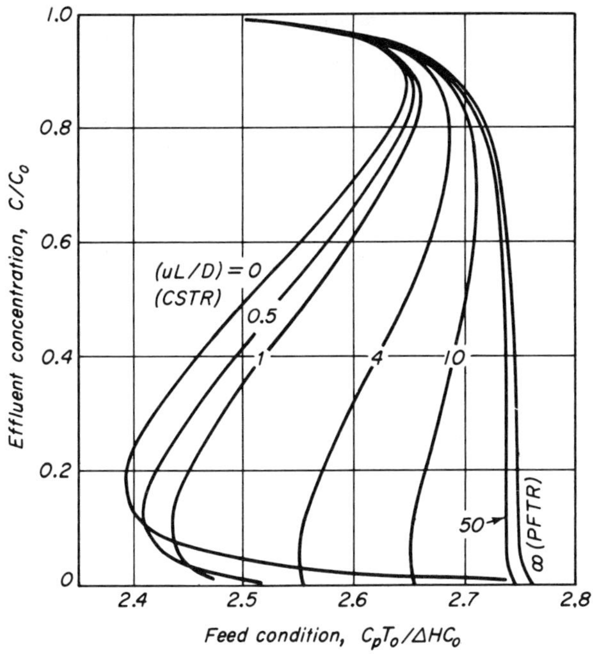

Figure 6-9b. Steady-state operating curves for the TRAM with $(uL/D = (uL/\alpha), (hL/uC_p) = 0.10$.

distinction may be noted by replotting Figure 6-9a on an expanded ordinate scale as in Figure 6-9c. The unusual shape of the curves in the range

$$0.18 < \frac{hL}{uC_p} < 0.49$$

suggests that as many as five steady states may exist for a given feed condition and fixed degree of axial dispersion, however the work of Hatfield and Aris (1969) on a related catalyst particle problem indicates that this result may be an artifact introduced by the particular choice of boundary conditions. In any case and whatever the degree of multiplicity, it should be understood that the various steady states do not exist simultaneously in a given reactor, but rather that when they are stable, they represent possible profiles that can develop from certain initial conditions.

Evidence of multiple steady states for a catalyst particle was also obtained by Raymond and Amundson (1965). Their computations were based on the same first-order kinetics in combination with linear relation (6-19) in the form

$$R(\eta) = k_0\left[C_0 + \frac{\alpha}{D}\eta_0 - \frac{\alpha}{D}\eta\right]\exp\left(-\frac{C_pQ}{\Delta H\eta}\right) \tag{6-39b}$$

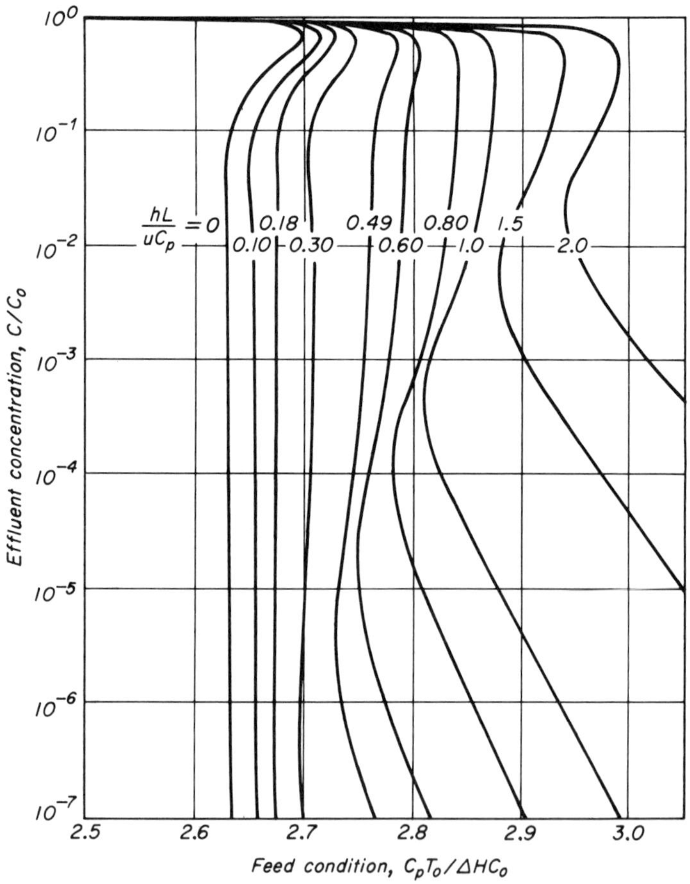

Figure 6-9c. Steady-state operating curves for the TRAM on expanded ordinate scale.

The result of numerically integrating equation (6-20) is reproduced for a multiple solution case in Figure 6-10, for the parameter groups:

$$\frac{\Delta H C_0 D}{C_p T_0 \alpha} = 0.7$$

$$\frac{L^2 k_0}{D} = 1.23 \times 10^7$$

$$\frac{Q}{T_0} = 20$$

That catalyst particle models can exhibit such behavior was in fact understood even earlier when computations of effectiveness factors by Carberry (1961), Tinkler and Metzner (1961), and Weisz and Hicks (1962) yielded

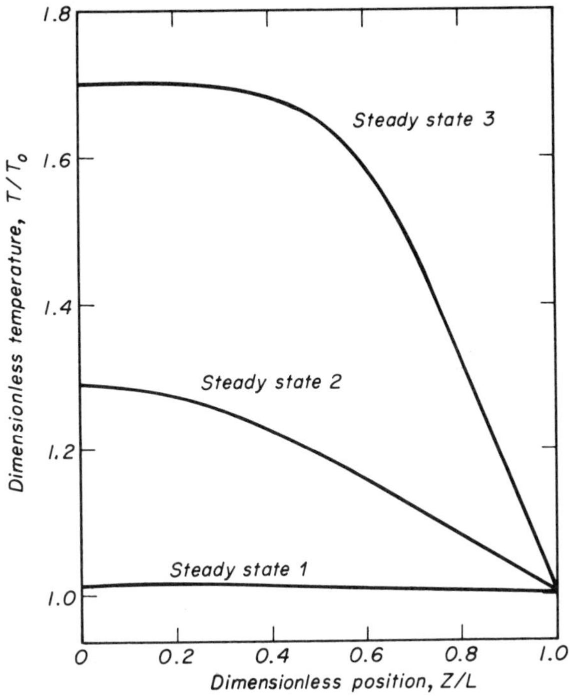

Figure 6-10. Steady-state profiles for a catalyst particle (after Raymond and Amundson, 1965).

multiple solutions for some ranges of the Thiele modulus. It remained for Roberts and Satterfield (1966), however, to demonstrate that the same can be true for the isothermal catalyst model

$$0 = D\frac{d^2C}{dZ^2} - R(C) \tag{6-40}$$

when the reaction rate is of the Langmuir–Hinshelwood form

$$R = \frac{kC(C + K_1)}{(1 + K_2C)^2} \tag{6-41}$$

More recent study has shown that although no more than three steady states are possible for the slab geometry, as many as five stationary states may be found for zero-order kinetics in a spherical particle [Hlavacek and Marek (1968)]. Computations have also been reported [Copelowitz and Aris (1970)] that show as many as fifteen steady states for the same equations. Most of these states are so close to one another as to be virtually indistinguishable in practice, but the theoretical implications are of interest in terms of what might be developed in more complex problems.

Yet another mode of thermal feedback was studied by Chen and Churchill

(1970), who recognized that steady-state multiplicity can arise in the PFTR when especially high temperatures cause large radiative energy movements upstream.

EXAMPLE 6-2. Solve equation (6-40) with boundary conditions (6-14) when the rate term is of the simple first-order type, $R = kC$. Are there any values of k for which multiple solutions are possible?

Solution: Substitution of the indicated rate term gives the linear constant coefficient equation

$$0 = D\frac{d^2C}{dZ^2} - kC \tag{6-42}$$

with boundary conditions

$$\frac{dC}{dZ} = 0, \qquad Z = 0 \tag{6-43a}$$

$$C = C_0, \qquad Z = \mathrm{L} \tag{6-43b}$$

If as is ordinarily understood to be the case, $k > 0$, equation (6-42) has the general solution

$$C = A_1 \exp\left(\sqrt{\frac{k}{D}}Z\right) + A_2 \exp\left(-\sqrt{\frac{k}{D}}Z\right) \tag{6-44}$$

The use of condition (6-43a) shows that $A_1 = A_2$ and reduces (6-44) to

$$C = 2A_1 \cosh\sqrt{\frac{k}{D}}Z \tag{6-45a}$$

Evaluating the coefficient A_1 from the remaining boundary condition leads to

$$C = C_0\frac{\cosh\sqrt{k/D}\,Z}{\cosh\sqrt{k/D}\,L} \tag{6-45b}$$

If on the contrary the mass balance is for a product of reaction rather than a reactant, $k < 0$ and the general solution is

$$C = A_1 \sin\sqrt{\frac{-k}{D}}Z + A_2 \cos\sqrt{\frac{-k}{D}}Z \tag{6-46}$$

For this case the same boundary conditions give in succession

$$C = A_2 \cos\sqrt{\frac{-k}{D}}Z \tag{6-47a}$$

and

$$C = C_0\frac{\cos\sqrt{-k/D}\,Z}{\cos\sqrt{-k/D}\,L} \tag{6-47b}$$

The remaining possibility is the circumstance of no reaction. Then $k = 0$, and the

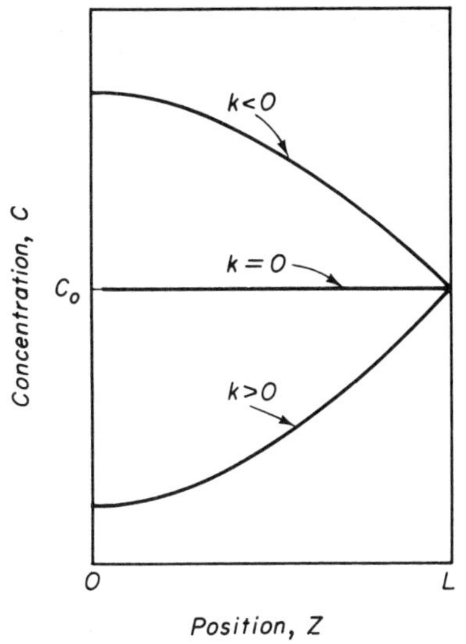

Figure 6-11. Possible solution forms for Example 6-2.

general solution

$$C = A_1 + A_2 Z \tag{6-48}$$

reduces for the specific boundary conditions to

$$C = C_0 \tag{6-49}$$

All three solutions are shown in Figure 6-11. There is no question but that a unique solution results for any choice of k.

6-6. Eigenvalues and Trivial Solutions

The solution to a differential equation system is of course as much dependent on the boundary conditions as on the equation. By emphasizing this aspect of the analysis, one can uncover a different kind of result which has a critical bearing on the question of multiple solutions.

Consider again the equation (6-42) but this time with boundary conditions

$$\frac{dC}{dZ} = 0, \qquad Z = 0 \tag{6-50a}$$

$$C = 0, \qquad Z = L \tag{6-50b}$$

Detailed examination will show that for $k > 0$, these boundary conditions are consistent with equation (6-44) only for $A_1 = A_2 = 0$. That is to say, the only possible solution to the system is the so-called *trivial* solution

$$C \equiv 0 \tag{6-51}$$

The same is true for $k = 0$, as is apparent for these boundary conditions from equation (6-48), but multiple solutions do arise for particular choices of $k < 0$. If for example $k = -D\pi^2/4L^2$, equation (6-46) becomes

$$C = A_1 \sin \frac{\pi Z}{2L} + A_2 \cos \frac{\pi Z}{2L} \tag{6-52}$$

which reduces by boundary condition (6-50b) to

$$C = A_2 \cos \frac{\pi Z}{2L} \tag{6-53}$$

Note that this is a solution to the problem for *every* value of the coefficient A_2. When the specific choice of a problem parameter produces such a solution, it is called an *eigenvalue*. Other eigenvalues for this problem are

$$k = -\frac{D}{L^2}\left[\frac{2i+1}{2}\pi\right]^2, \qquad i = 1, 2, \ldots \tag{6-54}$$

but they do not in this particular problem lead to physically acceptable answers because of the requirement that $C(Z)$ be everywhere positive. For any value of k that is not an eigenvalue, the trivial solution will be unique.

It would seem that this use of the term eigenvalue is different from that applicable to the matrix analysis of Chapter 3. That they are in fact related in an important sense can be seen from equation (3-26):

$$\lambda \mathbf{x} = \mathbf{A}\mathbf{x}$$

in which λ may be considered a parameter. In this earlier usage too, the eigenvalues are those choices of λ for which a solution exists other than the trivial one: $\mathbf{x} \equiv \mathbf{0}$.

While on the subject of eigenvalues for various catalyst particle problems, it is worth examining one further set of boundary conditions in anticipation of what will follow later in this chapter. Suppose that a mass transfer limitation exists at the particle surface, such that the flux reaching the surface is balanced by loss from the surface into a surrounding vacuum. Expressing the latter in terms of a mass transfer coefficient k_G gives the boundary conditions:

$$\frac{dC}{dZ} = 0, \qquad Z = 0 \tag{6-55a}$$

$$\frac{dC}{dZ} = -\frac{k_G}{D}C, \qquad Z = L \tag{6-55b}$$

Once again substitution of these conditions into equations (6-44), (6-46) and (6-48) produces the unique trivial solution for $k \geq 0$, as well as for negative k's that are not eigenvalues. Physical arguments support this result as to the sign of k, since a nonzero balance between reaction and transfer to the surroundings only makes sense if the reaction serves as a source term.

The more quantitative result regarding eigenvalues is not as intuitively obvious: that the sought for balance can only occur for specific discrete values of reaction rate.

EXAMPLE 6-3. Find the eigenvalues for the system of equations (6-42) and (6-55).

Solution: Exploring first the $k > 0$ case, it can be confirmed that the problem has no real eigenvalues by working with the partial solution (6-45a) which already accounts for boundary condition (6-55a). The remaining condition (6-55b) would be satisfied for any A_1 if

$$\sqrt{\frac{k}{D}} \sinh \sqrt{\frac{k}{D}} L = -\frac{k_G}{D} \cosh \sqrt{\frac{k}{D}} L \tag{6-56}$$

or

$$\sqrt{\frac{k}{D}} L \tanh \sqrt{\frac{k}{D}} L = -\frac{k_G}{D} L \tag{6-57}$$

but this is never true since the hyperbolic tangent always has the same sign as its argument.

Turning to the $k < 0$ case calls forth equation (6-47a). Now condition (6-55b) requires

$$\sqrt{\frac{-k}{D}} L \tan \sqrt{\frac{-k}{D}} L = \frac{k_G}{D} L \tag{6-58a}$$

That is to say, the values of k that satisfy equation (6-58a) are eigenvalues, and for such parameter choices multiple solutions of the form (6-47a) will result (for any A_2).

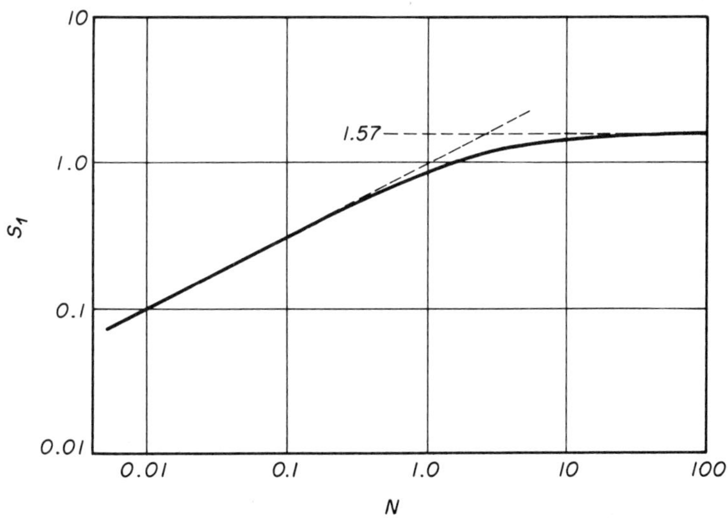

Figure 6-12. Solution of equation (6-58b).

The roots of equation (6-58a) are available in Carslaw and Jaeger (1959) in the form of solutions of

$$\mathscr{S} \tan \mathscr{S} = N \tag{6-58b}$$

where

$$\mathscr{S} = \sqrt{\frac{-k}{D}}L \tag{5-59a}$$

and

$$N = \frac{k_G L}{D} \tag{6-59b}$$

A concise criterion for unique solution may then be written as

$$k > -D\left(\frac{S_1}{L}\right)^2 \tag{6-60}$$

where S_1 is the smallest root of equation (6-58b). For numerical convenience values of S_1 are presented in Figure 6-12 as a function of the parameter N.

6-7. A Uniqueness Criterion

The CSTR results of Section 2-2, which showed how a uniqueness criterion could be developed in order to circumvent extensive numerical work, have a parallel in the realm of distributed systems. In fact, it was in this latter application that Luss and Amundson (1967) first showed the usefulness of the mean value theorem in such problems.

As a preliminary, consider the following statement of the mean value theorem with reference to the continuous function $R(C)$ shown in Figure 6-13: there always exists some C_* between C_1 and C_2 such that the slope of $R(C)$ at that location is parallel to the chord AB. Or algebraically:

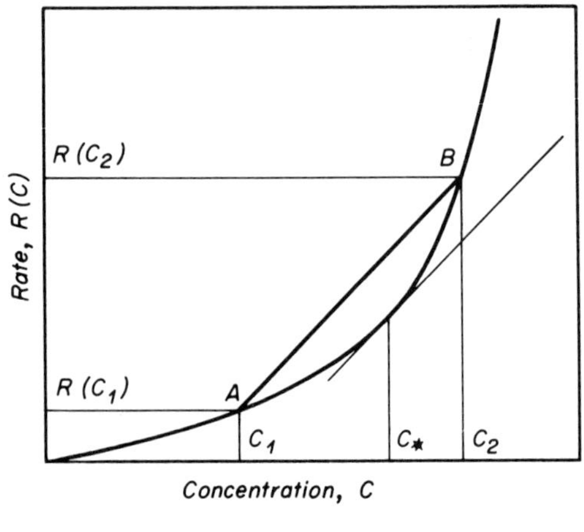

Figure 6-13. Illustration of the mean value theorem.

$$\left.\frac{dR}{dC}\right|_{C_*} = \frac{R(C_2) - R(C_1)}{C_2 - C_1} \tag{6-61}$$

Assume now that there are multiple solutions to equation (6-40) with boundary conditions (6-43). Call these solutions $C_1(Z)$ and $C_2(Z)$ and define

$$\Phi(Z) = C_2(Z) - C_1(Z) \tag{6-62}$$

Since any solution satisfies equation (6-40):

$$0 = D\frac{d^2C_1}{dZ^2} - R(C_1) \tag{6-63a}$$

$$0 = D\frac{d^2C_2}{dZ^2} - R(C_2) \tag{6-63b}$$

and by subtraction

$$0 = D\frac{d^2(C_2 - C_1)}{dZ^2} - [R(C_2) - R(C_1)] \tag{6-64}$$

Substituting equations (6-61) and (6-62) gives

$$0 = D\frac{d^2\Phi}{dZ^2} - \left.\frac{dR}{dC}\right|_{C_*}\Phi \tag{6-65}$$

with boundary conditions:

$$\frac{d\Phi}{dZ} = \frac{dC_2}{dZ} - \frac{dC_1}{dZ} = 0, \qquad Z = 0 \tag{6-66a}$$

$$\Phi = C_0 - C_0 = 0, \qquad Z = L \tag{6-66b}$$

If indeed the assumption of multiple solutions is correct, one should expect some $\Phi(Z)$ to exist that satisfied (6-65) and (6-66). Otherwise, $C_2(Z)$ and $C_1(Z)$ are indistinguishable and the only possible Φ is identically zero. Note however that this question has already been raised and answered in Section 6-6, since but for a change in symbols the following equivalences exist:

In terms of C	In terms of Φ
Equation (6-42)	Equation (6-65)
Boundary conditions (6-50)	Boundary conditions (6-66)
Parameter k	Parameter $(dR/dC)_{C_*}$

It may therefore be concluded that the system of equations (6-65) and (6-66) can have no solution other than $\Phi \equiv 0$, if

$$\left.\frac{dR}{dC}\right|_{C_*} > -\frac{D\pi^2}{4L^2} \tag{6-67a}$$

or

$$\left.\frac{dR}{dC}\right|_{C_*} \geq 0 \tag{6-67b}$$

and either of these inequalities is a sufficient condition for a unique solution of equation (6-40) with boundary conditions (6-14).

In some circumstances it is desirable to have available a criterion that is less conservative than inequality (6-67b), and Luss (1968) has marshalled the arguments that provide such a condition. He has developed a proof (see Appendix B for details) to show that uniqueness can be guaranteed if

$$(C - C_0)\frac{d\ln R}{dC} \leq 1 \tag{6-68}$$

over the full range of concentrations.

EXAMPLE 6-4. Reconsider equation (6-40) for specific choices of kinetic expressions, examining the system for possible unique solution.

Solution: Assume as case 1 that the isothermal rate expression is

$$R = kC^n \tag{6-69}$$

with $k > 0$ and $n > 0$. Then

$$\frac{dR}{dC} = nkC^{n-1} = \frac{nR}{C}$$

is positive for any feasible value of C, including C_*. Accordingly inequality (6-67b) is satisfied and the solution to equation (6-40) is unique for any choice of k and n.

Take as case 2 the Langmiur–Hinshelwood kinetics which Roberts and Satterfield (1965) identify as Type I:

$$R = \frac{kC(C + K_1)}{(1 + K_2C)} \tag{6-70}$$

Then

$$\frac{dR}{dC} = \frac{kC(2 + K_2C) + kK_1}{(1 + K_2C)^2}$$

which must be positive for k and K_1 positive. (That the signs of the terms in parentheses must also be positive is a consequence of positive rate R). As before, any solution is unique.

For a third case use Roberts and Satterfield (1966) Type II:

$$R = \frac{kC(C + K_1)}{(1 + K_2C)^2} \tag{6-71}$$

with k and $K_1 > 0$. Differentiating

$$\frac{dR}{dC} = \frac{k[K_1 + C(2 - K_1K_2)]}{(1 + K_2C)^3}$$

which has the required sign to guarantee a unique solution if

$$K_1K_2 \leq 2$$

or less conservatively if

$$K_1K_2 \leq 2 + \frac{K_1}{C} \tag{6-72}$$

at all possible values of C. Because $C \leq C_0$ for an irreversible reaction, inequality (6-72) may be replaced by

$$K_1 K_2 \leq 2 + \frac{K_1}{C_0}$$

Such an analysis was presented by Luss and Amundson (1967a) together with several comparisons to the numerical results of Roberts and Satterfield. If one seeks a still less conservative sufficient condition, it is necessary to use (6-68) rather than merely depend on the sign criterion (6-67b).

6-8. Extension to Nonisothermal Problems

There is no reason to limit the development of uniqueness criteria to isothermal systems. What is required is that the system model consist of only a single equation in order that the mean value theorem be applicable. A suitable candidate is equation (6-20) for the nonisothermal catalyst particle. Note the following equivalences:

In terms of η	In terms of C
Equation (6-20)	Equation (6-40)
Boundary conditions (6-14)	Boundary conditions (6-14)
Parameter: α	Parameter: D
Rate term: $R(\eta)$	Rate term: $-R(C)$

With changes only in symbols and a single sign, it may be concluded that no multiple solutions to equation (6-20) are possible if

$$\left.\frac{dR}{d\eta}\right|_{\eta*} < \frac{\alpha\pi^2}{4L^2} \tag{6-73a}$$

or

$$\left.\frac{dR}{dT}\right|_{\eta*} \leq 0 \tag{6-73b}$$

It is interesting to observe that this condition is virtually the same as inequality (2-40) derived for the CSTR.

Exploring the implication of this inequality for the specific kinetics of equation (6-39b):

$$\frac{dR}{d\eta} = R\left[\frac{QC_p}{\Delta H\eta^2} - \frac{n\alpha}{CD}\right] \tag{6-74}$$

which has the appropriate sign if

$$DQC_pC \leq n\alpha\Delta H\eta^2$$

or in terms of the temperature variable

$$DQ\Delta HC \leq n\alpha T^2 C_p$$

Finally noting that for this catalyst particle

$$C \leq C_0$$
$$T \geq T_0$$

for essentially the same reasons that the earlier inequalities (2-33) and (2-34) applied to the CSTR, a sufficient uniqueness condition is

$$\frac{Q\Delta H C_0}{nC_p T_0^2} \leq \frac{\alpha}{D} \tag{6-75}$$

As might perhaps be anticipated from recollections of the parallel CSTR results in Chapter 2, this criterion is rather conservative. It can be improved upon by making use of the temperature analog of inequality (6-68) for the adiabatic particle:

$$(T - T_0)\frac{d \ln R}{dT} = (\eta - \eta_0)\frac{1}{R}\frac{dR}{d\eta} \leq 1 \tag{6-76}$$

For kinetics of the nth-order Arrhenius form substitution of equation (6-74) provides

$$(\eta - \eta_0)\left[\frac{QC_p}{\Delta H \eta^2} - \frac{n\alpha}{CD}\right] \leq 1 \tag{6-77}$$

To remove the explicit concentration term, equation (6-19b) may be used in the form

$$C = \frac{\alpha}{D}(\eta_{\text{adi}} - \eta) \tag{6-78}$$

where η_{adi} is defined as the reduced temperature that corresponds to complete adiabatic reaction—i.e., the solution of equation (6-19b) for $C = 0$. By combining (6-77) and (6-78):

$$(\eta - \eta_0)\left[\frac{QC_p}{\Delta H \eta^2} - \frac{n}{\eta_{\text{adi}} - \eta}\right] \leq 1 \tag{6-79a}$$

or in terms of temperature

$$(T - T_0)\left[\frac{Q}{T^2} - \frac{n}{T_{\text{adi}} - T}\right] \leq 1 \tag{6-79b}$$

This relationship is in general cubic, but reduces for first-order kinetics to a quadratic in temperature:

$$(T_0 - T_{\text{adi}} - Q)T^2 + Q(T_{\text{adi}} + T_0)T - QT_0T_{\text{adi}} \leq 0$$

Equivalent to this inequality is the requirement of a zero or negative discriminant (repeated or complex roots). Thus

$$Q^2(T_{\text{adi}} - T_0)^2 \leq 4QT_0T_{\text{adi}}(Q + T_{\text{adi}} - T_0)$$

which simplifies to Luss' (1968) result:

$$\frac{Q(T_{\text{adi}} - T_0)}{T_0T_{\text{adi}}} \leq 4 \tag{6-80}$$

If one resubstitutes relations (6-19b) and (6-78) this inequality may also be written as

$$\frac{QD\Delta HC_0}{\alpha C_p T_0^2} \leq 4\left[1 + \frac{D\Delta HC_0}{\alpha C_p T_0}\right] \tag{6-81a}$$

but since we are dealing with a sufficient condition, it is acceptable to drop the final term in the bracket. This justifies the criterion

$$\frac{QD\Delta HC_0}{\alpha C_p T_0^2} \leq 4 \tag{6-81b}$$

The uniqueness conditions (6-81b), (6-81a), and (6-75) are all conservative but approach increasingly closer to the necessary condition in that order. They suffer, however, from an even greater fault from the designer's viewpoint in that the parameters involved are largely almost beyond adjustment, being fixed by either the system kinetics, the fluid properties, or the physical state of the catalyst. Remedy may be sought in the development of a criterion based on equation (6-73a):

$$\text{Max}\left[\frac{dR}{dT}\right] < \frac{\pi^2 \alpha C_p}{4L^2 \Delta H} \tag{6-82}$$

Further study of the specific kinetic equation (1-6) shows the derivative function to be identical to equation (2-51) provided that the parameter β is defined from the differentiation of (6-19):

$$\beta = -\frac{dC}{dT} = -\frac{C_p}{\Delta H}\frac{dC}{d\eta} = \frac{\alpha C_p}{D\Delta H} \tag{6-83}$$

If then follows directly that for first-order kinetics equation (2-52) leads to (2-55), where however ψ must be redefined as

$$\psi = \frac{\beta Q}{C_0 + \beta T_0} \tag{6-84}$$

which is identical in form to equation (2-54b). The final result can again be in one of two forms, depending on the test inequalities (2-59) and (2-60). For the parameter values that are commonly used in catalyst studies, the local maximum (dR/dT) will be appropriate rather than the boundary value. Combining equations (2-58) and (6-82) gives the uniqueness condition for a catalyst particle, which is parallel to condition (2-62) for the CSTR:

$$\frac{\pi^2 \alpha C_p}{4L^2 \Delta H} > \beta k_0 \left[\frac{\psi + 4}{\psi}\right] \exp(-\psi - 2) \tag{6-85a}$$

or

$$\frac{\pi^2 D}{4L^2 k_0} > \left[\frac{\psi + 4}{\psi}\right] \exp(-\psi - 2) = G(\psi) \tag{6-85b}$$

As before, Figure 2-5 may be used for numerical evaluations. Note that a designer could now guarantee a unique steady state by choosing a small enough catalyst particle dimension, if the other parameters are fixed. The report of Luss and Amundson (1967b) which first presented this result develops similar relationships for other geometries, and tabulates comparisons with prior integrations of earlier workers [Weisz and Hicks (1962)]. A complete parametric study by Drott and Aris (1969) has shown that the degree of conservatism associated with inequality (6-85) is not unduly large; nevertheless it may be reduced at the expense of some additional computations by adopting a procedure developed by Hlavacek, Marek, and Kubicek (1968). This latter study also includes provision for any mass or heat transfer boundary conditions when they are modeled by equations (1-13d and e).

Since the condition (6-85b) is sufficient but not necessary, it does not include the full range of parameter values for uniqueness. Not only does this produce the conservatism already discussed, but it also allows for uniqueness at the other end of the range of system constants. In his (1969) paper and subsequent (1971) amendment Luss attacked this question systematically and demonstrated that uniqueness can also be guaranteed by

$$\left.\frac{dR}{dT}\right|_{T_0} > \frac{\pi^2 \alpha C_p}{4L^2 \Delta H} \tag{6-86a}$$

for the catalyst pellet with slab geometry. By applying equations (2-55) and (6-83) this criterion reduces to the working inequality

$$\frac{\pi^2 D}{4L^2 k_0} < \left[\frac{Q}{T_0}\left(\frac{Q}{T_0 \psi} - 1\right) - 1\right] \exp\left(-\frac{Q}{T_0}\right) \tag{6-86b}$$

Thus, uniqueness can always be established by a choice of particle sufficiently large, as well as by a choice adequately small. It is to be noted that the group on the left side of (6-85b) and (6-86b) is proportional to the reciprocal of the commonly used Thiele modulus.

EXAMPLE 6-5. To estimate the range of catalyst particle sizes required to insure a unique steady-state solution, find the sufficient values of $(L^2 k_0/D)$, given the other parameter values listed with Figure 6-10.

Solution: From the data, equation (6-83) gives

$$\beta = \frac{\alpha C_p}{D \Delta H} = \frac{1}{0.7} \frac{C_0}{T_0}$$

and from equation (6-84)

$$\psi = \frac{\beta Q}{C_0 + \beta T_0} = \frac{Q/T_0}{1 + 0.7} = 11.8$$

The test inequality

$$\frac{Q}{\psi + 2} > T_0$$

may also be written as

$$\frac{(Q/T_0)}{\psi + 2} = 1.45 > 1$$

and shows that inequality (6-85b) is applicable. By rearrangement

$$\frac{L^2 k_0}{D} < \frac{\pi^2}{4G(\psi)} = 1.6 \times 10^6$$

To find the other limit in the range of large sizes, evaluate inequality (6-86b):

$$\frac{L^2 k_0}{D} > \frac{\pi^2}{4} \frac{\exp(Q/T_0)}{\left[\frac{Q}{T_0}\left(\frac{Q}{T_0 \psi} - 1\right) - 1\right]} = 94 \times 10^6$$

The numerical integration results for this problem produced multiple solutions for $(L^2 k_0/D) = 12.3 \times 10^6$, indicating by comparison that the computed result is only slightly conservative on each side. The estimate of particle size L differs from that corresponding to the triple steady state by a factor of only 2.8 on either side.

6-9. Extension to TRAM Problems

Because the mean value theorem was so successfully applied in finding uniqueness conditions for the CSTR and for the catalyst particle, it would not seem overly optimistic to hope that a related development should apply to a tubular reactor model. However, the investigation must as before be limited to single-equation models.

Accordingly consider the adiabatic TRAM modeled by equation (6-12) with boundary conditions (6-5c and d). The mean value theorem may now be written in terms of the reduced temperature variable as

$$\left.\frac{dR}{d\eta}\right|_{\eta_*} = \frac{R(\eta_2) - R(\eta_1)}{\eta_2 - \eta_1} \tag{6-87a}$$

Let the difference between two assumed solutions be

$$\Phi(Z) = \eta_2(Z) - \eta_1(Z) \tag{6-87b}$$

Then by subtraction

$$0 = \alpha \frac{d^2}{dZ^2}(\eta_2 - \eta_1) - u \frac{d}{dZ}(\eta_2 - \eta_1) + [R(\eta_2) - R(\eta_1)] \tag{6-88}$$

Substitution of (6-87a) and (6-87b) gives

$$0 = \alpha \frac{d^2\Phi}{dZ^2} - u \frac{d\Phi}{dZ} + \left.\frac{dR}{d\eta}\right|_{\eta_*} \Phi \tag{6-89}$$

with boundary conditions

$$\frac{d\Phi}{dZ} = \frac{d\eta_2}{dZ} - \frac{d\eta_1}{dZ} = \frac{u}{\alpha}\Phi, \qquad Z = 0 \tag{6-90a}$$

$$\frac{d\Phi}{dZ} = 0, \qquad Z = L \tag{6-90b}$$

In order to treat this problem by an eigenvalue analysis, it is desirable to recast it as one of those already examined in Section 6-6. This can be done quite readily by using a substitution suggested by Amundson (1965). Let

$$w = \Phi \exp\left(-\frac{uZ}{2\alpha}\right) \tag{6-91}$$

Substitution into (6-79) and (6-80) eliminates the first-derivative term to yield

$$0 = \alpha\frac{d^2w}{dZ^2} - \left[\frac{u^2}{4\alpha} - \frac{dR}{d\eta}\bigg|_{\eta_*}\right]w \tag{6-92}$$

with boundaries:

$$\frac{dw}{dZ} = \frac{u}{2\alpha}w, \qquad Z = 0 \tag{6-93a}$$

$$\frac{dw}{dZ} = -\frac{u}{2\alpha}w, \qquad Z = L \tag{6-93b}$$

But this problem is entirely symmetric with respect to distance about a center line at $Z = (L/2)$. Its boundary conditions may therefore be replaced by

$$\frac{dw}{dZ} = 0, \qquad Z = 0 \tag{6-94a}$$

$$\frac{dw}{dZ} = -\frac{u}{2\alpha}w, \qquad Z = (L/2) \tag{6-94b}$$

The entire problem is now identical in form to one solved previously, as shown by the following table:

In terms of C	In terms of w
Equation (6-42)	Equation (6-92)
Boundary conditions (6-55)	Boundary conditions (6-94)
Parameter k	Parameter $\left[\frac{u^2}{4\alpha} - \frac{dR}{d\eta}\Big\|_{\eta_*}\right]$
Parameter $\frac{k_G}{D}$	Parameter $\frac{u}{2\alpha}$
Parameter L	Parameter $\frac{L}{2}$
Parameter D	Parameter α

The conclusions must therefore parallel those of Example 6-3, with appropriate change of symbols. Except for the specific eigenvalues, the solution to (6-89) with the given boundary conditions is the trivial solution, and accordingly the solution to equation (6-12) is unique. The analog of equation (6-60) is the sufficient condition

$$\left[\frac{u^2}{4\alpha} - \frac{dR}{d\eta}\bigg|_{\eta_*}\right] > -\alpha\left(\frac{2S_1}{L}\right)^2 \tag{6-95a}$$

where S_1 may be read from Figure 6-11 provided that

$$N = \frac{uL}{4\alpha}$$

Note that in this context N is one-fourth of the Peclet number based on tube length. In order of increasing conservatism, equation (6-95a) may be reduced to

$$\left.\frac{dR}{d\eta}\right|_{\eta_*} \leq \frac{u^2}{4\alpha} \tag{6-95b}$$

and

$$\left.\frac{dR}{d\eta}\right|_{\eta_*} \leq 0 \tag{6-95c}$$

It remains to be seen whether these forms are appreciably poorer for practical numerical purposes. The result (6-95b) may also be found in the work of Gupalo and Ryazantsev (1969).

To relate these conditions to the first-order, Arrhenius-type kinetics as was done above, return to equation (2-52), this time interpreting β in terms of the derivative of equation (6-11):

$$\beta = -\frac{dC}{dT} = -\frac{C_p}{\Delta H}\frac{dC}{d\eta} = \frac{C_p}{\Delta H} \tag{6-96}$$

As twice before for the CSTR and catalyst particle, equation (2-52) leads directly to (2-55) where again

$$\psi = \frac{\beta Q}{C_0 + \beta T_0} \tag{6-97}$$

Finally, making use of equation (2-58) yet once more, the sufficient conditions (6-95a and b) produce

$$\frac{u^2}{4\alpha} > k_0 G(\psi) - \alpha\left(\frac{2S_1}{L}\right)^2 \tag{6-98a}$$

and

$$\frac{u^2}{4a} \geq k_0 G(\psi) \tag{6-98b}$$

Elaboration of condition (6-95c) for this kinetic form results in

$$\frac{Q\Delta H C_0}{nC_p T_0^2} \leq 1$$

an expression virtually identical to equation (6-81) on the assumption that $D = \alpha$. Further to this point, by applying the technique of Appendix B to equations of the form of (6-92) and (6-93) Luss (1968) has demonstrated that condition (6-76) is valid also for the adiabatic TRAM. For the case of first-order kinetics the result is analogous to equation (6-81b):

$$\frac{Q\Delta H C_0}{C_p T_0^2} \leq 4$$

a criterion also derived by Hlavacek and Hofmann (1970a) via another route. With this in mind it is interesting to examine the list prepared by the same authors (1970b) for some exothermic reactions of practical importance:

Reaction	$Q\Delta HC_0/C_pT_0^2$
Oxidation of SO_2 to SO_3	3.8
Water-gas shift reaction	3.9
Ethane oxidation	7.1
Methanol oxidation	7.6
Vinylacetate synthesis	7.8
CO oxidation	11.0
Ethylene oxidation	12.2
Ammonia synthesis	16.5
Methanol synthesis	25.6
Higher alcohols from $CO + H_2$	32.0
Napthalene oxidation	34.3

Evidently there are reasonable grounds to expect multiple steady states for many of these systems, the result in an individual case depending on the kinetic and geometric parameters. By numerical integrations of the adiabatic TRAM equations Hlavacek and Hofmann (1970b) concluded that uniqueness of steady state will always arise for sufficiently long reactors and sufficiently low conversions. Their results are presented in a series of graphs that are somewhat similar in form to Figures 6-9, but differ in that they emphasize the effect of changing Damkohler number (k_0L/u) for the adiabatic reactor, whereas McGowin's computations focussed on changing heat transfer and feed conditions.

The isothermal TRAM can also be modeled by a single conservation equation and is therefore amenable to the approach taken in this section. By comparing the isothermal equations (1-8a) and (1-9a and c) with the adiabatic equations (6-12) and (6-5c and d), results are obtained that are entirely analogous to the sufficient conditions (6-95). Thus, the isothermal system has a unique steady state if

$$\left.\frac{dR}{dC}\right|_{C_*} \geq -\frac{u^2}{4D} \tag{6-99a}$$

or

$$\left.\frac{dR}{dC}\right|_{C_*} \geq 0 \tag{6-99b}$$

The second of these conditions is identical to condition (6-67b) for the isothermal catalyst particle and specific kinetics may be examined as in Example 6-4.

EXAMPLE 6-6. Estimate the flow rate required to guarantee uniqueness of the steady state for the adiabate TRAM with parameters:

$$\begin{aligned} C_p &= 0.0009 && \text{cal/cm}^3{}^\circ\text{K} \\ \Delta H &= 3 \times 10^4 && \text{cal/mol} \\ k_0 &= 5 \times 10^{11} && \text{sec}^{-1} \\ T_0 &= 400 && {}^\circ\text{K} \\ C_0 &= 10^{-5} && \text{mol/cm}^3 \\ Q &= 1.3 \times 10^4 && {}^\circ\text{K} \\ (u/\alpha) &= 83.3 && \text{cm}^{-1} \end{aligned}$$

Is it feasible to achieve the same result by shortening the reactor if the flow rate were to fall 5% below the critical value?

Solution: From equations (6-96), (6-97), and (6-98a):

$$\beta = \frac{C_p}{\Delta H} = 3 \times 10^8 \text{ mol/cm}^3{}^\circ\text{K}$$

$$\psi = \frac{\beta Q}{C_0 + \beta T_0} = 17.7$$

$$\frac{u^2}{4\alpha} \geq k_0 G(\psi) = 1.8 \times 10^3 \text{ sec}^{-1}$$

Assuming the the flow rate and dispersion coefficient are approximately proportional [Kramers and Westerterp (1963)]:

$$u \geq \frac{4\alpha}{u} k_0 G(\psi) = 86.2 \text{ cm/sec}$$

If the flow rate were 5% lower:

$$u = 95\%(86.2) = 81.4 \text{ cm/sec}$$

$$\frac{u^2}{4\alpha} = \frac{u}{4} \cdot \frac{u}{\alpha} = \frac{81.4}{4}(83.3) = 1690 \text{ sec}^{-1}$$

From equation (6-98a):

$$\alpha\left(\frac{2S_1}{L}\right)^2 > -\frac{u^2}{4\alpha} + k_0 G(\psi) = 110 \text{ sec}^{-1}$$

leading to:

$$L < 2S_1\sqrt{\frac{\alpha}{110}} = 2S_1\sqrt{\frac{u}{(83.3)(110)}} = 0.188 S_1 \text{ cm}$$

In general, the solution would now require a trial-and-error procedure since S_1 is dependent on L, but in this case it can be avoided by observing from Figure 6-11 that S_1 can never exceed 1.57 for any value of L. As a result

$$L < 0.188(1.57) = 0.295 \text{ cm.}$$

which is a value so small in comparison to reasonable reactor lengths as to justify the rejection of condition (6-98a) for practically feasible results. Not that the equation is incorrect; merely that it does not in practical terms offer any less conservatism than the simpler (6-98b).

Having found that the eigenvalue computation was of little use in the TRAM calculation and recalling that the related computation for the catalyst particle Example 6-5 was most successful, leads one naturally to ask for an explanation of the difference. The answer is entirely to be found in a comparison of the numerical values that determine the respective eigenvalues:

$$-\alpha\left(\frac{2S_1}{L}\right)^2 \quad \text{for the TRAM}$$

and

$$-\alpha\left(\frac{\pi}{2L}\right)^2 \quad \text{for the catalyst particle}$$

Typically, the α's differ by perhaps 250 fold (in favor of the dispersion in the TRAM), and the lengths by about a factor of 100. Substitution shows that the TRAM eigenvalue will be smaller by at least an order of magnitude even if the S_1 parameter is taken at its maximum possible value. Accordingly the length term in (6-98a) has a relatively insignificant effect in this example, however it should be noted that the dispersion coefficients and length of a TRAM are greatly influenced by choice of flow rate. While it is true in general that it is always possible to insure uniqueness by use of a sufficiently short reactor, this does not seem feasible for typical numerical values of interest.

EXAMPLE 6-7. Estimate the feed concentration required to guarantee a unique steady state for the adiabatic TRAM with the parameters given by Raymond and Amundson (1964):

$$\begin{aligned} C_p &= 9.0 \times 10^{-4}\ \text{cal/cm}^3{}^\circ\text{K} & \alpha &= 0.6\ \text{cm}^2/\text{sec} \\ \Delta H &= 6.0 \times 10^4\ \text{cal/mol} & Q &= 1.5 \times 10^4\,{}^\circ\text{K} \\ u &= 50\ \text{cm/sec} & T_0 &= 500\ {}^\circ\text{K} \\ k_0 &= 5.0 \times 10^{11}\ \text{sec}^{-1} \end{aligned}$$

In the reference paper a value of $C_0 = 1.0 \times 10^{-5}$ produced a system with three steady states.

Solution:

$$\beta = \frac{C_p}{\Delta H} = 1.5 \times 10^{-8}\ \text{mol/cm}^3{}^\circ\text{K}$$

$$G(\psi) \leq \frac{u^2}{4k_0\alpha} = 2.04 \times 10^{-9}$$

From Figure 2-5, $\psi \geq 18.3$, and

$$C_0 \leq \beta\left[\left(\frac{Q}{18.3}\right) - T_0\right] = 3.8 \times 10^{-6}\ \text{mol/cm}^3$$

Comparing this result to the value of 10^{-5} chosen in the reference suggests again that the safe estimate is very conservative. The transition to multiple solutions occurs in the 38-fold range between the two values, probably much closer to the larger boundary.

6-10. Fixed-Bed Tubular Reactors

It would appear at first glance that uniqueness problems for distributed systems of more than one phase would be of extraordinary complexity, since so many more equations are to be considered. This is, however, not necessarily the case in spite of the fact that the numerical integration of such models is indeed an arduous and time consuming task.

Consider for example the fixed-bed tubular reactor modeled by equations (1-16) and (1-17), whose steady state is determined by:

$$0 = -u\frac{dC_1}{dZ} + k_P(C - C_1) \tag{6-100a}$$

$$0 = -uC_P\frac{dT_1}{dZ} + h_P(T - T_1) + h(T_w - T_1) \tag{6-100b}$$

$$0 = k_P(C_1 - C) - \theta R \tag{6-100c}$$

$$0 = h_P(T_1 - T) + \Delta H\theta R \tag{6-100d}$$

Because the PFTR description of the interstitial fluid always leads to a unique profile, any nonuniqueness must be the result of multiplicity in the solution of the particle equations (6-100c and d). But these relations are algebraic, identical in form to equations (2-71c and d) for the fluidized bed particles and may be analyzed by a van Heerden diagram as was done in Chapter 2. The uniqueness condition for this fixed-bed model is

$$\frac{\partial Q_R}{\partial T} > \text{Max}\left[\frac{\partial Q_G}{\partial T}\right] \tag{6-101}$$

where the partial derivatives are used to emphasize that the analysis considers only a catalyst particle in a given (T_1, C_1) fluid environment, and depends on the fact that overall reactor multiplicity can only occur in the presence of multiple particle states. Working from the terms of equation (6-100d) and carrying through the indicated differentiations gives

$$h_P > \Delta H\theta\frac{\partial R}{\partial T} \tag{6-102}$$

Further steps depend on specified kinetics. The usual first choice of rate equation (1-6b) gives from rearranged (6-100c):

$$C = \left[\frac{C_1}{1 + \frac{k_0}{k_P}\exp\left(-\frac{Q}{T}\right)}\right] \tag{6-103}$$

Noting the similarity in functional form to equation (2-76) makes possible solution by inspection of the prior results. Instead of equations (2-80) and (2-81), the fixed-bed analysis gives

$$\frac{\partial R}{\partial T} = R\frac{Q}{T^2}\frac{C}{C_1} = \frac{k_P QC(C_1 - C)}{C_1 T^2} \tag{6-104}$$

This function has a maximum at approximately

$$T = \left[\frac{Q}{\ln\left(\frac{k_0}{k_P}\right)}\right] \tag{6-105}$$

which is similar to equation (2-86). As a result

$$\text{Max}\left[\frac{\partial R}{\partial T}\right] = \frac{k_P C_1}{4Q}\left[\ln\left(\frac{k_P}{k_0}\right)\right]^2 \tag{6-106}$$

Using this result in equation (6-102) gives the criterion

$$\frac{4Qh_P}{k_P \Delta H \theta C_1} > \left[\ln \frac{k_P}{k_0}\right]^2 \tag{6-107}$$

Since $C_1 < C_0$ for irreversible reaction, this inequality will be satisfied if:

$$\frac{4Qh_P}{\Delta H k_P \theta C_0} > \left[\ln \frac{k_P}{k_0}\right]^2 \tag{6-108}$$

Since this result is based entirely on the particle equations, it is equally valid for adiabatic and nonadiabatic systems. This observation was first made by Liu, Aris, and Amundson (1963). The numerical integrations of Liu and Amundson (1963) suggest that the same is probably true if axial dispersion terms are added to the equations of the fluid phase, although some reservation may be held in view of the fact that TRAM models can have multiple solutions without packing.

EXAMPLE 6-8. Use the following data (adapted from the Liu and Amundson 1962 paper) to estimate the maximum feed concentration that will assure a unique steady state for a fixed-bed reactor. How conservative is the estimate of inequality (6-108) in view of the fact that the system is reported to have multiple steady states for $C_0 = 2.18 \times 10^{-4}$ mol/ft³?

$$\frac{k_P}{k_0 \theta} = \exp(-12.98)$$

$$\frac{k_P}{h_P} = 34.3°\text{F ft}^3/\text{Btu}$$

$$Q = 22{,}000°\text{R}$$

$$\theta = 0.74$$

$$\Delta H = 1.2 \times 10^5 \text{ Btu/mol}$$

Solution: Substitution of the given values into inequality (6-108) gives:

$$C_0 < \frac{4Qh_p}{k_p \theta \Delta H[\ln k_p/k_0)]^2} = 1.21 \times 10^{-4} \text{ mol/ft}^3$$

This result is only mildly conservative, differing from the given value by less than a factor of two.

That the fixed-bed reactor model could be analyzed with such relative ease is due to the fact that the system of four equations could be reduced to two for the particles. The same logic applies also for any variation in the particle equations, so long as the equations describing the interstitial fluid are of the PFTR type. Allowing for finite mass and heat diffusion rates within the particles retains equations (6-100a and b) unchanged, but modifies (6-100c and d) to:

$$0 = D\frac{d^2C}{dZ^2} - R \tag{6-109a}$$

$$0 = \alpha\frac{d^2\eta}{dZ^2} + R \tag{6-109b}$$

with boundary conditions:

$$\frac{dC}{dZ} = \frac{d\eta}{dZ} = 0, \qquad Z = 0 \tag{6-109c}$$

$$\frac{dC}{dZ} = \frac{k_G}{D}(C_0 - C), \qquad Z = L \tag{6-109d}$$

$$\frac{d\eta}{dZ} = \frac{h_G}{\alpha}(\eta_0 - \eta), \qquad Z = L \tag{6-109e}$$

In this formulation, the symbols refer back to Section 6-2 to avoid extensive subscripting. The state variables are concentration and reduced temperature for the catalyst particle, and similarly for the distance variable and various parameters. The subscripted C_0 and η_0 refer to the concentration and temperature of the interstitial fluid, and as before, they are treated as constant ambient conditions.

Following the procedure of Section 6-2, the substitution of equation (6-16) reduces equations (6-109) to

$$w(Z) = A + BZ \tag{6-110a}$$

with boundaries:

$$\frac{dw}{dZ} = 0, \qquad Z = 0 \tag{6-110b}$$

$$\frac{dw}{dZ} = k_G(C_0 - C) + h_G(\eta_0 - \eta), \qquad Z = L \tag{6-110c}$$

Luss and Amundson (1967b) observed that for the special case when

$$\frac{k_G}{h_G} = \frac{D}{\alpha}$$

equation (6-110c) becomes

$$\frac{dw}{dZ} = \frac{k_G}{D}(w_0 - w), \qquad Z = L \tag{6-110d}$$

and the set of equations (6-110) reduces to the same linear result as equation (6-19).

Accordingly, a uniqueness condition for the catalyst particle (and hence the fixed-bed reactor) may be derived by following the well-trodden path:

1. Use (6-19) to reduce (6-109b) to a single-variable equation
2. Substitute equations (6-86) and (6-87) to obtain

$$0 = \alpha\frac{d^2\Phi}{dZ^2} + \frac{dR}{d\eta}\bigg|_{\eta_*}\Phi \tag{6-111a}$$

$$\frac{d\Phi}{dZ} = 0, \qquad Z = 0 \tag{6-111b}$$

$$\frac{d\Phi}{dZ} = -\frac{h_G}{\alpha}\Phi, \qquad Z = L \tag{6-111c}$$

3. Note that except for changes in symbols equations (6-111) are the same as equation (6-42) with boundary conditions (6-55).

The uniqueness condition is therefore

$$\frac{dR}{d\eta}\bigg|_{\eta_*} < \alpha\left(\frac{S_1}{L}\right)^2 \tag{6-112}$$

where S_1 is determined from Figure 6-11 in terms of

$$N = \frac{k_G L}{D} = \frac{h_G L}{\alpha} \tag{6-113}$$

The interpretation of condition (6-112) for specific kinetics follows the development already detailed in Sections 6-8 and 6-9, but whatever the kinetics comparison of (6-112) with equation (6-77a) shows that the fixed-bed condition is more restrictive than that developed for the catalyst particle alone, since S_1 is always less than $(\pi/2)$.

Finally, while on the subject of fixed-bed tubular reactors, mention should be made of the contributions of McGreavy and Thornton (1970) and Cresswell (1970), who took advantage of a practical observation to derive a very simple algebraic uniqueness criterion. Their development follows from equations (6-110) but does not depend on any special ratio between the transfer coefficients and the mass and thermal diffusivities. Instead, equations (6-110a, b, and c) are used to establish the relationship between fluid temperature and particle temperature at any reactor position:

$$\eta_0 = \eta + \frac{k_G}{h_G}(C - C_0) \tag{6-114}$$

The essential simplification arises from prior studies of temperature profiles within catalyst pellets and their surrounding fluid boundary layers. The evidence [McGreavy and Cresswell (1969)] suggests that a catalyst particle is approximately isothermal over the entire practical range of heat transfer parameters and may therefore be modeled by equation (6-109a) with boundary conditions (6-109c and d). This permits an algebraic link to be established between the surface concentration and temperature in the catalyst particle and the reactant concentration in the fluid. If, for example, first-order kinetics are applicable: $R = kC$; the system of equations (6-109a, c, and d) is linear and easily integrated to give

$$C(Z) = C_0 \frac{\cosh(\sqrt{k/D}\,Z)}{\cosh(\sqrt{k/D}\,L) + (\sqrt{kD}/k_G)\sinh(\sqrt{k/D}\,L)} \tag{6-115a}$$

from which

$$C(L) = C_0\left[\frac{1}{1 + (\sqrt{kD}/k_G)\tanh(\sqrt{k/D}\,L)}\right] \tag{6-115b}$$

Substitution of this result into equation (6-114) gives

$$\eta_0 = \eta - \frac{k_G C_0}{h_G}\left[\frac{(\sqrt{kD}/k_G)\tanh(\sqrt{k/D}\,L)}{1 + (\sqrt{kD}/k_G)\tanh(\sqrt{k/D}\,L)}\right] \tag{6-116}$$

from which it is possible to compute the fluid temperature-pellet temperature interdependence at any chosen value of C_0, given the various system parameters (k is a function of η). McGreavy and Thornton have shown in this manner that multiple steady states occur over certain parameter ranges; that is, three particle temperatures can exist for a given fluid condition. As is commonly the case, the intermediate steady state of the three is unstable, the others stable. Further numerical details for a spherical pellet geometry may be found in the reference cited.

The approximate nature of this approach should be emphasized, since it rests on the prospect of an isothermal particle. This assumption is a reasonable one, but it implies that $(dC/dZ) = 0$ because

$$\frac{dw}{dZ} = \alpha\frac{d\eta}{dZ} + D\frac{dC}{dZ} = 0 \tag{6-117}$$

at all $0 \leq Z \leq L$, and this is in flat contradiction to an integrated result such as equation (6-115). In effect this treatment of the fixed-bed considers the catalyst pellets to be sufficiently isothermal to ignore the energy balance requirements, but not so isothermal as to prevent the development of a concentration gradient. Cresswell (1970) has pointed out that this is in essence the case over a reasonable range of industrially important conditions.

REFERENCES

Amundson, N. R., *Can. J. Chem. Eng.*, **43**, 49 (1965).

Aris, R., *Introduction to the Analysis of Chemical Reactors*, Englewood Cliffs, New Jersey: Prentice-Hall (1965).

Barkelew, C. H., *Chem. Eng. Prog. Symp. Series No. 25*, **55**, 37 (1959).

Bilous, O. and N. R. Amundson, *AIChE Journal*, **2**, 117 (1956).

BOYNTON, D. E., W. B. NICHOLS, and H. M. SPURLIN, *Ind. Eng. Chem.*, **51**, No. 4, 489 (1959).

CARBERRY, J. J., *AIChE Journal*, **7**, 350 (1961).

CARBERRY, J. J. and D. WHITE, *Ind. Eng. Chem.*, **61**, No. 7, 27 (1969).

CARSLAW, H. S. and J. C. JAEGER, *Conduction of Heat in Solids*, Oxford University Press (1959), p. 377.

CHEN, J. L. P., Ph. D. Thesis, Univ. of Pennsylvania, Philadelphia. (1970).

COPELOWITZ, I. and R. ARIS, *Chem. Eng. Sci.*, **25**, 906 (1970).

CRESSWELL, D. L. *Chem. Eng. Sci.*, **25**, 267 (1970).

DOUGLAS, J. M. and L. C. EAGLETON, *Ind. Eng. Chem. Fund.*, **1**, 116 (1962).

DROTT, D. W. and R. ARIS, *Chem. Eng. Sci.*, **24**, 541 (1969).

FROMENT, G. F., *Ind. Eng. Chem.*, **59**, No. 2, 19 (1967).

GUPALO, Y. P. and Y. S. RYAZANTSEV, *Chem. Eng. Sci.*, **24**, 1441 (1969).

HANRATTY, T. J., *Chem. Eng. Sci.*, **3**, 209 (1954).

HARRIOT, P., *Chem. Eng.*, **68**, No. 10, 165 (May 15, 1961).

HATFIELD, B. and R. ARIS, *Chem. Eng. Sci.*, **24**, 1213 (1969).

HLAVACEK, V. and M. MAREK, *Chem. Eng. Sci.*, **23**, 865 (1968).

HLAVACEK, V., M. MAREK, and M. KUBICEK, *Chem. Eng. Sci.*, **23**, 1083 (1968).

HLAVACEK, V. and H. HOFMANN, *Chem. Eng. Sci.*, **25**, 173 (1970a).

HLAVACEK, V. and H. HOFMANN, *Chem. Eng. Sci.*, **25**, 187 (1970b).

KRAMERS, H. and K. R. WESTERTERP, *Elements of Chemical Reactor Design and Operation*, New York: Academic Press (1963).

LIU, S-L. and N. R. AMUNDSON, *Ind. Eng. Chem. Fund.*, **1**, 200 (1962).

LIU, S-L. and N. R. AMUNDSON, *Ind. Eng. Chem. Fund.*, **2**, 183 (1963).

LIU, S-L., R. ARIS, and N. R. AMUNDSON, *Ind. Eng. Chem. Fund.*, **2**, 12 (1963).

LUSS, D. and N. R. AMUNDSON, *Ind. Eng. Chem. Fund.*, **6**, 457 (1967a)

LUSS, DAN, *Chem. Eng. Sci.*, **23**, 1249 (1968).

LUSS, DAN, *Chem. Eng. Sci.*, **24**, 879 (1969).

LUSS, DAN, *Chem. Eng. Sci.*, **26**, to appear (1971).

LUSS, D. and N. R. AMUNDSON, *Chem. Eng. Sci.*, **22**, 253 (1967b).

MCGOWIN, C. R. and D. D. PERLMUTTER, *AIChE Journal*, **17**, 831 (1971a).

MCGOWIN, C. R. and D. D. PERLMUTTER, *Ibid*, 837 (1971b).

MCGREAVY, C. and D. L. CRESSWELL, *Can. J. Ch.E.*, **47**, 583 (1969).

MCGREAVY, C. and J. M. THORNTON,, *Can. J. Ch. E.*, **48**, 178 (1970).

PADBERG, G. and E. WICKE, *Chem. Eng. Sci.*, **22**, 1035 (1967).

PRATER, C. D., *Chem. Eng. Sci.*, **8**, 284 (1958).

RAYMOND, L. R. and N. R. AMUNDSON, *Can J. Ch. E.*, **42**, 173 (1964).

RAYMOND, L. R. and N. R. AMUNDSON, *AIChE Journal*, **11**, 339 (1965).

REILLY, M. J. and R. A. SCHMITZ, *AIChE Journal*, **12**, 153 (1966).

ROBERTS, G. W. and C. N. SATTERFIELD, *Ind. Eng. Chem. Fund.*, **4**, 228 (1965).

ROBERTS, G. W. and C. N. SATTERFIELD, *Ind. Eng. Chem. Fund.*, **5**, 317 (1966).

TINKLER, J. D. and A. B. METZNER, *Ind. Eng. Chem.*, **53**, 663 (1961).

VAN HEERDEN, C., *Chem. Eng. Sci.*, **8**, 133 (1958).

WANG, F. S. and D. D. PERLMUTTER, *AIChE Journal*, **14**, 934 (1968).

WEISZ, P. B. and J. S. HICKS, *Chem. Eng. Sci.*, **17**, 265 (1962).

WICKE, E., G. PADBERG and H. ARENS, *Fourth European Symp. on Chem. Reaction Eng., Brussels* (1968).

WICKE, E. and D. VORTMEYER, *Z. Elektrochem.*, **63**, 145 (1959).

WICKE, E., *Z. Elektrochem.*, **65**, 267 (1961).

WILSON, K. B., *Tr. I. Ch. E.* (*London*), **24**, 77 (1946).

Local Stability of Distributed-Parameter Systems

7

7-1. Linear Systems With Constant Coefficients

Just as the linear, constant coefficient equation played a special role among lumped-parameter systems, so the corresponding features in partial differential equations are basic in that they permit relatively simple solutions which point the way to techniques applicable to more complex problems.

Considering as an example the model for an isothermal catalyst particle in one spacial dimension

$$\frac{\partial C}{\partial t} = D\frac{\partial^2 C}{\partial Z^2} - R \tag{7-1a}$$

with boundary conditions to insure symmetry

$$\frac{\partial C}{\partial Z} = 0, \qquad Z = 0 \tag{7-1b}$$

and physical equilibrium at the outer surfaces

$$C = C_0, \qquad Z = L \tag{7-1c}$$

The steady-state profile $C_s(Z)$ is the solution of the ordinary differential equation

$$0 = D\frac{d^2C}{dZ^2} - R(C) \tag{7-2}$$

with the same boundary conditions. Having found a steady-state profile, it is convenient to express the dynamics of equations (7-1) transformed in terms of the deviation variables

$$x(Z, t) = C(Z, t) - C_s(Z) \tag{7-3a}$$

$$r = R - R_s \tag{7-3b}$$

to give:

$$\frac{\partial x}{\partial t} = D\frac{\partial^2 x}{\partial Z^2} - r \tag{7-4a}$$

$$\frac{\partial x}{\partial Z} = 0, \qquad Z = 0 \tag{7-4b}$$

$$x = C_0 - C_0 = 0, \qquad Z = \mathrm{L} \tag{7-4c}$$

In the special simplest-of-all case, the reaction rate is first-order in concentration:

$$R = kC \tag{7-5a}$$

$$r = R - R_s = k(C - C_s) = kx \tag{7-5b}$$

Then equation (7-4a) reduces to

$$\frac{\partial x}{\partial t} = D\frac{\partial^2 x}{\partial Z^2} - kx \tag{7-6}$$

which together with boundary conditions (7-4b and c) and some as yet unspecified initial profiles can be solved for the concentration deviation.

This linear system has the well-known solution

$$x(Z, t) = \sum_{i=1}^{\infty} A_i \left[\cos\left(i - \frac{1}{2}\right)\frac{\pi Z}{L}\right] \exp(\lambda_i t) \tag{7-7a}$$

where the constants A_i are determined such that the initial condition is satisfied, and

$$\lambda_i = -k - D\left[\frac{(i - \frac{1}{2})\pi}{L}\right]^2, \qquad i = 1, 2, \ldots, \infty \tag{7-7b}$$

the eigenvalues of the problem. It is obvious by inspection of the solution equation (7-7a) that the system stability will depend on the signs of λ_i. Since equation (7-7b) shows all the eigenvalues are real and negative, the steady state must be stable regardless of initial condition chosen. This most straightforward argument developed from the system linearity, a simple and rather special case.

A second example that will be useful is the model for isothermal zero-order reaction in a TRAM:

$$\frac{\partial C}{\partial t} = D\frac{\partial^2 C}{\partial Z^2} - u\frac{\partial C}{\partial Z} - k \tag{7-8}$$

which in terms of the deviation variables (7-3) becomes

$$\frac{\partial x}{\partial t} = D\frac{\partial^2 x}{\partial Z^2} - u\frac{\partial x}{\partial Z} \tag{7-9a}$$

with boundary conditions

$$D\frac{\partial x}{\partial Z} = ux, \qquad Z = 0 \tag{7-9b}$$

$$\frac{\partial x}{\partial Z} = 0, \qquad Z = L \tag{7-9c}$$

The analytic series solution to equations (7-9) is available from Carslaw and Jaeger (1959):

$$x(Z, t) = \sum_{i=1}^{\infty} A_i\left[\frac{2D}{uL}\beta_i \cos\left(\frac{\beta_i Z}{L}\right) + \sin\left(\frac{\beta_i Z}{L}\right)\right]\left[\exp(\lambda_i t)\right]\left[\exp\left(\frac{uZ}{2D}\right)\right] \tag{7-10a}$$

where

$$\lambda_i = -\frac{D}{4uL}\left[4\beta_i^2 + \left(\frac{uL}{D}\right)^2\right] \tag{7-10b}$$

and the β_i are the roots of

$$\beta \cot \beta = \frac{D}{4uL}\left[4\beta^2 - \left(\frac{uL}{D}\right)^2\right] \tag{7-10c}$$

Once again, the system is stable for all initial conditions, since λ_i computed from equation (7-10b) is always real and negative.

7-2. Linearization without Proof

Remembering the great successes that emerged from the linearization of nonlinear differential equations, it is very tempting to apply analogous arguments and computations to the analysis of local stability in distributed systems. This approach has in fact yielded very satisfactory results in the cases so far tested, but it should be noted that there is no proof known that justifies linearization in the same rigorous way that Liapunov's Fundamental theorem did for lumped-parameter systems. Judging from the results known so far, this restriction is not of much practical importance for the common variety of well-behaved systems. It suggests a note of caution in out-of-the-ordinary models or disturbances, however.

The mechanics of linearization are straightforward, and may be illustrated most simply by considering the more general (nonisothermal) catalyst particle model:

$$\frac{\partial \eta}{\partial t} = \alpha\frac{\partial^2 \eta}{\partial Z^2} + R \tag{7-11a}$$

$$\frac{\partial C}{\partial t} = D\frac{\partial^2 C}{\partial Z^2} - R \tag{7-11b}$$

with boundary conditions:

$$\frac{\partial C}{\partial Z} = \frac{\partial \eta}{\partial Z} = 0, \qquad Z = 0 \tag{7-11c}$$

$$C = C_0, \eta = \eta_0, \qquad Z = L \tag{7-11d}$$

which in terms of the deviation variables

$$x_1(Z, t) = \eta(Z, t) - \eta_s(Z) \tag{7-12a}$$

$$x_2(Z, t) = C(Z, t) - C_s(Z) \tag{7-12b}$$

$$r = R - R_s \tag{7-12c}$$

becomes

$$\frac{\partial x_1}{\partial t} = \alpha \frac{\partial^2 x_1}{\partial Z^2} + r \tag{7-13a}$$

$$\frac{\partial x_2}{\partial t} = D \frac{\partial^2 x_2}{\partial Z^2} - r \tag{7-13b}$$

and

$$\frac{\partial x_1}{\partial Z} = \frac{\partial x_2}{\partial Z} = 0, \qquad Z = 0 \tag{7-14a}$$

$$x_1 = x_2 = 0, \qquad Z = L \tag{7-14b}$$

Linearization about the steady-state profiles now requires only substitution for $r(x_1, x_2)$ from the truncated Taylor-series expansion

$$r(x_1, x_2) = \left.\frac{\partial r}{\partial x_1}\right|_s x_1 + \left.\frac{\partial r}{\partial x_2}\right|_s x_2 = \left.\frac{\partial R}{\partial \eta}\right|_s x_1 + \left.\frac{\partial R}{\partial C}\right|_s x_2 = R_\eta x_1 + R_c x_2 \tag{7-15}$$

but note that the partial derivatives must be evaluated as functions of the position variable Z. As a result, the linearized form is more complicated than equation (7-6) not only because the number of equations is larger, but also because the model is no longer one with constant coefficients. Accordingly, alternative solution methods will be needed, even when an uncoupling of the paired equations is possible, as in Section 7-3.

7-3. Uncoupling the Transient Balances

The key to the uncoupling of the reactor steady-state relationships in Chapter 6 lay in the observation that concentration and temperature were linearly (algebraically) related. One might ask whether a similar simplification is possible for dynamic equations such as (7-13) and (7-14). The answer is yes, but only for a rather restricted case [Wei (1965)]; nevertheless, the results are useful in several ways.

In particular if $\alpha = D$, equations (7-13) may be added to produce

$$\frac{\partial w}{\partial t} = \alpha \frac{\partial^2 w}{\partial Z^2} \tag{7-16a}$$

with boundary conditions

$$\frac{\partial w}{\partial Z} = 0, \qquad Z = 0 \tag{7-16b}$$

$$w = 0, \qquad Z = L \tag{7-16c}$$

The new variable defined by

$$w = x_1 + x_2 \tag{7-17}$$

is proportional to the *residual enthalpy* as defined by Luss and Lee (1968). In general the solution to equations (7-16) will be some $w(Z, t)$ depending on the initial disturbance condition; however, for the special case of

$$w = 0, \qquad t = 0 \tag{7-18}$$

the solution reduces to the trivial one

$$w = x_1(Z) + x_2(Z) \equiv 0 \tag{7-19}$$

which is the dynamic counterpart of the steady-state equation (6-10). The physical restriction that only disturbances are considered which arise from prior steady states leads to initial condition (7-18). That is to say, one imagines that the transient under study starts from what was a steady state before some major parameter change occurred. It moves to a new steady state as $t \longrightarrow \infty$ (assuming it is stable), and it is the latter set of profiles that has been identified as $C_s(Z)$ and $\eta_s(Z)$ for this analysis.

An entirely parallel argument applies if one starts with the adiabatic TRAM transient equations:

$$\frac{\partial x_1}{\partial t} = \alpha \frac{\partial^2 x_1}{\partial Z^2} - u \frac{\partial x_1}{\partial Z} + r \tag{7-20a}$$

$$\frac{\partial x_2}{\partial t} = D \frac{\partial^2 x_2}{\partial Z^2} - u \frac{\partial x_2}{\partial Z} - r \tag{7-20b}$$

and boundary conditions:

$$\frac{\partial x_1}{\partial Z} = \frac{u}{\alpha} x_1, \qquad Z = 0 \tag{7-20c}$$

$$\frac{\partial x_2}{\partial Z} = \frac{u}{D} x_2, \qquad Z = 0 \tag{7-20d}$$

$$\frac{\partial x_1}{\partial Z} = \frac{\partial x_2}{\partial Z} = 0, \qquad Z = L \tag{7-20e}$$

The same restrictions as before, that the Lewis number equal unity and that the disturbances be of the special sort described, will lead again to conclusion (7-19). Under these conditions, single equations may be used to model the catalyst particle

$$\frac{\partial x_1}{\partial t} = \alpha \frac{\partial^2 x_1}{\partial Z^2} + r(x_1, Z) \tag{7-21}$$

or the adiabatic TRAM:

$$\frac{\partial x_1}{\partial t} = \alpha \frac{\partial^2 x_1}{\partial Z^2} - u\frac{\partial x_1}{\partial Z} + r(x_1, Z) \tag{7-22}$$

but note that in either equation the space variable Z has been introduced by the substitution of equation (7-19) to remove x_2.

7-4. Galerkin's Method

Having uncoupled the transient mass and energy balance equations, we may simplify the result by linearization and seek suitable techniques for analysis of the stability questions. As will turn out presently, the method to be outlined is in fact applicable even without prior uncoupling, but it is initially simpler to follow the argument using only a single equation.

Specifically, consider the catalyst particle model (7-21), linearized by the substitution

$$r(x_1, Z) = \left.\frac{dr}{dx_1}\right|_s x_1 = \left.\frac{dR}{d\eta}\right|_s x_1 = R_\eta(Z)x_1$$

Make the space variable dimensionless by defining

$$z = \frac{Z}{L} \tag{7-23}$$

to give

$$\frac{\partial x_1}{\partial t} = \frac{\alpha}{L^2}\frac{\partial^2 x_1}{\partial z^2} + R_\eta(z)x_1 \tag{7-24}$$

Since this equation has a close resemblance to the constant coefficient equation (7-6), it appears reasonable to suppose a solution related to (7-7). For trial purposes consider the approximation

$$x_1(z, t) = \sum_{i=1}^{n} a_i(t)\left[\cos\left(i - \frac{1}{2}\right)\pi z\right] \tag{7-25}$$

which immediately satisfied boundary conditions (7-14), but leaves the $a_i(t)$ as yet undetermined. Then by substitution into (7-24):

$$\begin{aligned}\sum_{i=1}^{n} \frac{da_i}{dt} \cos\left(i - \frac{1}{2}\right)\pi z = &-\frac{\alpha}{L^2}\sum_{i=1}^{n} a_i\left[\left(i - \frac{1}{2}\right)\pi\right]^2 \cos\left(i - \frac{1}{2}\right)\pi z \\ &+ R_\eta(z)\sum_{i=1}^{n} a_i \cos\left(i - \frac{1}{2}\right)\pi z + \Gamma(z)\end{aligned} \tag{7-26}$$

where $\Gamma(z)$ is the residual by which at any time the approximation (7-25) fails to exactly satisfy the differential equation.

A major simplification of equation (7-26) is obtainable by making use of the observation that the members of the above family of cosine functions are *orthogonal* to one another, by which is meant that

$$2\int_0^1 \left[\cos\left(i-\frac{1}{2}\right)\pi z\right]\left[\cos\left(j-\frac{1}{2}\right)\pi z\right] dz = \begin{cases}1, & i=j\\ 0, & i\neq j\end{cases} \tag{7-27}$$

To utilize this convenient relationship, multiply equation (7-26) by each of the n functions

$$\cos\left(j-\frac{1}{2}\right)\pi z, \qquad j = 1, 2, \ldots, n$$

respectively, and integrate with respect to reactor length from inlet to outlet. The orthogonality of the cosines will remove many of the summation signs, since in each sum containing an integral of the form (7-27) only the single term for which $i = j$ makes a finite contribution, the other integrals being exactly zero. The result will be the set of n equations

$$\frac{da_j}{dt} = -\alpha\left[\left(j-\frac{1}{2}\right)\frac{\pi}{L}\right]^2 a_j + \sum_{i=1}^{n} I(i,j)a_i + 2\int_0^1 \Gamma(z)\cos\left(j-\frac{1}{2}\right)\pi z\, dz \tag{7-28}$$

where

$$I(i,j) = I(j,i) = 2\int_0^1 R_n(z)\left[\cos\left(i-\frac{1}{2}\right)\pi z\right]\left[\cos\left(j-\frac{1}{2}\right)\pi z\right] dz \tag{7-29}$$

is a vector of n components for each choice of i. The removal of the final integral in (7-28) may be justified by considering such a step a part of the definition of $a_i(t)$. Since this time function has in no way been restricted until now, it is permissible to consider it to be orthogonal to the residual for all i. With this last modification, equation (7-28) becomes a set of linear ordinary differential equations with constant coefficients

$$\frac{d\mathbf{a}}{dt} = \mathbf{A}\mathbf{a} \tag{7-30}$$

where the matrix $\mathbf{A}$ is symmetric and composed of the elements:

$$\mathbf{A} = \begin{bmatrix} I_{11} - \alpha\left(\dfrac{\pi}{2L}\right)^2 & I_{12} & I_{13} & \cdots & I_{1n} \\ I_{12} & I_{22} - \alpha\left(\dfrac{3}{2}\dfrac{\pi}{L}\right)^2 & I_{23} & \cdots & I_{2n} \\ \cdots & & & & \\ \cdots & & & & \\ I_{1n} & I_{2n} & I_{3n} & \cdots & I_{nn} - \alpha\left[\left(n-\dfrac{1}{2}\right)\dfrac{\pi}{L}\right]^2 \end{bmatrix} \tag{7-31}$$

The problem of distributed parameter stability has in this way been reduced to the familiar linear, constant coefficient problem discussed in Chapter 3, but note that this is only a study of *local* stability—a consequence of the linearization. The only significant new numerical difficulty (aside from that of finding the steady state) arises from the integrations needed in the evaluation of equations (7-29). The technique of using orthogonalization to average the residual by integration is named for Galerkin.

In the special case where

$$R_n(z) = -k$$

is a constant, it may be removed from the integral. Then from the orthogonality definition of equation (7-27):

$$I(i, j) = 0$$

$$I(i, i) = -k$$

and matrix **A** reduces to the diagonal form

$$\mathbf{A} = \begin{bmatrix} -k - \alpha\left(\frac{\pi}{2L}\right)^2 & 0 & \cdots & 0 \\ 0 & -k - \alpha\left(\frac{3\pi}{2L}\right)^2 & \cdots & 0 \\ & & \cdots & 0 \\ 0 & 0 & \cdots & -k - \alpha\left[\left(n - \frac{1}{2}\right)\frac{\pi}{L}\right]^2 \end{bmatrix}$$

which has eigenvalues

$$\lambda_i = -k - \alpha\left[\frac{(i - \frac{1}{2})\pi}{L}\right]^2 \qquad i = 1, 2, \ldots, n \tag{7-32}$$

By comparison with equation (7-7b), it is clear that the Galerkin method has extracted the first n eigenvalues of the infinite set. An increase in n corresponds in trial solution (7-25) to an improved approximation, but such a move would not yield any advantage in this problem, since the additional eigenvalues are increasingly negative in value and have no bearing on the stability question per se. In general one will want in a Galerkin stability analysis to use a sufficiently large n to detect with some confidence at least the sign of the largest eigenvalue. This usually requires a series of tests with increasing n.

EXAMPLE 7-1. How large a value of n is needed in the study of the local stability of the catalyst particle modeled by equation (7-21), if the kinetics are of the first-order Arrhenius form and the parameter values are those given with Figure 6-10, the three steady-state case.

Solution: For the kinetics specified

$$R_\eta(z) = \frac{dR}{d\eta}\bigg|_s = k_0 \exp\left(-\frac{C_p Q}{\Delta H \eta_s(z)}\right)\left[\frac{C_s(z) Q C_p}{\Delta H \eta_s^2(z)} - 1\right] \tag{7-33}$$

Substitution in (7-29) makes possible numerical integration, and subsequent solution for the eigenvalues of **A**. The necessary computations were carried through by McGowin (1969) for different choices of n for each of the steady states, with the results shown in Table 7-1. Although a trial solution with six terms was needed to determine the largest eigenvalues to three significant figures for *all* steady states, it is clear that $n = 2$ is adequate if only information on the signs is sought. The same result was obtained by Wei (1965) by a slightly different but essentially equivalent route.

TABLE 7-1
MAXIMUM EIGENVALUES OF MATRIX (7-31)
FOR PARAMETERS SPECIFIED IN EXAMPLE 7-1
[FROM MCGOWIN (1969)]

Number of Terms in Trial Solution	Steady State 1	Steady State 2	Steady State 3
2	−2.078	4.148	−29.35
3	−2.078	4.151	−18.82
4	−2.078	4.151	−18.72
5			−18.36
6			−18.27
7			−18.27
8			−18.26
9			−18.26
Conclusion	Stable	Unstable	Stable

7-5. The Collocation Method

The technique of the previous section for reducing a partial differential equation to an approximately equivalent set of ordinary differential equations is only one of a family of such *weighted residual* methods. As described with great clarity in Finlayson and Scriven's review (1966), the members of this family differ essentially in the way that the residual term is eliminated, and there is no reason to suppose that Galerkin's alternative is the best.

Consider, for example, that the residual in equation (7-26) is set equal to zero at n selected points over the interval: $0 \leq z \leq 1$. Then at these so-called *collocation* points z_j, equation (7-26) becomes the set of ordinary differential equations

$$\sum_{i=1}^{n} \frac{da_i}{dt} \cos\left(i - \frac{1}{2}\right)\pi z_j = -\sum_{i=1}^{n} a_i\left[\alpha\left(i - \frac{1}{2}\right)^2 \frac{\pi^2}{L^2} - R_\eta(z)_j\right] \cos\left(i - \frac{1}{2}\right)\pi z_j$$

$$j = 1, 2, \ldots, n \tag{7-34}$$

The approach to numerical solution becomes apparent when this equation is put into a vector-matrix notation as

$$\mathbf{M}\frac{d\mathbf{a}}{dt} = \mathbf{B}\mathbf{a} \tag{7-35}$$

where the elements of the square matrices **M** and **B** are:

$$M(j, i) = \cos\left(i - \frac{1}{2}\right)\pi z_j \tag{7-36a}$$

$$B(j, i) = -\left[\alpha\left(i - \frac{1}{2}\right)^2\frac{\pi^2}{L^2} - R_\eta(z_j)\right]\cos\left(i - \frac{1}{2}\right)\pi z_j \tag{7-36b}$$

for now the obvious identification of

$$\mathbf{A} = \mathbf{M}^{-1}\mathbf{B} \tag{7-37}$$

will reduce the analysis to the standard form of equation (7-30).

The computations required by this collocation method are for low-order matrices appreciably less than those called for by Galerkin's method, the additional matrix inversion being more than offset by the savings in integrations. Note that the integrals of equation (7-29) do not even appear in the collocation statements.

EXAMPLE 7-2. In view of the satisfactory result in Example 7-1 when n was only two, assume that this low-order approximation will also be adequate in the collocation procedure, and derive stability conditions for the same problem.

Solution: For the given degree of approximation $n = 2$, the matrix determined by equation (7-36a) will be

$$\mathbf{M} = \begin{bmatrix} \cos\frac{\pi z_1}{2} & \cos\frac{3\pi z_1}{2} \\ \cos\frac{\pi z_2}{2} & \cos\frac{3\pi z_2}{2} \end{bmatrix} \tag{7-38}$$

Since we are at liberty to choose the collocation points, let $z_1 = \frac{1}{3}$ and $z_2 = \frac{2}{3}$ in order to simplify **M** to

$$\mathbf{M} = \begin{bmatrix} \frac{\sqrt{3}}{2} & 0 \\ \frac{1}{2} & -1 \end{bmatrix}$$

Then

$$\mathbf{B} = \begin{bmatrix} \frac{\sqrt{3}}{2}\left(R_\eta(z_1) - \alpha\frac{\pi^2}{4L^2}\right) & 0 \\ \frac{1}{2}\left(R_\eta(z_2) - \alpha\frac{\pi^2}{4L^2}\right) & \left(\frac{9\alpha\pi^2}{4L^2} - R_\eta(z_2)\right) \end{bmatrix}$$

and

$$\mathbf{A} = \mathbf{M}^{-1}\mathbf{B} = \begin{bmatrix} \left(R_\eta(z_1) - \alpha\dfrac{\pi^2}{4L^2}\right) & 0 \\ \dfrac{1}{2}\left(R_\eta(z_1) - R_\eta(z_2)\right) & \left(R_\eta(z_2) - \dfrac{9\alpha\pi^2}{4L^2}\right) \end{bmatrix}$$

The eigenvalues of this matrix will be the diagonal elements (regardless of where z_2 has been placed). For local stability:

$$\lambda_1 = R_\eta(z_1) - \alpha\frac{\pi^2}{4L^2} < 0 \tag{7-39a}$$

and

$$\lambda_2 = R_\eta(z_2) - \alpha\frac{9\pi^2}{4L^2} < 0 \tag{7-39b}$$

To connect these conditions with the uniqueness criteria of Chapter 6, it may be observed that

$$R_\eta(z_j) = \left.\frac{dR}{d\eta}\right|_{s,\,z_j} \leq \text{Max}\left[\frac{dR}{d\eta}\right]$$

As a consequence the condition (6-82) which is sufficient to guarantee uniqueness for this system will also assure local stability through equations (7-39), with the reservation, however, that the analysis is provisional until tests are run for larger values of n.

7-6. Modified Collocation

A feature common to both the Galerkin method and the standard collocation approach is that they recast a problem into terms of the $a_i(t)$ variables of the trial solution. This is not a serious disadvantage so long as one asks only about the signs of eigenvalues, but it leaves much to be desired if one looks further toward details of the solutions more directly related to the state variables. To correct this deficiency, Villadsen and Stewart (1967) have developed a modified collocation technique which is built around the fact that the assumption of a power series trial solution allows derivative terms in a system model to be expressed as a linear combination of the state variables.

Consider for example the trial solution

$$x(z, t) = \sum_{i=1}^{n} a_i(t)z^{2i-2} = a_1(t) + a_2(t)z^2 + a_3(t)z^4 + \ldots + a_n(t)z^{2n-2} \tag{7-40}$$

At any collocation point z_j the derivative

$$\left.\frac{\partial x}{\partial z}\right|_{z_j} = 2a_2(t)z_j + 4a_3(t)z_j^3 + \ldots + 2(n-1)a_n(t)z_j^{2n-3} \tag{7-41}$$

can also be written as a component of the vector

$$\frac{\partial x}{\partial \mathbf{z}} = \mathbf{N}_1\mathbf{N}_0^{-1}x(\mathbf{z}) \tag{7-42}$$

where

$$\mathbf{N}_1 = \begin{bmatrix} 0 & 2z_1 & 4z_1^3 & \cdots & 2(n-1)z_1^{2n-3} \\ 0 & 2z_2 & 4z_2^3 & \cdots & 2(n-1)z_2^{2n-3} \\ \cdots \\ \cdots \\ 0 & 2z_n & 4z_n^3 & \cdots & 2(n-1)z_n^{2n-3} \end{bmatrix} \tag{7-43a}$$

$$\mathbf{N}_0 = \begin{bmatrix} 1 & z_1^2 & z_1^4 & \cdots & z_1^{2n-2} \\ 1 & z_2^2 & z_2^4 & \cdots & z_2^{2n-2} \\ \cdots \\ \cdots \\ 1 & z_n^2 & z_n^4 & \cdots & z_n^{2n-2} \end{bmatrix} \tag{7-43b}$$

and

$$x(\mathbf{z}) = \begin{bmatrix} x(z_1) \\ x(z_2) \\ \cdot \\ \cdot \\ \cdot \\ x(z_n) \end{bmatrix} \tag{7-43c}$$

Similarly, the second derivative at z_j is

$$\left.\frac{\partial^2 x}{\partial z^2}\right|_{z_j} = 2a_2(t) + 12a_3(t)z_j^2 + \ldots + (2n-2)(2n-3)a_n(t)z_j^{2n-4} \tag{7-44}$$

but can be more advantageously formulated as a component of the vector

$$\frac{\partial^2 x}{\partial \mathbf{z}^2} = \mathbf{N}_2\mathbf{N}_0^{-1}x(\mathbf{z}) \tag{7-45}$$

where

$$\mathbf{N}_2 = \begin{bmatrix} 0 & 2 & 12z_1^2 & \ldots & (2n-2)(2n-3)z_1^{2n-4} \\ 0 & 2 & 12z_2^2 & \ldots & (2n-2)(2n-3)z_2^{2n-4} \\ \cdots \\ \cdots \\ 0 & 2 & 12z_n^2 & \ldots & (2n-2)(2n-3)z_n^{2n-4} \end{bmatrix} \tag{7-46}$$

To illustrate an application of this modified collocation, suppose a trial solution of the form (7-40) were to be attempted for the catalyst particle problem given as equation (7-24). It should first be noted that the proposed trial solution is acceptable as regards the boundary condition at $z = 0$, since it has a zero derivative for this position, but the trial form does not in itself restrict the value of x at any position z. It is necessary at some stage of the analysis to assure that it is consistent with the remaining boundary condition at $z = 1$. This may be most conveniently done by choosing the last collocation point at this boundary; i.e., let

$$z_n = 1$$

$$x(z_n) = x(1) = 0$$

With these definitions, substitution into equation (7-24) gives the ordinary differential equation set

$$\frac{d\mathbf{x}}{dt} = \frac{\alpha}{L^2}\mathbf{N}_2\mathbf{N}_0^{-1}\mathbf{x} + \mathbf{R}_\eta\mathbf{x} \tag{7-47a}$$

where $\mathbf{R}_\eta$ is the diagonal matrix

$$\mathbf{R}_\eta = \begin{bmatrix} \frac{dR}{d\eta}(z_1) & 0 & \cdots & 0 \\ 0 & \frac{dR}{d\eta}(z_2) & \cdots & 0 \\ \cdots & & & \\ \cdots & & & \\ 0 & 0 & \cdots & \frac{dR}{d\eta}(z_n) \end{bmatrix} \tag{7-47b}$$

This is identically the standard linear form

$$\dot{\mathbf{x}} = \mathbf{A}\mathbf{x} \tag{7-48a}$$

with

$$\mathbf{A} = \frac{\alpha}{L^2}\mathbf{N}_2\mathbf{N}_0^{-1} + \mathbf{R}_\eta \tag{7-48b}$$

The eigenvalues of $\mathbf{A}$ determine stability in the usual way.

In placing a collocation point at the boundary location where $x(1)$ is fixed, the dimensionality of the problem is in effect reduced by one. To see why this is the case write equations (7-48) in the expanded form:

$$\frac{dx}{dt}(z_1) = a_{11}x(z_1) + a_{11}x(z_2) + \ldots + a_{1n}x(z_n) \tag{7-49a}$$

$$\frac{dx}{dt}(z_2) = a_{21}x(z_1) + a_{22}x(z_2) + \ldots + a_{2n}x(z_n) \tag{7-49b}$$

$$\cdots$$

$$\cdots$$

$$\frac{dx}{dt}(z_n) = a_{n1}x(z_1) + a_{n2}x(z_2) + \ldots + a_{nn}x_n(z_n) \tag{7-49c}$$

and note that with $x(z_n)$ fixed, the system has been reduced to one with $(n - 1)$ variables. From the viewpoint of stability analysis, one can ignore entirely the last row and column of the $\mathbf{A}$ matrix. This should not be used as a justification for avoiding a third collocation point, however, since each element of the $\mathbf{N}_0$ and $\mathbf{N}_2$ matrices makes a contribution to the result before the simplification. More to the point, this reduction in dimensionality should be thought of as a degenerate case of the broader problem, arising from the particular case of a constant condition at $z_n = 1$. The full $n \times n$ matrix would be needed if for example the second boundary condition were a constant derivative value at $z = 1$.

EXAMPLE 7-3. Find local stability conditions for the catalyst particle by analyzing equation (7-48) for the choice of $z_1 = \frac{1}{3}, z_2 = \frac{2}{3}, z_3 = 1$.

Solution: At the collocation points specified:

$$\mathbf{N}_0 = \begin{bmatrix} 1 & \frac{1}{9} & \frac{1}{81} \\ 1 & \frac{4}{9} & \frac{16}{81} \\ 1 & 1 & 1 \end{bmatrix}$$

$$\mathbf{N}_2 = \begin{bmatrix} 0 & 2 & \frac{4}{3} \\ 0 & 2 & \frac{16}{3} \\ 0 & 2 & 12 \end{bmatrix}$$

$$\mathbf{N}_2\mathbf{N}_0^{-1} = \frac{3}{20}\begin{bmatrix} -35 & 32 & 3 \\ 55 & -112 & 57 \\ 205 & -352 & 147 \end{bmatrix}$$

Since the third row and column may be neglected:

$$\mathbf{A} = \begin{bmatrix} R_\eta\left(\frac{1}{3}\right) - 35\left(\frac{3\alpha}{20L^2}\right) & 32\left(\frac{3\alpha}{20L^2}\right) \\ 55\left(\frac{3\alpha}{20L^2}\right) & R_\eta\left(\frac{2}{3}\right) - 112\left(\frac{3\alpha}{20L^2}\right) \end{bmatrix}$$

Necessary and sufficient conditions for stability may be obtained from equation (4-38):

$$R_\eta\left(\frac{1}{3}\right) + R_\eta\left(\frac{2}{3}\right) < 22.0\left(\frac{\alpha}{L^2}\right) \tag{7-50a}$$

$$R_\eta\left(\frac{1}{3}\right)R_\eta\left(\frac{2}{3}\right) > 22.0\left(\frac{\alpha}{L^2}\right) - 48.6\left(\frac{\alpha}{L^2}\right)^2 \tag{7-50b}$$

One should keep in mind of course that the validity of these conditions is associated with the two-term approximation of the trial solution and the restricted disturbance forms used in the uncoupling, but that each of these constraints can be relaxed at the expense of added computation.

The contribution of Villadsen and Stewart to modified collocation is not restricted to trial solutions of the form of (7-40). The equations (7-42) and (7-45) are valid for any trial solution of the form

$$x(z, t) = \sum_{i=1}^{n} a_i(t)\Phi_i(z) \tag{7-51a}$$

provided the **N** matrices are generalized appropriately to:

$$\mathbf{N}_0 = \begin{bmatrix} \Phi_{11} & \Phi_{21} & \cdots & \Phi_{n1} \\ \Phi_{12} & \Phi_{22} & \cdots & \Phi_{n2} \\ \cdots & & & \\ \cdots & & & \\ \Phi_{1n} & \Phi_{2n} & \cdots & \Phi_{nn} \end{bmatrix} \tag{7-51b}$$

$$\mathbf{N}_1 = \begin{bmatrix} \Phi'_{11} & \Phi'_{21} & \cdots & \Phi'_{n1} \\ \Phi'_{12} & \Phi'_{22} & \cdots & \Phi'_{n2} \\ \cdots & & & \\ \cdots & & & \\ \Phi'_{1n} & \Phi'_{2n} & \cdots & \Phi'_{nn} \end{bmatrix} \tag{7-51c}$$

$$\mathbf{N}_2 = \begin{bmatrix} \Phi''_{11} & \Phi''_{21} & \cdots & \Phi''_{n1} \\ \Phi''_{12} & \Phi''_{22} & \cdots & \Phi''_{n2} \\ \cdots & & & \\ \cdots & & & \\ \Phi''_{1n} & \Phi''_{2n} & \cdots & \Phi''_{nn} \end{bmatrix} \tag{7-51d}$$

where the primes refer to first and second derivatives with respect to z and the subscripts identify the $\Phi_i(z)$ and the collocation point in that order. The entry Φ''_{12}, for example, refers to the second derivative of Φ_1 with respect to z, evaluated at the collocation point z_2.

In particular, the same investigators have emphasized the convenience of using for $\Phi_i(z)$ functions of the series of Jacobi polynomials:

$$\begin{aligned} \Phi_1 &= (1 - z^2)P_o = (1 - z^2) \\ \Phi_2 &= (1 - z^2)P_1 = (1 - z^2)(1 - 5z^2) \\ \Phi_3 &= (1 - z^2)P_2 = (1 - z^2)(1 - 4z^2 + 21z^4) \\ &\cdots \\ \Phi_i(z^2) &= (1 - z^2)P_{i-1}(z^2) \end{aligned} \tag{7-52}$$

where the Jacobi polynomials are generated by

$$P_i(z^2) = \sum_{j=0}^{i} \frac{[(-i)(1-i)(1-2i)\ldots(-1)][(2i+3)(2i+5)\ldots(2i+2j+i)]}{(i)!(1)(3)(5)\ldots(-2i+2j+1)} z^{2j}$$

One advantage of the choice of Jacobi polynomials in Φ_i lies in its applicability to boundary conditions such as govern the catalyst particle model. Since the boundary conditions at $z = 0$ and $z = 1$ are both satisfied by equation (7-51a), it is not necessary to use $z_n = 1$ as one of the collocation points. On the other hand, if one is dealing with a model that calls for a non-zero boundary condition at $z = 1$, the same functions are still applicable with the addition of a constant to the trial solution:

$$x(z, t) = x(1) + \sum_{i=1}^{n} a_i(t)\Phi_i(z) \tag{7-53}$$

but now the equations (7-51) are no longer correct and one must return to equations (7-43) and (7-46) to find $\mathbf{N}_0$, $\mathbf{N}_1$, and $\mathbf{N}_2$. This may be done because the trial solution (7-53) is in fact just a special case of equation (7-40); it can be so written by expanding the series and collecting coefficients of like terms, as follows:

$$\begin{aligned} x(z, t) &= x(1) + a_1\Phi_1 + a_2\Phi_2 + \ldots \\ &= x(1) + a_1(1 - z^2) + a_2(1 - z^2)(1 - 5z^2) + \ldots \\ &= [x(1) + a_1 + a_2 + \ldots] + [-a_1 - 6a_2 - \ldots]z^2 \\ &\qquad + [5a_2 + \ldots]z^4 + \ldots \end{aligned}$$

The choice of collocation points has to this point still been left open, but a useful hint may be found in the orthogonality property of the Jacobi polynomials

$$\int_0^1 \Phi_i(z^2)P_n(z^2)dz = \int_0^1 (1 - z^2)P_{i-1}P_n dz = \begin{cases} 0, & (i-1) \neq n \\ K_n, & (i-1) = n \end{cases} \tag{7-54}$$

The founders of this technique proposed using for the n collocation points the zeros of the nth Jacobi polynomial, that is the roots of

$$P_n(z^2) = 0$$

With this selection, the residual $\Gamma(z)$ will be zero at the same values of z for which P_n is zero. As a result

$$\int_0^1 \Gamma(z)\Phi_i(z^2)dz$$

is closely approximated by

$$\int_0^1 P_n(z^2)\Phi_i(z^2)dz = 0$$

With this choice of collocation points the residual is restricted in two ways. First, as in any collocation it is made zero at the selected points; secondly, its weighted integral is reduced close to zero by a judicious orthogonalization (thus the name *orthogonal collocation*).

A modest increase in arithmetic detail results from this particular choice of z_j points, but since the computation need only be done once and for all, Villadsen and Stewart have presented numerical values of $\mathbf{N}_1\mathbf{N}_0^{-1}$ and $\mathbf{N}_2\mathbf{N}_0^{-1}$ for various n. For uniformity they included a collocation point at $z = 1$ in all analyses thus avoiding the need for a separate decision in each case. This does not significantly affect the analysis if the boundary condition at $z = 1$ is a constant [compare the discussion of equations (7-49)], but makes possible the treatment of other cases from the same tabulation. Their results are recorded in Appendix C for easy reference.

EXAMPLE 7-4. Reconsider the previous Example 7-3 in order to compare the earlier result with two other approaches based on Jacobi polynomials. Find $\mathbf{N}_2\mathbf{N}_0^{-1}$ via (a) equation (7-51) and (b) Appendix C based on orthogonal collocation.

Solution: Part (a): Following equations (7-51) for $n = 2$:

$$\mathbf{N}_0 = \begin{bmatrix} \Phi_{11} & \Phi_{22} \\ \Phi_{12} & \Phi_{22} \end{bmatrix} = \begin{bmatrix} (1 - z_1^2) & (1 - z_1^2)(1 - 5z_1^2) \\ (1 - z_2^2) & (1 - z_2^2)(1 - 5z_2^2) \end{bmatrix}$$

Since the choice of collocation points is open, choose $z_1 = \sqrt{1/5} = 0.447$ to simplify $\mathbf{N}_0$ and $\mathbf{N}_2$ to

$$\mathbf{N}_0 = \begin{bmatrix} 0.80 & 0 \\ (1 - z_2^2) & (1 - z_2^2)(1 - 5z_2^2) \end{bmatrix}$$

$$\mathbf{N}_2 = -2\begin{bmatrix} 1 & 0 \\ 1 & 6(1 - 5z_2^2) \end{bmatrix}$$

Then

$$\mathbf{N}_2\mathbf{N}_0^{-1} = 2.5\begin{bmatrix} -1 & 0 \\ 5 & -\dfrac{4.8}{(1 - z_2^2)} \end{bmatrix}$$

The $\mathbf{A}$ matrix is diagonal from which the conditions of stability are

$$R_\eta(z_1) < 2.5\frac{\alpha}{L^2} \tag{7-56a}$$

$$R_\eta(z_2) < \frac{12}{(1 - z_2^2)}\frac{\alpha}{L^2} \tag{7-56b}$$

If for a specific $z_2 = \sqrt{\dfrac{1}{2}} = 0.707$, the second of these is

$$R_\eta(z_2) < 24\frac{\alpha}{L^2} \tag{7-56c}$$

Part (b) From Appendix C for $n = 2$:

$$\mathbf{N}_2\mathbf{N}_0^{-1} = \begin{bmatrix} -4.74 & 5.68 & -0.937 \\ 8.32 & -23.3 & 14.9 \\ 19.07 & -47.1 & 28 \end{bmatrix}$$

Since this problem has a constant boundary condition fixed at $z = 1$, the third row and column of $\mathbf{N}_2\mathbf{N}_0^{-1}$ are to be ignored and the $\mathbf{A}$ matrix degenerates to the 2×2:

$$\mathbf{A} = \begin{bmatrix} R_\eta(z_1) - 4.74\dfrac{\alpha}{L^2} & 5.68\dfrac{\alpha}{L^2} \\ 8.32\dfrac{\alpha}{L^2} & R_\eta(z_2) - 23.3\dfrac{\alpha}{L^2} \end{bmatrix}$$

where

$$z_1 = 0.285$$
$$z_2 = 0.765$$

are the roots of $P_2(z^2) = 1 - 14z^2 + 21z^4 = 0$. The stability conditions analogous to inequalities (7-50) are:

$$R_\eta(z_1) + R_\eta(z_2) < 28.0\frac{\alpha}{L^2} \tag{7-57a}$$

$$R_\eta(z_1)R_\eta(z_2) > 28.0\left(\frac{\alpha}{L^2}\right) - 64.6\left(\frac{\alpha}{L^2}\right)^2 \tag{7-57b}$$

The various results are all roughly comparable.

7-7. Coupled Equations

It is worth repeating here the point made in introducing the weighted residual methods: the techniques are not dependent on prior uncoupling. The arguments presented are as valid for the paired equations (7-20) as for the single equation (7-22), and similarly for the catalyst particle models. Because of the rather severe assumptions that were necessary to justify the uncoupling, it is especially satisfying to observe that no restrictions on initial conditions or adiabatic transients limit either Galerkin's method or the various collocation approaches. The only penalty for the use of paired equations is in an increase in system dimensionality, showing up eventually in the computational load associated with the manipulation of higher-order matrices.

This may be demonstrated by combining equations (7-13), (7-15) and (7-23) to obtain the linearized catalyst particle equations:

$$\frac{\partial x_1}{dt} = \frac{\alpha}{L^2}\frac{\partial^2 x_1}{\partial z^2} + R_\eta x_1 + R_c x_2 \tag{7-58a}$$

$$\frac{\partial x_2}{\partial t} = \frac{D}{L^2}\frac{\partial^2 x_2}{\partial z^2} - R_\eta x_1 - R_c x_2 \tag{7-58b}$$

To apply Galerkin's method, one now needs a trial solution for each state variable:

$$x_1(z, t) = \sum_{i=1}^{n} a_i(t) \cos\left(i - \frac{1}{2}\right)\pi z \tag{7-59a}$$

$$x_2(z, t) = \sum_{i=1}^{n} b_i(t) \cos\left(i - \frac{1}{2}\right)\pi z \tag{7-59b}$$

Substitution into (7-58) produces two equations of the form of equations (7-26) which simplify through the relationship (7-27) to:

$$\frac{da_j}{dt} = -\alpha\left[\left(j - \frac{1}{2}\right)\frac{\pi}{L}\right]^2 a_j + \sum_{i=1}^{n} I(i, j)a_i + \sum_{i=1}^{n} J(i, j)b_i \tag{7-60a}$$

$$\frac{db_j}{dt} = -D\left[\left(j - \frac{1}{2}\right)\frac{\pi}{L}\right]^2 b_j - \sum_{i=1}^{n} I(i, j)a_i - \sum_{i=1}^{n} J(i, j)b_i \tag{7-60b}$$

where:

$$I(i, j) = 2\int_0^1 R_n(z)\left[\cos\left(i - \frac{1}{2}\right)\pi z\right]\left[\cos\left(j - \frac{1}{2}\right)\pi z\right] dz \tag{7-61a}$$

$$J(i, j) = 2\int_0^1 R_c(z)\left[\cos\left(i - \frac{1}{2}\right)\pi z\right]\left[\cos\left(j - \frac{1}{2}\right)\pi z\right] dz \tag{7-61b}$$

These equations may be recognized to be in the same form as equations (7-30), but now

$$\mathbf{a} = \begin{bmatrix} a_1 \\ a_2 \\ \cdot \\ \cdot \\ \cdot \\ a_n \\ b_1 \\ b_2 \\ \cdot \\ \cdot \\ \cdot \\ b_n \end{bmatrix} \tag{7-62a}$$

and the dimension of the square **A** matrix has doubled to $2n$:

$$\mathbf{A} = \left[\begin{array}{c:c:c:c:c:c} I_{11} - \alpha\left(\frac{\pi}{2L}\right)^2 & I_{12} & \cdots & J_{11} & J_{12} & \cdots \\ I_{12} & I_{22} - \alpha\left(\frac{3\pi}{2L}\right)^2 & & J_{12} & J_{22} & \\ \hdashline \cdots & & & & & \\ \hdashline -I_{11} & -I_{12} & \cdots & J_{11} - D\left(\frac{3\pi}{2L}\right)^2 & -J_{12} & \cdots \\ -I_{12} & -I_{22} & & J_{12} & -J_{22} - D\left(\frac{3\pi}{2L}\right)^2 & \\ \hdashline \cdots & & & & & \end{array}\right] \tag{7-62b}$$

Further numerical difficulty arises in the eigenvalue determinations because matrix (7-62b) is not in general symmetric, as was matrix (7-31). The eigenvalues of a symmetric matrix are always real and simpler computational methods are applicable. Equations which produce symmetric **A** matrices are called *self-adjoint.*

The essential question in this or any weighted residual method is that of determining the degree of approximation needed; that is, of finding the n small enough to make computation easy but large enough to retain the accuracy or qualitative behavior sought. Evidence to date suggests that surprisingly small n is satisfactory for stability study of the catalyst particle system. Kuo and Amundson (1967) reported results for accurate examination of four steady-state profiles by Galerkin's method. In each case the stability conclusion was correct at $n = 1$, and in no case did it require more than $n = 3$ to get eigenvalues correct to three significant figures. A study of the same system by McGowin (1969) also used Galerkin's method, but focussed especially on the affect of variation in Lewis number over a wide range. The particular example chosen was the one with three steady states shown as Figure 6-10. The same profiles are valid for all Lewis numbers if the other parameters are held at

$$\frac{\Delta H C_0 D}{C_p T_0 \alpha} = 0.70$$

$$\frac{L^2 k_0}{D} = 1.23 \times 10^7$$

$$\frac{Q}{T_0} = 20$$

The analysis showed that the low conversion steady state is locally stable and the intermediate conversion steady state is unstable for all positive Lewis numbers, while the high conversion steady state is locally stable for $(D/\alpha) > 0.265$ and is unstable otherwise.

In every case, the Galerkin's method estimate of the dominant eigenvalue of the linearized equations converged monotonically in the positive or negative direction, requiring from three to ten terms in the trial solutions for the dominant eigenvalue to converge to a constant value. Monotonic convergence occurred in the negative direction for the low conversion steady state for $(D/\alpha) < 1$ and for the high conversion steady state for (D/α) between approximately 0.50 and 1.0; it occurred in the positive direction in the remaining cases. Figure 7-1 shows the convergence patterns of the estimated dominant eigenvalues for the high conversion steady state for Lewis numbers of 0.25 and 0.50, respectively. In the former case, the monotonic convergence is in the positive direction and the unstable nature of the steady state is established for $n = 3$ before convergence occurs. In the latter case, monotonic convergence is positive, and the stable nature of the high conversion

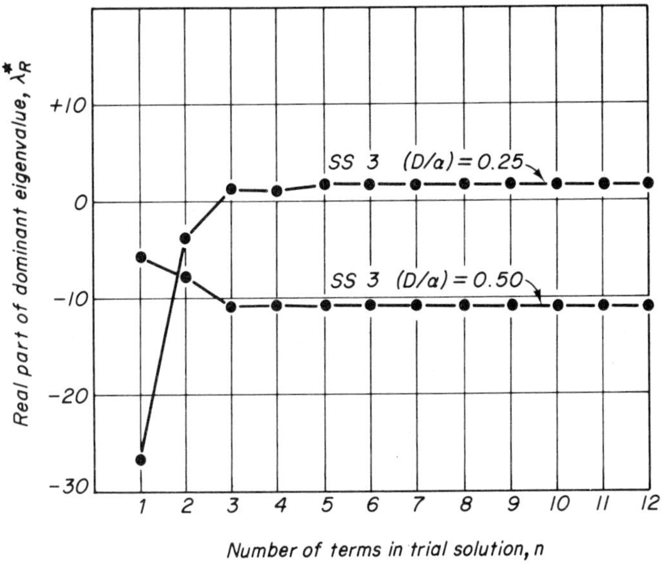

Figure 7-1. Convergence pattern of dominant eigenvalue of the catalyst particle problem estimated using Galerkin's method.

steady state is established for $n = 2$ terms in the trial solutions. Similarly, the unstable character of the intermediate steady state and the locally stable character of the low conversion steady state for $(D/\alpha) < 1$ are established before the dominant eigenvalue converges to a constant value.

Results for the same problem in spherical geometry were obtained independently by Lee and Luss (1970) using a seven-term Galerkin expansion for moderate values of the Lewis number. In agreement with McGowin's findings, instabilities showed up for small values of the Lewis number. Below $(D/\alpha) = 0.5$ the system transients tended to oscillation and for $(D/\alpha) = 0.1$ cases of unique but unstable steady states were obtained. As expected from the earlier proof of Gavalas (1968), a steady state is unstable for any Lewis number if it is unstable for $(D/\alpha) = 1$. Fortunately, most practical catalysts have Lewis numbers large enough to avoid the extraordinary behavior predicted by these computations.

As an alternative to Galerkin's method, the coupled equations (7-58) may be dealt with by collocation. Then equation (7-35) will again describe the dynamics of the linearized problem, where now

$$M(j, i) = \cos\left(i - \frac{1}{2}\right)\pi z_j$$

$$M(n + j, n + i) = M(j, i)$$

$$M(j, n + i) = M(n + j, i) = 0 \tag{7-63a}$$

and

$$B(j, i) = -\left[\alpha\left(i - \frac{1}{2}\right)^2 \frac{\pi^2}{L^2} - R_\eta(z_j)\right] \cos\left(i - \frac{1}{2}\right)\pi z_j$$

$$B(n + j, n + i) = -\left[D\left(i - \frac{1}{2}\right)^2 \frac{\pi^2}{L^2} + R_c(z_j)\right] \cos\left(i - \frac{1}{2}\right)\pi z_j$$

$$B(j, n + i) = R_c(z_j) \cos\left(i - \frac{1}{2}\right)\pi z_j$$

$$B(n + j, i) = -R_\eta(z_j) \cos\left(i - \frac{1}{2}\right)\pi z_j \tag{7-63b}$$

This approach produces a nonsymmetric **A** as can be verified by inspection of equations (7-36) or (7-63).

The same problem yields also to analysis by orthogonal collocation. For trial solutions of the form:

$$x_1(z, t) = x_1(1) + \sum_{i=1}^{n} a_i(t)\Phi_i(z) \tag{7-64a}$$

$$x_2(z, t) = x_2(1) + \sum_{i=1}^{n} b_i(t)\Phi_i(z) \tag{7-64b}$$

the counterpart of equation (7-48) is

$$\dot{\mathbf{x}} = \mathbf{A}\mathbf{x} \tag{7-65a}$$

with

$$\mathbf{x} = \begin{bmatrix} x_1(z_1) \\ x_1(z_2) \\ \cdot \\ \cdot \\ \cdot \\ x_1(z_n) \\ x_2(z_1) \\ x_2(z_2) \\ \cdot \\ \cdot \\ \cdot \\ x_2(z_n) \end{bmatrix} \tag{7-65b}$$

and in partitioned form

$$\mathbf{A} = \left[\begin{array}{c|c} \frac{\alpha}{L^2}\mathbf{N}_2\mathbf{N}_0^{-1} + \mathbf{R}_\eta & \mathbf{R}_c \\ \hline -\mathbf{R}_\eta & \frac{D}{L^2}\mathbf{N}_2\mathbf{N}_0^{-1} - \mathbf{R}_c \end{array}\right] \tag{7-65c}$$

It should be recalled that the subscripts on x refer to the temperature and concentration state variables, respectively, whereas the subscripting on z

refers to the collocation points. The subscripted R's refer in this context to *partial* derivatives with respect to the subscript variable.

EXAMPLE 7-5. Derive stability conditions for the coupled system of equations (7-58) to compare (a) collocation for $n = 2$ at $z_1 = \frac{1}{3}$ and $z_2 = \frac{2}{3}$, based on the trial solution (7-59) with (b) orthogonal collocation.

Solution: Part (a) With the specifications and trial function given:

$$\mathbf{M} = \left[\begin{array}{cc|cc} \sqrt{\frac{3}{2}} & 0 & 0 & 0 \\ \frac{1}{2} & -0 & 0 & 0 \\ \hline 0 & 0 & \sqrt{\frac{3}{2}} & 0 \\ 0 & 0 & \frac{1}{2} & -1 \end{array}\right]$$

$$\mathbf{B} = \left[\begin{array}{c|c|c|c} \sqrt{\frac{3}{2}}\left[R_\eta(z_1) - \alpha\frac{\pi^2}{4L^2}\right] & 0 & \sqrt{\frac{3}{2}}[R_c(z_1)] & 0 \\ \hline \frac{1}{2}\left[R_\eta(z_2) - \alpha\frac{\pi^2}{4L^2}\right] & \alpha\frac{9\pi^2}{4L^2} - R_\eta(z_2) & \frac{1}{2}[R_c(z_2)] & -R_c(z_2) \\ \hline \sqrt{\frac{3}{2}}[-R_\eta(z_1)] & 0 & -\sqrt{\frac{3}{2}}\left[D\frac{\pi^2}{4L^2} + R_c(z_1)\right] & 0 \\ \hline \frac{1}{2}[-R_\eta(z_2)] & R_\eta(z_2) & -\frac{1}{2}\left[D\frac{\pi^2}{4L^2} + R_c(z_2)\right] & R_c(z_2) + D\frac{9\pi^2}{4L^2} \end{array}\right]$$

$$\mathbf{A} = \left[\begin{array}{c|c|c|c} R_\eta(z_1) - \frac{\alpha\pi^2}{4L^2} & 0 & R_c(z_1) & 0 \\ \hline \frac{1}{2}[R_\eta(z_1) - R_\eta(z_2)] & R_\eta(z_2) - \alpha\frac{9\pi^2}{4L^2} & \frac{1}{2}[R_c(z_1) - R_c(z_2)] & R_c(z_2) \\ \hline -R_\eta(z_1) & 0 & -D\frac{\pi^2}{4L^2} - R_c(z_1) & 0 \\ \hline \frac{1}{2}[R_\eta(z_2) - R_\eta(z_1)] & -R_\eta(z_2) & \frac{1}{2}[R_c(z_2) - R_c(z_1)] & -D\frac{9\pi^2}{4L^2} - R_c(z_2) \end{array}\right] \tag{7-66}$$

The most general conditions for local stability depend of course on the signs of the eigenvalues of $\mathbf{A}$ but a simplified sufficient condition may be obtained by applying Gerschgorin's criteria to the columns of $\mathbf{A}$. Asuming, as is usually the case, that all the partial derivatives of R are positive, the eigenvalues of **A** will have negative real parts if:

$$\begin{aligned} 2R_\eta(z_1) + |R_\eta(z_2) - R_\eta(z_1)| &< \frac{\alpha\pi^2}{4L^2} \\ 2R_\eta(z_2) &< \frac{9\alpha\pi^2}{4L^2} \\ |R_c(z_2) - R_c(z_1)| &< \frac{D\pi^2}{4L^2} \end{aligned} \tag{7-67}$$

Part (b): As was the case in Example 7-4, Part (b), the last row and column of $\mathbf{N}_2\mathbf{N}_0^{-1}$ are ignored because $x_1(1) = x_2(1) = 0$. This reduces equation (7-65c) to

$$\mathbf{A} = \left[\begin{array}{c:c:c:c} R_\eta(z_1) - 4.74\frac{\alpha}{L^2} & 5.68 & R_c(z_1) & 0 \\ \hdashline 8.32 & R_\eta(z_2) - 23.3\frac{\alpha}{L^2} & 0 & R_c(z_2) \\ \hdashline -R_\eta(z_1) & 0 & -R_c(z_1) - 4.74\frac{D}{L^2} & 5.68 \\ \hdashline 0 & -R_\eta(z_2) & 8.32 & -R_c(z_2) - 23.3\frac{D}{L^2} \end{array}\right] \tag{7-68}$$

Again, a sufficient condition is available from the columns of **A** via Gerschgorin's theorem:

$$\begin{aligned} 2R_\eta(z_1) &< 4.74\frac{\alpha}{L^2} - 8.32 \\ 2R_\eta(z_2) &< 23.3\frac{\alpha}{L^2} - 5.68 \\ 0 &< 4.74\frac{D}{L^2} - 8.32 \end{aligned} \tag{7-69}$$

These results appear roughly equivalent to conditions (7-67), but are perhaps a bit simpler to apply since the computations of R_c are not required.

7-8. Self-Adjoint Equations

The numerical experience cited above suggests that the weighted residual methods lead to correct stability results for relatively modest choices of n. When this is the case, a symmetric **A** matrix is convenient but not essential; however, for matrices of larger dimensions, the advantage of a symmetric **A** becomes increasingly attractive. Of the weighted residual methods described above, only Galerkin's variation can lead to symmetry, but even this is lost in the coupled equation analysis that developed to matrix (7-62).

A partial relief for this difficulty has been found by Kuo and Amundson (1967) in the form of an auxiliary equation set:

$$\frac{\partial W_1}{\partial t} = \frac{\alpha}{L^2}\frac{\partial W_1}{\partial z^2} + R_\eta W_1 + \frac{1}{2}(R_c - R_\eta)W_2 \tag{7-70a}$$

$$\frac{\partial W_2}{\partial t} = \frac{D}{L^2}\frac{\partial^2 W_2}{\partial z^2} + \frac{1}{2}(R_c - R_\eta)W_1 - R_c W_2 \tag{7-70b}$$

which since it is self-adjoint, has real eigenvalues determined from a symmetric **A**. It is proved in the reference cited that the largest eigenvalue of the auxiliary equation (7-70) is a conservative bound on the real parts of the eigenvalues of the original system (7-58), with the same boundary conditions. Consequently a sufficient condition for stability of (7-58) is that the largest eigenvalue of (7-70) is negative.

7-9. Application to the Tubular Reactor

If the deviation variables defined by equations (7-12) are used to reduce the TRAM equations (1-8), the resulting set is

$$\frac{\partial x_1}{\partial t} = \alpha\frac{\partial^2 x_1}{\partial Z^2} - u\frac{\partial x_1}{\partial Z} + r - \frac{h}{C_p}x_1 \tag{7-71a}$$

$$\frac{\partial x_2}{\partial t} = D\frac{\partial^2 x_2}{\partial Z^2} - u\frac{\partial x_2}{\partial Z} - r \tag{7-71b}$$

By linearization and normalization as per equations (7-15) and (7-23):

$$\frac{\partial x_1}{\partial t} = \frac{\alpha}{L^2}\frac{\partial^2 x_1}{\partial z^2} - \frac{u}{L}\frac{\partial x_1}{\partial z} + \left(R_\eta - \frac{h}{C_p}\right)x_1 + R_c x_2 \tag{7-72a}$$

$$\frac{\partial x_2}{\partial t} = \frac{D}{L^2}\frac{\partial^2 x_2}{\partial z^2} - \frac{u}{L}\frac{\partial x_2}{\partial z} - R_\eta x_1 - R_c x_2 \tag{7-72b}$$

and the boundary conditions corresponding to equations (1-9) are

$$\frac{\alpha}{L}\frac{\partial x_1}{\partial z} = ux_1, \qquad z = 0 \tag{7-73a}$$

$$\frac{D}{L}\frac{\partial x_2}{\partial z} = ux_2, \qquad z = 0 \tag{7-73b}$$

$$\frac{\partial x_1}{\partial z} = \frac{\partial x_2}{\partial z} = 0, \qquad z = 1 \tag{7-73c}$$

Galerkin's method can be applied to this system by defining a suitable trial function. Comparison with equation (7-10), the solution of (7-8), suggests the form:

$$x_1(z, t) = \sum_{i=1}^{n} a_i(t)\left[\frac{2\alpha}{uL}\beta_i \cos\beta_i z + \sin\beta_i z\right]\left[\exp\left(\frac{uLz}{2\alpha}\right)\right] \tag{7-74a}$$

$$x_2(z, t) = \sum_{i=1}^{n} b_i(t)\left[\frac{2D}{uL}\beta'_i \cos\beta'_i z + \sin\beta'_i z\right]\left[\exp\left(\frac{uLz}{2D}\right)\right] \tag{7-74b}$$

For algebraic simplicity in the next steps, consider that $\alpha = D$, from which $\beta_i = \beta'_i$. Substitution into (7-72) followed by the usual rearrangement then leads to:

$$\frac{da_j}{dt} = -\left[\frac{D}{L^2}\beta_j^2 + \frac{u^2}{4D} + \frac{h}{C_p}\right]a_j + \sum_{i=1}^{n} I(i,j)a_i + \sum_{i=1}^{n} J(i,j)b_i \tag{7-75a}$$

$$\frac{db_j}{dt} = -\left[\frac{D}{L^2}\beta_j^2 + \frac{u^2}{4D}\right]b_j - \sum_{i=1}^{n} I(i,j)a_i - \sum_{i=1}^{n} J(i,j)b_j \tag{7-75b}$$

where:

$$I(i,j) = \frac{1}{K_i}\int_0^1 R_\eta(z)\left[\frac{2D}{uL}\beta_i \cos \beta_1 z + \sin \beta_i z\right]\left[\frac{2D}{uL}\beta_j \cos \beta_j z + \sin \beta_j z\right]dz \tag{7-76a}$$

$$J(i,j) = \frac{1}{K_i}\int_0^1 R_c(z)\left[\frac{2D}{uL}\beta_i \cos \beta_i z + \sin \beta_i z\right]\left[\frac{2D}{uL}\beta_j \cos \beta_j z + \sin \beta_j z\right]dz \tag{7-76b}$$

and the constants K_i are defined by the orthogonality relation

$$\int_0^1 \left[\frac{2D}{uL}\beta_i \cos \beta_i z + \sin \beta_i z\right]\left[\frac{2D}{uL}\beta_j \cos \beta_j z + \sin \beta_j z\right]dz = \begin{cases} 0, & j \neq i \\ K_i, & j = i \end{cases} \tag{7-76c}$$

The parallelism between equations (7-75) for the TRAM and equations (7-60) for the catalyst particle is so great as to require scant comment. The stability will be determined by the signs of the eigenvalues of a matrix quite similar in form to (7-62b).

The details of such a calculation were carried through by McGowin (1971a and 1971b). He demonstrated that this method yields the most rapid convergence of four common techniques tested. Depending on the steady state being studied, as few as four or as many as twenty-four terms were needed in the trial solution to achieve three significant figures in the dominant eigenvalue; however, since convergence occurred monotonically in the positive direction in every case, the local stability character of unstable states could be determined well before this. Two examples are shown in Figure 7-2. Curve (*a*) results from the eigenvalue calculations for the kinetics given with Figure 6-4 and the parameters

$$\frac{uL}{D} = \frac{uL}{\alpha} = 10$$

$$\frac{hL}{uC_p} = 0.10$$

$$\text{Effluent } (C/C_0) = 0.65$$

$$\text{Feed condition } (C_pT_0/\Delta HC_0) = 2.71$$

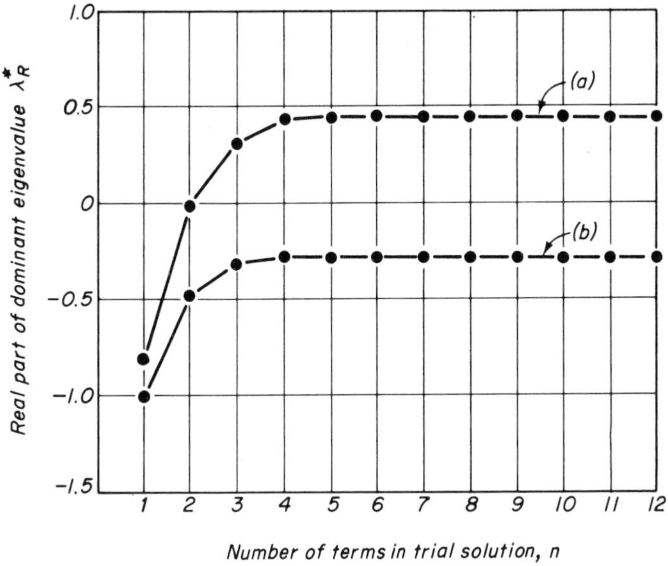

Figure 7-2. Convergence pattern of dominant eigenvalues of the TRAM problem, estimated using Galerkin's method.

The unstable character of this steady state is established for $n = 3$ terms in the trial solutions. Curve (*b*) for the same system with Effluent $(C/C_0) = 0.70$ illustrates the convergence pattern for a locally stable steady state. In this case it is necessary to obtain convergence to establish local stability. After enough steady states have been analyzed in this way a region of instability can be located on the plane of the steady state operating curves. Such a result is given as Figure 7-3 for numerical values identical to those of Figure 6-9a.

EXAMPLE 7-6. In the course of introducing Galerkin's method in Section 7-4 it was mentioned that a trial solution should be chosen to satisfy the problem boundary conditions. Is this the case for the trial functions (7-74) with reference to the TRAM conditions (7-73)?

Solution: By formal differentiation of equation (7-74a)

$$\frac{\partial x_1}{\partial z} = \sum_{i=1}^{n} a_i(t) \exp\left(\frac{uLz}{2\alpha}\right)\left[\left(\frac{uL}{2\alpha} - \frac{2\alpha\beta_i^2}{uL}\right)\sin\beta_i z + 2\beta_i \cos\beta_i z\right]$$
$$= \sum_{i=1}^{n} 2a_i(t) \sin\beta_i z \exp\left(\frac{uLz}{2\alpha}\right)\left[\left(\frac{uL}{4\alpha} - \frac{\alpha\beta_i^2}{uL}\right) + \beta_i \cot\beta_i z\right]$$

Comparison with equation (7-10c) shows that the term in brackets reduces to zero at $z = 1$, thus satisfying condition (7-73c). The other boundary conditions are also satisfied since at $z = 0$:

$$\left.\frac{\partial x_1}{\partial t}\right|_0 = 2\sum_{i=1}^{n} a_i(t)\beta_i$$

and

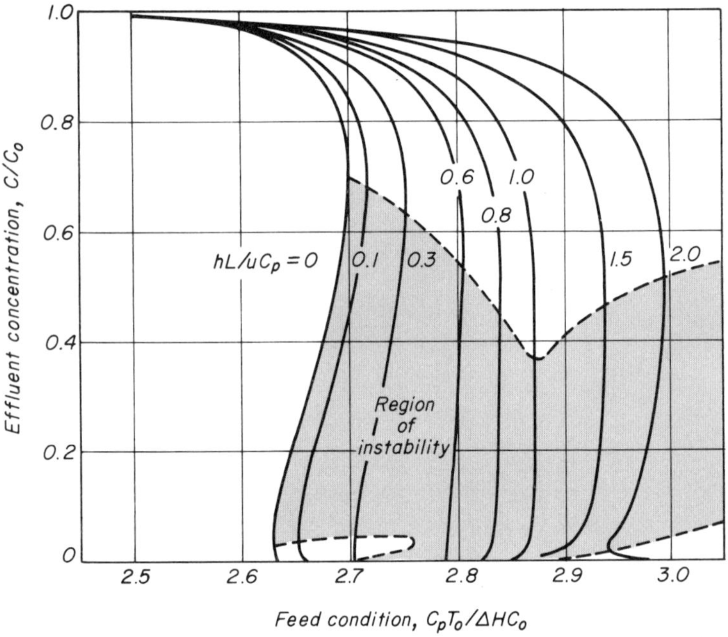

Figure 7-3. Steady-state operating curves for the TRAM [$\alpha = D$, $(uL/D = 10]$ showing a region of instability.

$$x_1(0, t) = \frac{2\alpha}{uL} \sum_{i=1}^{n} a_i(t)\beta_i$$

and similarly for the x_2 variable.

7-10. Liapunov Functionals

The formal extension of Liapunov methods to the study of distributed systems may be found in books by Zubov (1957) and Hahn (1963) and papers by Wang (1963a, 1963b, 1964) and Movchan (1959, 1960, 1963). A fine summary of Movchan's work is by Knops and Wilkes (1966). The definitions and arguments parallel to a great extent the earlier treatment of lumped-parameter models, but must be pursued with some care, since the addition of a spacial distribution introduces particular complications.

To deal effectively with distributed systems a norm is needed which takes the space-coordinate z into account. This distance measure is usually defined as the so-called L_2-norm

$$||\mathbf{x}(z)|| = \left\{\int_0^1 [x_1^2(z) + x_2^2(z)]dz\right\}^{1/2} \tag{7-77}$$

Such a relationship is called a *functional* (rather than a function), since it defines a scalar for every value of the function $\mathbf{x}(z)$. Another way of stating this is to describe it as a function of a function (as opposed to a function of a variable). This norm can be generalized for any number of state variables and spacial dimensions, but equation (7-77) is adequate for the discussion of this chapter.

The usual Liapunov definition of local stability may be applied to distributed systems by using the L_2-norm. The familiar ϵ and $\delta(\epsilon)$ are needed, such that

$$\|\mathbf{x}(0, z)\| \leq \delta \tag{7-78a}$$

implies

$$\|\mathbf{x}(t, z)\| \leq \epsilon \tag{7-78b}$$

If in addition

$$\|\mathbf{x}(t, z)\| \longrightarrow 0, \quad t \longrightarrow \infty \tag{7-78c}$$

the system is said to be asymptotically stable. Recalling from Example 5-5 and the discussion of Section 5-8 that the Liapunov function may also be thought of as a norm, it is reasonable to expect that the functional

$$v = \int_0^1 \mathbf{x}^T(z)\mathbf{x}(z)\,dz \tag{7-79}$$

would be useful for distributed systems, but the decision is subject to the requirements of a specific model as to feasibility and convenience.

The especially simple model of the uncoupled catalyst particle has been successfully analyzed by Wei (1965) and Berger and Lapidus (1968). Since equation (7-21) is in terms of only a single state variable, the Liapunov functional (7-79) reduces to

$$v = \int_0^1 x^2(z)\,dz \tag{7-80}$$

where x is the temperature deviation as defined by equation (7-12a) with the unnecessary subscript dropped for convenience of notation. Differentiation with respect to time gives

$$\dot{v} = 2\int_0^1 x\frac{\partial x}{\partial t}\,dz \tag{7-81}$$

and by substitution of equations (7-21) and (7-23):

$$\dot{v} = 2\int_0^1 x\left(\frac{\alpha}{L^2}\frac{\partial^2 x}{\partial z^2} + r\right)dz \tag{7-82}$$

Integration by parts with boundary conditions (7-14) converts this into

$$\dot{v} = -\frac{2\alpha}{L^2}\int_0^1 \left(\frac{\partial x}{\partial z}\right)^2 dz + 2\int_0^1 xr\,dz \tag{7-83}$$

from which it is apparent that

$$\int_0^1 xr(x, z)dz \leq 0 \tag{7-84}$$

is sufficient to ensure that $\dot{v}$ will be negative.

One way to satisfy inequality (7-84) is to find conditions for which the integrand $xr(x, z)$ is zero or negative at all positions. This is the case when

$$\frac{dr}{dx} = \frac{\Delta H}{C_p}\frac{dR}{dT} \leq 0, \qquad 0 \leq z \leq 1 \tag{7-85}$$

because as shown in Figure 7-4, such a condition places the r versus x curve in the second and fourth quadrants of the plane. The similarity of equations

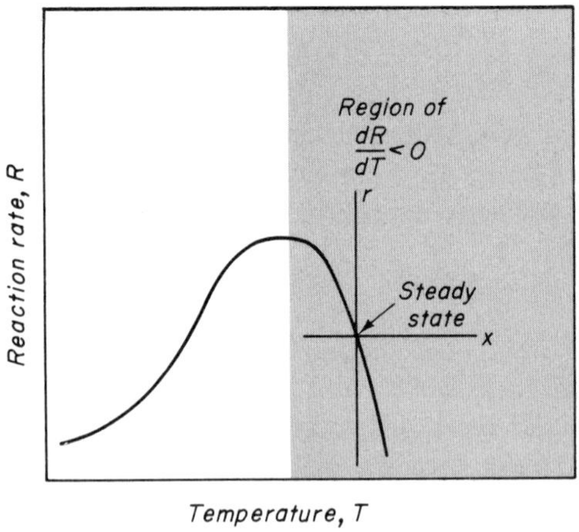

Figure 7-4. A typical function $R(T)$, showing that $(dR/dT) < 0$ implies that $xr < 0$.

(7-85) and (6-77b) indicates that this stability criterion also guarantees a unique steady state. For the standard first-order Arrhenius kinetics this is given in terms of the various system parameters by equation (6-81). As emphasized before, this condition is quite conservative. A sharper estimation technique was developed by Berger and Lapidus who were able to exploit an inequality of Gavalas (1966) as follows. By definition:

$$\int_z^1 \frac{\partial x}{\partial z} dz = x(t, z) + \text{const.} \tag{7-86}$$

but boundary condition (7-14b) shows the constant to be zero. As a result

$$x^2 = \left[\int_z^1 \frac{\partial x}{\partial z}\, dz\right]^2 \tag{7-87}$$

Applying the Schwartz inequality (3-13f) to the norm defined by

$$\|\cdot\| = \int_z^1 (\cdot)\, dz \tag{7-88}$$

produces

$$\int_z^1 \left(\frac{\partial x}{\partial z}\right)^2 dz \leq \int_z^1 \left(\frac{\partial x}{\partial z}\right)^2 dz \int_z^1 (1) dz = (1 - z) \int_z^1 \left(\frac{\partial x}{\partial z}\right)^2 dz$$
$$\leq (1 - z) \int_0^1 \left(\frac{\partial x}{\partial z}\right)^2 dz \tag{7-89a}$$

and also

$$\left[\int_z^1 \left(\frac{\partial x}{\partial z}\right) dz\right]^2 = \int_z^1 \left(\frac{\partial x}{\partial z}\right) dz \int_z^1 \left(\frac{\partial x}{\partial z}\right) dz \leq \int_z^1 \left(\frac{\partial x}{\partial z}\right)^2 dz \tag{7-89b}$$

Combining inequalities (7-89) with (7-87) gives

$$x^2 \leq (1 - z) \int_0^1 \left(\frac{\partial x}{\partial z}\right)^2 dz \tag{7-90}$$

By using the linearized model (7-24), equation (7-83) can be approximated as

$$\dot{v} = -\frac{2\alpha}{L^2} \int_0^1 \left(\frac{\partial x}{\partial z}\right)^2 dz + 2 \int_0^1 x^2 R_\eta \, dz \tag{7-91}$$

Examination of typical kinetics shows as for example in Figure 2-4 that the function R_η can change sign over the range of interest. By including in the analysis only the positive contribution and ignoring the ranges $z < z_1$ and $z > z_2$ over which R_η is negative, equation (7-90) becomes the inequality

$$\dot{v} \leq -\frac{2\alpha}{L^2} \int_0^1 \left(\frac{\partial x}{\partial z}\right)^2 dz + 2 \int_{z_1}^{z_2} x^2 R_\eta \, dz \tag{7-92}$$

which in combination with inequality (7-90) gives

$$\dot{v} \leq -\frac{2\alpha}{L^2} \int_0^1 \left(\frac{\partial x}{\partial z}\right)^2 dz + 2 \int_0^1 \left(\frac{\partial x}{\partial z}\right)^2 dz \int_{z_1}^{z_2} (1 - z) R_\eta \, dz$$
$$\leq 2 \int_0^1 \left(\frac{\partial x}{\partial z}\right)^2 dz \left[-\frac{\alpha}{L^2} + \int_{z_1}^{z_2} (1 - z) R_\eta \, dz\right] \tag{7-93}$$

A sufficient condition for stability (i.e., for $\dot{v} < 0$) is therefore

$$\int_{z_1}^{z_2} (1 - z) R_\eta \, dz < \frac{\alpha}{L^2} \tag{7-94}$$

Berger and Lapidus applied this criterion to each of the three steady states of the particular example given as Figure 6-10, and were able to confirm that steady states one and three are stable. No conclusion was obtained for the intermediate steady state, since violation of a sufficient condition leaves the question unresolved. The authors were able to extend the technique to include more general cases including coupled equations for various Lewis numbers, but they demonstrated that numerical results on sample problems were not conclusive.

An attempt to analyze the adiabatic TRAM by the same approach led to the stability criterion

$$\int_{z_1}^{z_2}\left(R_\eta - \frac{u^2}{4\alpha}\right)z\,dz < \frac{\alpha}{L^2} \tag{7-95}$$

but once again numerical work on an example led to results that were unsatisfactory in their inconclusive nature.

Other investigators have also looked at distributed parameter systems in the light of Liapunov functional theory [Nishimura and Matsubara (1970); Murphy and Crandall (1970); Clough and Ramirez (1970); Kugelman (1969)]. The various findings demonstrate that this method can be used to establish local stability, but the computational load is generally not at all trivial. A particularly simple sufficient condition was obtained by Clough and Ramirez for the adiabatic TRAM:

$$R_\eta < \frac{u^2}{\alpha} \tag{7-96}$$

EXAMPLE 7-7. The stability criteria (7-94) and (7-95) can be simplified (though made more conservative) by replacing R_η by its maximum value. Show that this results in inequalities comparable to the uniqueness conditions developed in Chapter 6.

Solution: Carrying out the suggested substitution inequality (7-94) becomes

$$\left[\text{Max}\,\frac{dR}{d\eta}\right]\int_{z_1}^{z_2}(1-z)dz < \frac{\alpha}{L^2}$$

The remaining integral will always be less than 0.5, since this is the value when $z_1 = 0$, $z_2 = 1$ (the largest possible range). As a result

$$\text{Max}\,\frac{dR}{d\eta} < \frac{2\alpha}{L^2}$$

is sufficient to insure stability. This may be compared to inequality (6-77a) from which

$$\text{Max}\,\frac{dR}{d\eta} < \frac{\pi^2}{4}\frac{\alpha}{L^2} = 2.46\frac{\alpha}{L^2}$$

guarantees uniqueness. Evidently, a steady state whose stability is established in this way will also be unique. It should also be observed that this stability condition compares closely to inequalities (7-39a) and (7-56a) obtained by Galerkin's method and collocation, respectively.

The corresponding treatment of (7-95) leads to

$$\left[\text{Max}\,\frac{dR}{d\eta}\right] - \frac{u^2}{4\alpha} < \frac{2\alpha}{L^2} \tag{7-97}$$

for the stability of the adiabatic TRAM. The appropriate comparison is with inequality (6-95a), but the details depend on the value of S_1, which is in turn a function of the Peclet number via Figure 6-11. If the right-side term in inequality (7-97) is replaced by zero:

$$\text{Max}\,\frac{dR}{d\eta} \le \frac{u^2}{4\alpha} \tag{7-98}$$

is virtually the same as the sufficient stability condition developed by Gupalo and Ryazantsev (1969). It is worth noting also that inequality (7-96) is sufficient to establish condition (7-97) in the event that the Peclet number

$$(uL/\alpha) < \sqrt{8/3}.$$

REFERENCES

BERGER, A. J. and L. LAPIDUS, *AIChE Journal*, **14**, 558 (1968).

CARSLAW, H. S. and J. C. JAEGER, *Conduction of Heat in Solids*, 2nd Ed., Oxford Univ. Press (1959).

CLOUGH, D. E. and W. F. RAMIREZ, Paper at Session No. 35, A.I.Ch.E. Meeting, Denver, Colo. (Sept., 1970).

FINLAYSON, B. A. and L. E. SCRIVEN, *Appl. Mech. Rev.*, **19**, 735 (1966).

GAVALAS, G. R., *Chem. Eng. Sci.*, **21**, 477 (1966).

GAVALAS, G. R., *Nonlinear Differential Equations of Chemically Reacting Systems*, Springer Verlag, Germany (1968), p. 91.

GUPALO, Y. P. and Y. S. RYAZANTSEV, *Chem. Eng. Sci.*, **24**, 1441 (1969).

HAHN, W., *Theory and Application of Liapunov's Direct Method*, Englewood Cliffs, N. J.: Prentice-Hall (1963).

KNOPS, R. J. and E. W. WILKES,, *Int. J. Engng. Sci.*, **4**, 303 (1966).

KUGELMAN, A. M., Ph. D. Thesis, Univ. of Pennsylvania, Philadelphia. (1969).

KUO, J. C. W. and N. R. AMUNDSON, *Chem. Eng. Sci.*, **22**, 1185 (1967).

LEE, J. C. M. and DAN LUSS, *AIChE Journal*, **16**, 620 (1970).

MCGOWIN, C. R., Ph. D. Thesis, Univ. of Pennsylvania, Philadelphia. (1969).

MCGOWIN, C. R. and D. D. PERLMUTTER, *AIChE Journal*, **17**, 831 (1971a).

MCGOWIN, C. R. and D. D. PERLMUTTER, *Chem. Eng. Journal*, **2**, 125 (1971b).

MOVCHAN, A. A., *Appl. Math. Mech.*, **23**, 483 (1959).

MOVCHAN, A. A., *Ibid.*, **24**, 988 (1960).

MOVCHAN, A. A., *Archiom. Mech. stosow.*, **15**, 659 (1963).

MURPHY, E. N. and E. D. CRANDALL, *J. Basic Eng.*, **92**, 265 (1970).

NISHIMURA, Y. and M. MATSUBARA, *Chem. Eng. Sci.*, **24**, 1427 (1970).

VILLADSEN, J. V. and W. E. STEWART, *Chem. Eng. Sci.*, **22**, 1483 (1967).

WANG, P. K. C., *Trans. ASME: J. Appl. Mech.*, **85**, 500 (1963a).

WANG,, P. K. C., *Advances in Control Systems*, **1**, N. Y.: Academic Press (1964).

WANG, P. K. C., and M. L. BANDY, *J. Elec. and Control*, **15**, No. 4, 343 (1963b).

WEI, J., *Chem. Eng. Sci.*, **20**, 729 (1965).

ZUBOV, V. I., *Methods of A. M. Lyapunov and Their Application*, Leningrad (1957); available as AEC Translation 4439 (1961).

Regions of Stability for Distributed-Parameter Systems

8

8-1. A Composite Phase-Plane

Since distributed systems differ from lumped-parameter models in their dependence on spacial variables, it is to be expected that ordinary phase-plane arguments are not applicable. On the other hand, it was noted in Chapter 6 that a fluid section of the PFTR can be treated as a small batch reactor sliding along the tube axis. Wang (1968a) has shown that this property of the PFTR model is not limited to the steady state, and is a basis for developing a special kind of phase-plane useful in computing regions of stability. By using it the frame of reference for the conservation equations is shifted to allow a viewpoint which keeps account of transfer across the boundaries of the moving plug.

The argument can be presented formally by reducing the partial differential equations (1-7) to an equivalent pair of ordinary differential equations

$$\frac{dC}{d\tau'} = -R \tag{8-1a}$$

$$\frac{d\eta}{d\tau'} = R - \frac{h}{C_p}(\eta - \eta_w) \tag{8-1b}$$

by defining the so-called substantial derivative [Bird, Stewart, and Lightfoot, (1960)]:

$$\frac{d}{d\tau'} = \frac{\partial}{\partial t} + u\frac{\partial}{\partial Z} = \frac{\partial}{\partial t} + \frac{\partial}{\partial \tau} \tag{8-2}$$

Because of their similar forms, it is tempting to simply identify equations (8-1) with those for the steady-state profiles, equations (6-23); however, for clarity it is worthwhile retaining a distinction between them. In particular the difference lies in the boundary conditions, for equations (6-23) lead only to $C_s(Z)$ and $\eta_s(Z)$, the steady-state profiles for a given input feed; whereas (8-1) are generally valid for the dynamic change from any initial condition, however it may arise at reactor entrance.

By introducing the deviation variables defined in equations (7-12), the dynamic PFTR is modeled by

$$\frac{dx_1}{d\tau'} = r - \frac{h}{C_p}x_1 \tag{8-3a}$$

$$\frac{dx_2}{d\tau'} = -r \tag{8-3b}$$

suggesting that bounds on possible transients might be obtainable on an (x_1, x_2) *composite phase-plane* (CPP). The modifier *composite* is included to emphasize that if τ' is to be treated as an implicit variable, it includes both time and position; that is to say, trajectories on a composite phase-plane show the variations of concentration and temperature at different time and position. In effect, it is as if the ordinary phase-planes at all the positions along the reactor were superimposed on one another, collapsed so as to keep the origins in place. This is illustrated in Figure 8-1.

As a fluid section moves along the reactor, its transients will trace a trajectory on the CPP that is unique to the initial conditions. Such a trajectory may appear on the CPP to be a loop, a contour crossing itself, or a

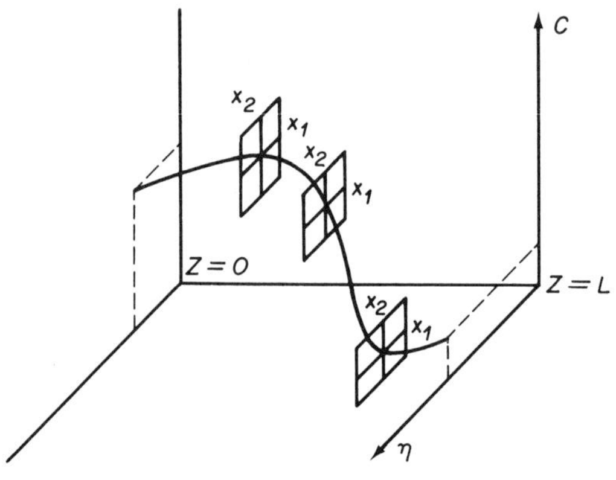

Figure 8-1. Construction of composite phase-plane by projection along a steady-state line in space.

simple point other than the origin. This peculiarity arises from the fact that the origin of the CPP corresponds to a pair of varying parameters. A trajectory that appears to be a single point on the CPP implies that the transient profiles traced by a fluid section as it moves down the reactor run parallel to the steady profile in (x_1, x_2, Z) space. A trajectory that crosses itself on the CPP means nothing more than that the differences between the trajectory of the fluid section and the steady-state profile are identical at two or more positions. Such characteristics on the CPP do not in any way indicate the nonuniqueness of the transient solutions of equations (8-3). Some of the possible trajectories of a fluid section are shown in the composite phase-plane of Figure 8-2.

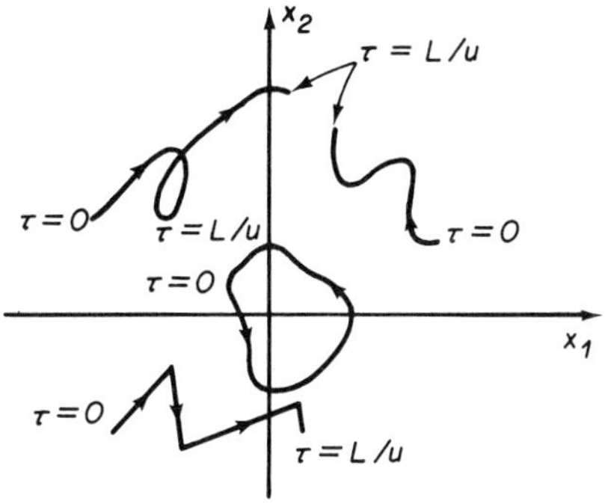

Figure 8-2. Trajectories on the composite phase-plane.

It is important to add that the lines of the CPP have a dual character. Seen as trajectories, they describe the states of a fluid element moving along the tube, but viewed as a profile, a line on the CPP is not a sequence of states but rather a state at a fixed time. The evolution of a distributed system from its initial to final states will appear as a set of lines, and if the transients move in such a way that the system approaches its steady state, the profile trajectories will appear to collapse into the origin of the composite phase-plane.

8-2. Bounds on CPP Trajectories

The simplest constraint that can be placed on trajectories in the CPP arises from an adiabatic assumption. By adding equations (8-3) with h set equal to zero

$$\frac{d}{d\tau'}(x_1 + x_2) = 0 \tag{8-4}$$

from which

$$x_1 + x_2 = \text{const.} = C_0 + \eta_0 = K \tag{8-5}$$

This result is entirely analogous to equation (6-25) and leads to a conclusion that the composite phase plane transient for adiabatic reaction follows the pattern observed earlier for the steady-state adiabatic reaction path of Figure 6-5: the changes must follow lines of -1 slope. The direction of this change can vary, however, depending on the particular deviations from the steady-state profiles (the origin of the CPP).

Examination of equation (8-3b) shows that the movement will be in the direction of increased x_2 if $r < 0$, and vice versa. With this motivation it is desirable to find the locus of $r = 0$ on the CPP, but since

$$r = R(x_1, x_2) - R_s(Z) = 0 \tag{8-6}$$

is a function of the steady-state profile which fixes R_s at each position, there is an $r = 0$ line for each position in the reactor, and the $r = 0$ locus is a band on the CPP. Lacking an identification of specific lines within the band with their appropriate positions, the sign of r is indeterminate; however, for the rest of the CPP: $r > 0$ always applies in the region above and to the right, $r < 0$ below and to the left.

A typical result of a numerical calculation by Wang is reproduced as Figure 8-3. Several constant K lines are drawn together with the extremes of

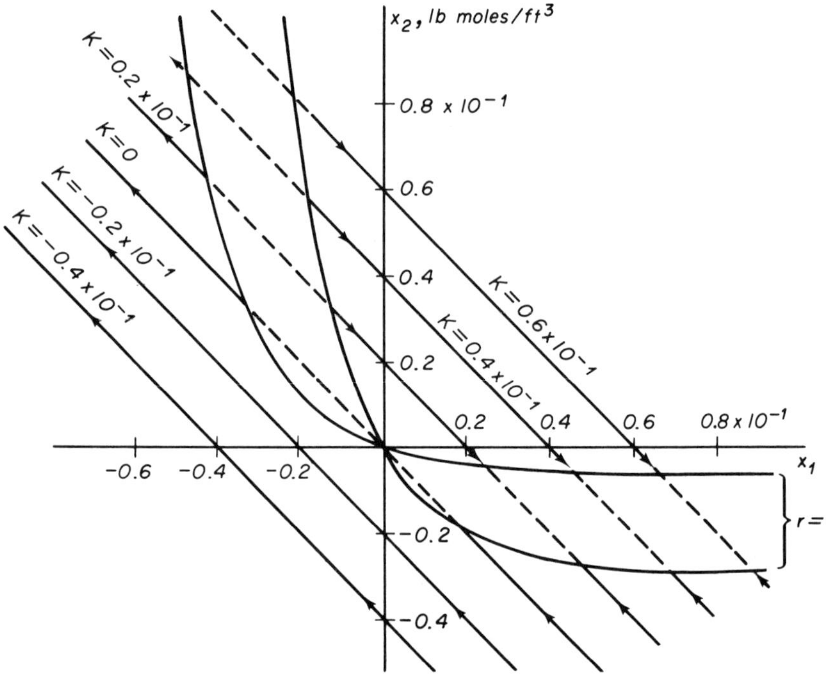

Figure 8-3. Adiabatic reaction paths on the composite phase-plane.

the $r = 0$ family for a particular steady-state curve. There are thus two regions in the CPP which are separated by the $r = 0$ band, for which opposite signs of $(dx_2/d\tau')$ will be found. At any point within the region $r = 0$, this derivative can be either positive, zero, or negative depending on which point of the steady state curve it refers to. The arrows in Figure 8-3 indicate the tendencies of transients at different sections of the constant K lines. Anywhere to the left and below the $r = 0$ family the movement is to decreasing temperature, whereas anywhere to the right and above the $r = 0$ family the movement is to increasing temperature along 45 degree lines. Within the $r = 0$ region the tendency is not defined without further detailed calculation.

Recall that initial conditions of a series of consecutive fluid plugs are the inlet conditions of the reactor for a period of time. Seen in this light each of the constant K lines of Figure 8-3 is a trace through the reactor of a single fluid section, or it can be the connection between any inlet condition to the PFTR and a corresponding effluent condition. Such a calculation would most conveniently be accomplished by integration of equation (6-23a) which in combination with adiabatic condition (6-25) is in terms of only a single dependent variable and parameters:

$$\frac{dC}{d\tau} = -R(C, C_0, \eta_0) \tag{8-7a}$$

or

$$-\int_{C_0}^{C} \frac{dC}{R(C, C_0, \eta_0)} = \int_0^{(L/u)} d\tau = \frac{L}{u} \tag{8-7b}$$

For a given inlet condition and known residence time in the reactor, the computation produces C, the upper integration limit and effluent concentration level. Taking this a step further, suppose that such inlet conditions are confined within a δ-region in the (x_1, x_2) plane. Then by this direct integration, one and only one effluent condition can be determined for each point on the δ-boundary. The group of effluent conditions found in this way enclose an ϵ-region, related to δ by a point-to-point mapping. If the inlet conditions of a PFTR are limited within a CPP region δ, the reactor exit conditions must lie in the CPP within a corresponding ϵ region. Moreover, because the transient curves generated from initial conditions within δ are confined within a CPP region between δ and ϵ, it is possible by the same construction to find bounds to the transient curves.

Several numerical studies of this technique were reported by Wang (1968a), who also extended the method to allow for realistic inequality constraints on the state variables. A sample result is shown in Figure 8-4 for the adiabatic PFTR specified by the kinetics equation (1-6b) and the parameters:

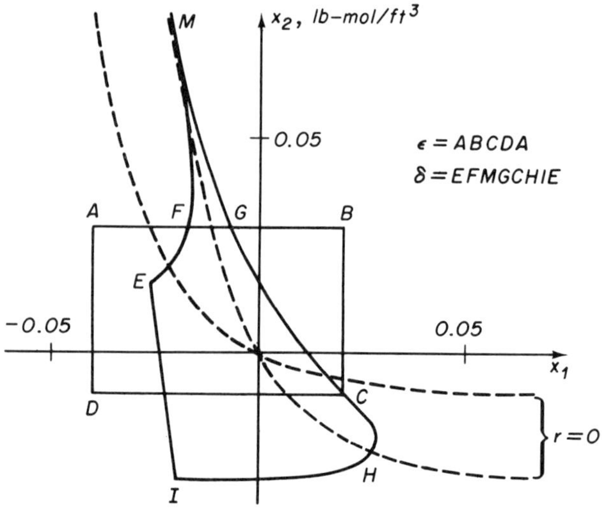

Figure 8-4. Example of δ-and ϵ-regions for a PFTR.

$$
\begin{aligned}
k_0 &= 1 \times 10^{10}\ \text{hr}^{-1} \\
Q &= 9{,}400^\circ\text{R} \\
\Delta H &= 1 \times 10^5\ \text{Btu/lb-mol} \\
C_p &= 50\ \text{Btu/}^\circ\text{F ft}^3 \\
T_0 &= 580^\circ\text{F} \\
C_0 &= 0.03\ \text{lb-mol/ft}^3 \\
L &= 5\ \text{ft} \\
u &= 7{,}200\ \text{ft/hr}
\end{aligned}
$$

One can think of the ϵ-region found in this way as a region of stability in the CPP, but clearly it is different than the concept applied to lumped-parameter systems. Since the ϵ and δ are both to be chosen by the designer to meet predecided specifications on likely disturbances and allowable effluent deviations, the regions are closest to the idea of practical stability, as presented in Chapter 5.

EXAMPLE 8-1. Apply the idea of a constrained integration to the isothermal PFTR modeled by:

$$
\frac{\partial C_A}{\partial t} = -u\frac{\partial C_A}{\partial Z} - R(C_A, C_B)
$$

$$
\frac{\partial C_B}{\partial t} = -u\frac{\partial C_B}{\partial Z} + \gamma R
$$

to compute the δ-region for a square ϵ-region centered about the origin of the composite (x_1, x_2) phase-plane and having sides 0.2 lb-mol/ft^3 in magnitude. Assume the kinetics

$$R = k_1 C_A - k_2 C_B$$

with the parameter values:

$$\gamma = 2$$
$$k_1 = 2 \text{ min}^{-1}$$
$$k_2 = 1 \text{ min}^{-1}$$
$$u = 4 \text{ ft/min}$$
$$L = 1 \text{ ft}$$

Solution: In terms of the deviation variables:

$$x_1 = C_A - C_{A_s}$$
$$x_2 = C_B - C_{B_s}$$
$$r = R - R_s$$

and the substantial derivatives, the isothermal PFTR becomes:

$$\frac{dx_1}{d\tau'} = -r \tag{8-8a}$$

$$\frac{dx_2}{d\tau'} = \gamma r \tag{8-8b}$$

and

$$r = k_1 x_1 - k_2 x_2 \tag{8-8c}$$

By addition and integration the lines of constrained integration have slope $-\gamma$ and follow

$$\gamma x_1 + x_2 = K$$

The equivalent of equations (8-7) is then

$$\frac{dx_1}{d\tau} = -k_1 x_1 + k_2 x_2 = -\lambda x_1 + k_2 K$$

where $\lambda = (k_1 + k_2 \gamma)$. By integration with the appropriate boundary condition at $\tau = (L/u)$:

$$\frac{\lambda x_1(\epsilon) - k_2 K}{\lambda x_1(\delta) - k_2 K} = \exp\,[-\lambda L/u]$$

which upon substitution of the parameters reduces to

$$\frac{4x_1(\delta) - K}{4x_1(\epsilon) - K} = \exp\,(+1) = 2.71$$

with

$$K = 2x_1(\epsilon) + x_2(\epsilon)$$

This relationship is the mapping with which every point on a chosen ϵ-contour is connected to a point on the δ-contour, The result of this computation for the given ϵ-square is the δ-parallelogram shown in Figure 8-5. The family of parallel lines is

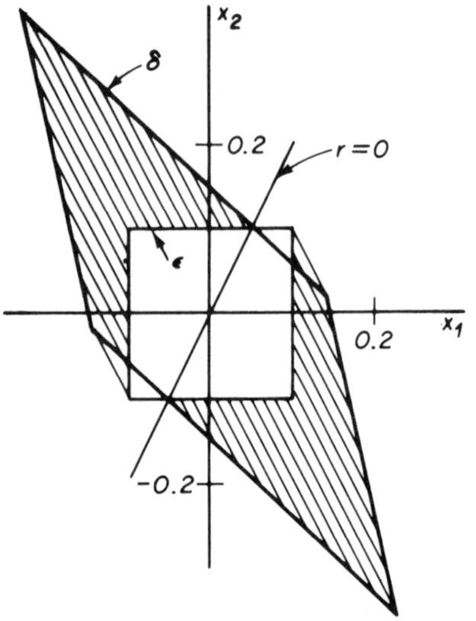

Figure 8-5. Stability region for isothermal PFTR of Example 8-1.

included to show the mappings of slope $-\gamma$ that connect the individual points. No disturbance wholly contained within the δ-region will ever generate a transient exceeding the ϵ-bound during the residence time in the reactor.

A weakness of the (δ, ϵ) procedure just presented is that it requires integration of equation (8-7). Furthermore, it is restrictive in treating only the adiabatic reactor, or one whose model can be reduced to a single equation. Both these difficulties can be avoided if one is willing to settle for sufficient (and therefore more conservative) bounds, obtained by applying the tracking function method that was so successfully used in Chapter 5 for second-order phase-plane problems.

Consider the tracking functions

$$E_1 = x_1 + x_2 \tag{8-9a}$$

$$E_2 = x_2 \tag{8-9b}$$

with regard to the CPP and the system equations (8-3). The lines of constant E_1 are of slope -1, and those of constant E_2 are horizontal. By differentiation:

$$\frac{dE_1}{d\tau'} = \frac{dx_1}{d\tau'} + \frac{dx_2}{d\tau'} = -\frac{h}{C_p} x_1 \tag{8-10a}$$

$$\frac{dE_2}{d\tau'} = \frac{dx_2}{d\tau'} = -r \tag{8-10b}$$

Since x_1 and r are both positive in the first quadrant, any trajectory in that region must move downward and in a generally leftward or only slightly rightward direction, as indicated on Figure 8-6 by the usual arrow-angle sym-

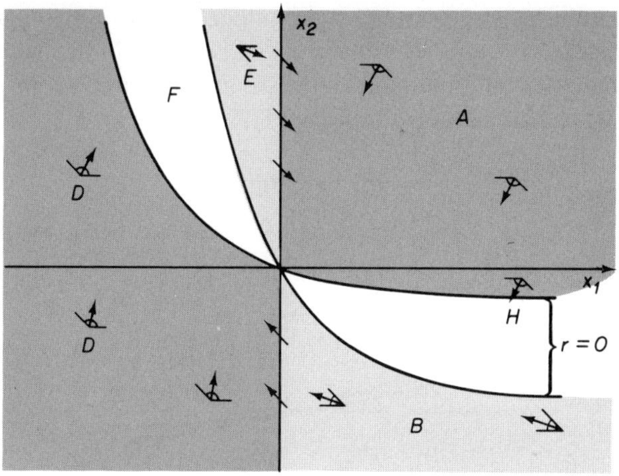

Figure 8-6. Constraints from tracking functions on the CPP.

bols. The same reasoning may be applied to motions in the third quadrant, but because x_1 and r are necessarily negative in this region, the directions of limits are reversed as shown. The treatment of the second and fourth quadrants is a bit more complicated because the sign of r changes on either side of the $r = 0$ band and is ambiguous within the band.

Nevertheless, careful study of the two tracking functions does yield useful information in most parts of the CPP. The result is that an almost-closed region of practical stability can be constructed as for example in Figure 8-7a. If as is often the case, a designer is concerned only with possible excursions in the direction of temperature increase (hotspots), these bounds can be a reasonable basis for design. If, on the other hand, there is danger in the low temperature, high concentration direction, a means must be found to

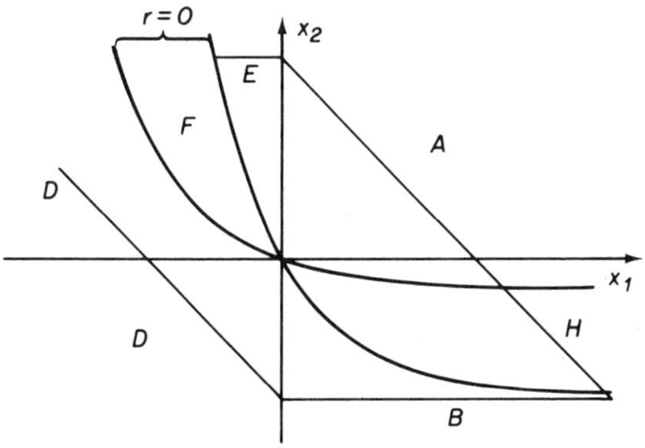

Figure 8-7a. A nearly-closed region on the CPP.

close the gap in regions D and F. A technique for this is explored in Section 8-4, but it should be noted here that no gap will arise in the first place, when the $r = 0$ band is shallow enough in the second quadrant. A closed parallelogram for such a case is shown in Figure 8-7b.

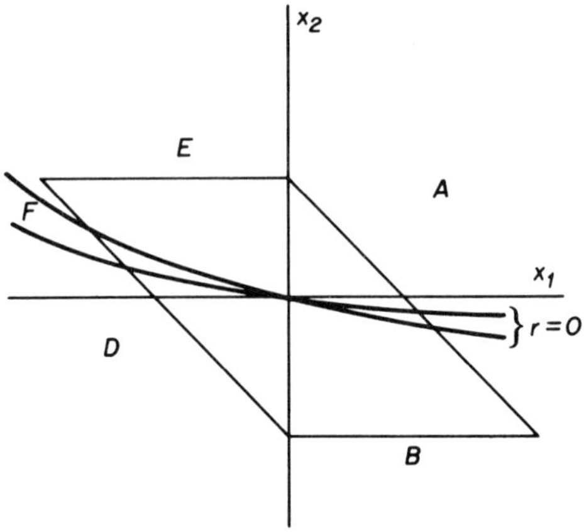

Figure 8-7b. A closed $\epsilon = \delta$ region on the CPP.

EXAMPLE 8-2. Use tracking functions on the CPP to develop a closed bounding region for the isothermal PFTR running the consecutive irreversible reaction sequence

$$A \xrightarrow{R_1} B \xrightarrow{R_2} P$$

Solution: This system can be modeled by

$$\frac{\partial C_A}{\partial t} = -u\frac{\partial C_A}{\partial Z} - R_1$$

$$\frac{\partial C_B}{\partial t} = -u\frac{\partial C_B}{\partial Z} + R_1 - R_2$$

which becomes in the reduced deviation coordinates:

$$\frac{\partial x_1}{\partial t} = -u\frac{\partial x_1}{\partial Z} - r_1$$

$$\frac{\partial x_2}{\partial t} = -u\frac{\partial x_2}{\partial Z} + r_1 - r_2$$

or, in terms of the substantial derivative:

$$\frac{dx_1}{d\tau'} = -r_1$$

$$\frac{dx_2}{d\tau'} = +r_1 - r_2$$

Introducing the tracking functions

$$E_1 = x_1 + x_2$$
$$E_2 = x_1$$

leads to

$$\frac{dE_1}{d\tau'} = \frac{dx_1}{d\tau'} + \frac{dx_2}{d\tau'} = -r_2 \tag{8-11a}$$

$$\frac{dE_2}{d\tau'} = \frac{dx_1}{d\tau'} = -r_1 \tag{8-11b}$$

from which a parallelogram bounding region may be constructed in the CPP, as follows. Assuming that (i) the rates R_1 and R_2 are dependent only on the concentrations of A and B respectively, and (ii) the functions are both monotonic assures that

$$r(x) > 0, \qquad x > 0 \tag{8-12a}$$
$$r(x) = 0, \qquad x = 0 \tag{8-12b}$$
$$r(x) < 0, \qquad x < 0 \tag{8-12c}$$

for both $r_1(x_1)$ and $r_2(x_2)$. Comparison of this observation with equations (8-11) shows that for this problem the loci of $(dE/d\tau') = 0$ coincide with the axes of the (x_1, x_2) composite phase-plane. The details are summarized in Figure 8-8 by the directional arrow symbols that bound possible changes, and a selection of two of the many parallelogram regions that meet these standards. It is noteworthy that this result was obtained without numerical specification of system parameters and with only moderate restrictions on kinetic forms.

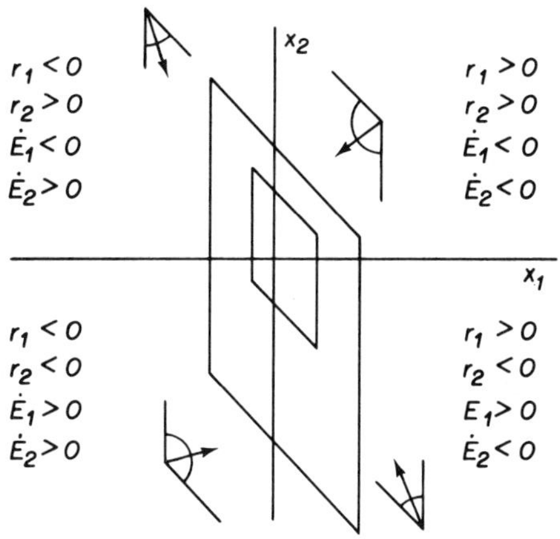

Figure 8-8. Stability regions for Example 8-2.

8-3. Finite-Time Stability

Weiss and Infante (1965) have developed rigorously the definitions and concepts that are appropriate to the study of lumped-parameter models, when the nature of the system is such that stability is a meaningful notion only over a finite-time interval. The PFTR is a fine example of such a system, for since the residence time is finite, it is only of interest to estimate the bounds on transients while any fluid section remains in the reactor. It is rather immaterial whether the profiles are stable or unstable for *all* time. With particular reference to equations (8-3) the interval of concern is for τ from zero to (L/u). In such a case it is unnecessarily conservative to demand stability or trajectory bounds for *all* time. The cost of such a guarantee is too high, for it could serve to reject system designs that are in fact workable and satisfactory.

A more realistic stability requirement can be built upon the definitions of Weiss and Infante which apply to a finite time interval and to two regions of state space that are specified in terms of a prechosen norm. Formally, a system is *stable over a finite-time interval* if for every ϵ there is a $\delta(\epsilon)$ such that

$$\|\mathbf{x}(0)\| \leq \delta \tag{8-13a}$$

implies

$$\|\mathbf{x}(\tau)\| \leq \epsilon \tag{8-13b}$$

for

$$0 \leq \tau \leq \frac{L}{u} \tag{8-13c}$$

This formulation is in essence a relaxation of the Liapunov definition to make it easier for the defender (in the two-man game of Section 4-2) to find useful δ's. The same idea can of course be carried over to a corresponding practical viewpoint by again asking the challenger and defender to cooperate. Then a system will be classed as *practically stable over a finite-time interval* when the two designers can find *any* δ-ϵ pair that suitably satisfy their engineering purposes as well as condition (8-13). Moreover, as previously noted in Section 5-6, the relatively more relaxed requirements of practical stability make it possible to ignore the usual norms and use instead the more convenient graphically derived regions. The tracking function again becomes attractive, as does indeed any other technique that can establish practical (δ, ϵ) regions.

To establish useful design regions for the PFTR, consider the positive-definite tracking function

$$E = x_1^2 + x_2^2 \tag{8-14}$$

for which a growth or decay coefficient may be defined by analogy to equation (5-46):

$$\xi = \text{Max}\left[\frac{(dE/d\tau')}{E}\right] \tag{8-15}$$

The significance of this function for the problem of interest here can be seen by separating three subcases, according to the sign of ξ:

(1) If $\xi < 0$, E will decay with time over the finite interval, $0 < \tau < (L/u)$. Then, if a trajectory of equations (8-3) is initially within the region

$E \leq \delta$, it will always remain within that region, and will end up at reactor exit bounded by some smaller region $E \leq \epsilon$. Such a system is called *contractively stable* with respect to $[\delta, \epsilon, (L/u), E]$ by Weiss and Infante.

(2) If $\xi > 0$, E may grow with time over the range from zero to (L/u). However, since this growth occurs for only a finite time, it is still possible for the trajectory to be bounded by $E \leq \epsilon$ at $\tau = (L/u)$. If this is in fact the case, the system is stable according to the Weiss–Infante definition, but not contractively stable.

(3) If $\xi = 0$, E will decay or remain constant. In either case, the test of stability with respect to $[\delta, \epsilon, (L/u), E]$ remains the same.

The formulations of Weiss and Infante are broad enough to include other norms, but for the purposes of this application it is convenient to use the circular E-function. It is important to emphasize that this E-function may have positive as well as negative time derivative. It is therefore not a Liapunov function according to the definition given by LaSalle and Lefschetz (1961). Because ξ is simply the bound governing the growth or decay rate of the proposed positive-definite function, E-functions which cannot guarantee stability of a given system defined for all time, may still be used to establish the practical stability of the same system defined over a finite-time interval. In this respect less restrictive E-functions are required than for a stability analysis which relies on Liapunov functions. Whenever the choice of δ and ϵ takes into account the region in the state space which is physically feasible and practically allowable for a particular system, the result may be called a region of practical stability in the composite phase-plane.

For the particular E-function at hand, differentiation gives

$$\frac{dE}{d\tau'} = 2x_1\frac{dx_1}{d\tau'} + 2x_2\frac{dx_2}{d\tau'} \tag{8-16}$$

and by substitution from equations (8-3)

$$\frac{dE}{d\tau'} = 2\left[(x_1 - x_2)r - \frac{h}{C_p}x_1^2\right] \tag{8-17}$$

Evaluation of defining equation (8-15) then produces

$$\xi = 2\,\text{Max}\left[\frac{(x_1 - x_2)r - (h/C_p)x_1^2}{x_1^2 + x_2^2}\right] \tag{8-18}$$

Before searching this function for its maximum in the (x_1, x_2) plane at each position, it is worthwhile to observe that the answer must lie in a region for which

$$(x_1 - x_2)r > 0 \tag{8-19}$$

These regions are outlined in Figure 8-9, and Wang (1966) has shown that the maximum sought is always found in the right-side region.

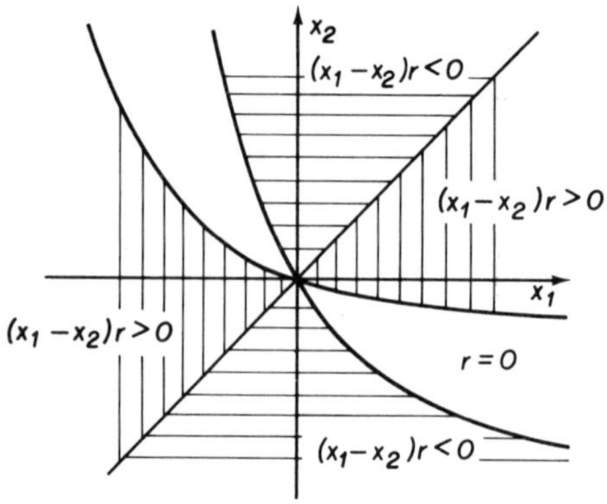

Figure 8-9. Regions in the CPP showing the signs of $(x_1 - x_2)r$.

Once ξ is known, a region $E \leq \epsilon$ is specified that is consistent with practical limitations on the reactor effluent, and equation (8-15) is used in the form suggested by (5-43):

$$E(L/u) \leq E(0) \exp(\xi L/u) \tag{8-20a}$$

or

$$\epsilon \leq \delta \exp(\xi L/u) \tag{8-20b}$$

to determine the permissible inlet disturbance range: $E \leq \delta$. Typical results of such calculations, taken from the reference cited, are shown in Figures 8-10 and 8-11, in the first instance as E profiles for various ϵ's, and in the

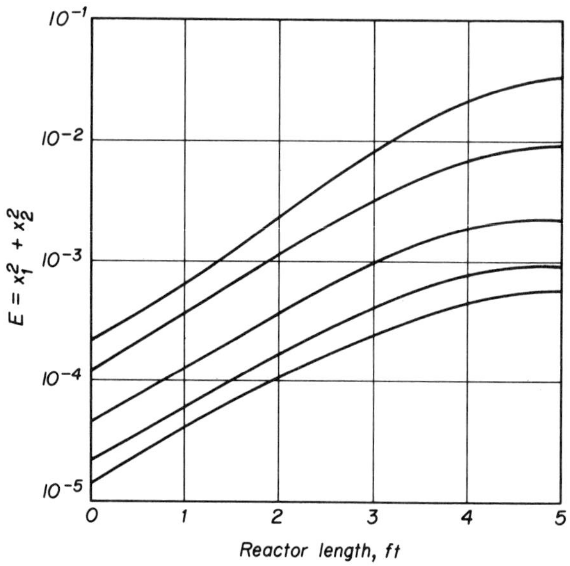

Figure 8-10. Profiles of E along the reactor.

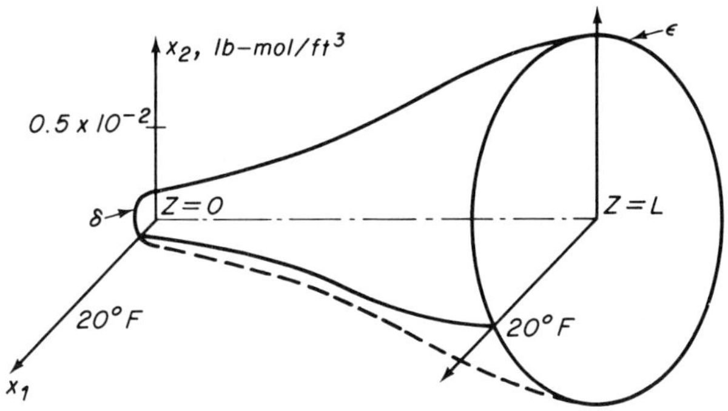

Figure 8-11. An isometric view of a typical bounding region.

latter figure as an isometric view at a given ϵ corresponding to $\pm 20°$F. The conical shape forms a region of stability for this PFTR, since it bounds all possible transients of this dynamic system. The range of ϵ's studied for Figure 8-10 corresponds to temperatures in the range from $\pm 20°$F to $\pm 120°$F; the associated δ's are from $\pm 3°$F to $\pm 9.7°$F. The following data were used:

$$
\begin{aligned}
k_0 &= 1.2 \times 10^{10}\ \text{hr}^{-1} \\
Q &= 9{,}400°\text{R} \\
\Delta H &= 1.0 \times 10^5\ \text{Btu/mol} \\
u &= 7{,}200\ \text{ft/hr} \\
h &= 7{,}000\ \text{Btu/ft}^3\,°\text{F hr} \\
C_p &= 50\ \text{Btu/ft}^3\,°\text{F} \\
T_w &= 530°\text{F} \\
T_0 &= 600°\text{R} \\
C_0 &= 0.035\ \text{mol/ft}^3 \\
L &= 5\ \text{ft}
\end{aligned}
$$

EXAMPLE 8-3. Reconsider the PFTR described in Example 8-1, this time using the ξ-bound method to compute a δ-circle for a given ϵ-circle of radius 0.1 (mol/ft^3). Compare with the earlier result.

Solution: By differentiating equation (8-14) and substituting the system equations (8-8):

$$\frac{\dot{E}}{E} = \frac{2r(\gamma x_2 - x_1)}{x_1^2 + x_2^2} = \frac{2(k_1 x_1 - k_2 x_2)(\gamma x_2 - x_1)}{x_1^2 + x_2^2}$$

Evaluating the parameters by the numerical values given in Example 8-1:

$$\frac{\dot{E}}{E} = \frac{2(5x_1x_2 - 2x_1^2 - 2x_2^2)}{x_1^2 + x_2^2} = \frac{10x_1x_2}{x_1^2 + x_2^2} - 4$$

This function has a maximum at $x_1 = x_2$ for which from definition (8-15): $\xi = 1$. Application of equation (8-20) then gives

$$\delta = \epsilon \exp\left(-\frac{\xi L}{u}\right) = 0.1 \exp(-0.25) = 0.0607 \text{ mol/ft}^3$$

Comparison with the (δ, ϵ) regions of Figure 8-5 illustrates the essential conservatism associated with the ξ-bound method.

8-4. A Combination Analysis

Detailed examination of the last example and of the regions of stability reported as Figures 8-10 and 8-11 show that the inlet deviations allowed are rather small. Such conservative δ's are partly due to the exponential dependence of E on the temperature of the system and partly due to the inherent nature of the ξ-bound method. In contrast, consider the almost-closed region in Figure 8-7 that was developed with relative ease by using tracking functions. Such results are less conservative, but are weak to a greater or lesser extent when the regions are not entirely closed. It is clear in comparison that a combination of both approaches can produce results with the best features of each, less conservative, easily computed, closed regions.

Referring to the composite phase-plane (CPP) diagram in Figure 8-7, it is evident that the gap in regions D and F is the one previously left open. Because it is open, no assurance can yet be given that transients in the region can be bounded. In order to close this open gap, the ξ-bound method cited before may be adopted just in regions D and F to utilize part of a closed E-contour. By using such a contour to close a gap an ϵ and δ pair may be obtained. Since the open gap is always in regions D and F, only that part of an E-contour that lies in these regions need be considered. In other words, the range in which Max $(\dot{E}/E)$ is sought can be restricted within regions D and F, consequently reducing considerably the necessary computations. Because the largest $(\dot{E}/E)$ in regions D and F is smaller than that in region A, the estimation of δ from ϵ should be less conservative than would be found by the ξ-bound method alone.

Since the slopes of transients in region D and possibly in region F are upward to the left, an elliptical E-contour is proposed in order to cover most of the transients in these regions:

$$E = (a^2 + b^2)(x_1^2 + x_2^2) + 2(b^2 - a^2)x_1x_2 \tag{8-21}$$

where a and b are the minor and major axes of the ellipse. By differentiating and substituting equations (8-3):

$$\frac{\dot{E}}{E} = -\frac{h}{C_p} - \frac{h}{C_pE}(a^2 + b^2)(x_1^2 - x_2^2) + \frac{4a^2}{E}(x_1 - x_2)r$$

The search for ξ is along that part of the ellipse that is in regions D and F and above the 45 degree line. The result for the particular example is shown by the closed (δ, ϵ) pair of Figure 8-12.

One caution must be observed in combining regions from the two techniques. If an E-contour at the inlet is larger than that at the exit $(\delta > \epsilon)$, there is no guarantee that all possible disturbances within δ will lead to transients within a smaller ϵ, because the boundary of δ is based on isoclines and tracking functions which give only limits in slopes and directions of the transients rather than a bound on their rate of growth or decay. For such cases a compromise is reached by considering δ as equal to ϵ. This single region will then be a region of practical stability for the PFTR.

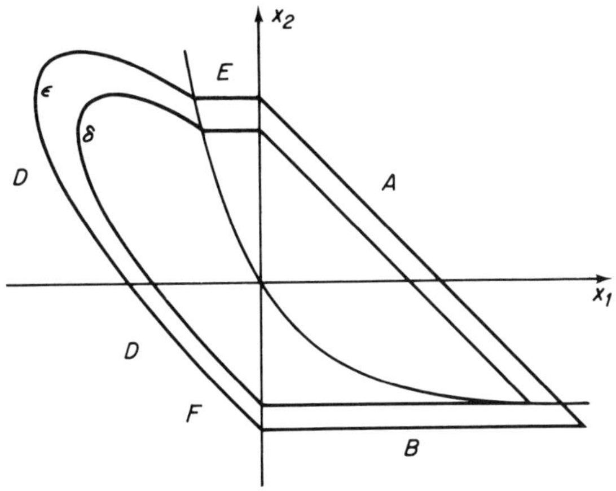

Figure 8-12. A typical ϵ and δ pair for a nonadiabatic PFTR.

8-5. A Conjecture on TRAM Transients

The essential difference between the TRAM and PFTR models is of course that the former includes diffusion-like terms for mass and energy. This can change the qualitative behavior of the system, even when the quantitative effect is small in magnitude, but because diffusion is by nature a smoothing phenomenon it is a reasonable expectation that the introduction of the second-derivative terms in equations (7-71) will not ordinarily cause greater instability. This observation led Wang (1971) to the conjecture that a graphical superposition of different bounds on the CPP might serve to fix limits on the possible transients of a TRAM. The argument is most conveniently pursued in two parts.

First consider the pseudo-PFTR that is produced when the dispersion terms of equations (7-71) are deleted:

$$\frac{\partial x_1}{\partial t} = -u\frac{\partial x_1}{\partial Z} + r - \frac{h}{C_p}x_1 \tag{8-22a}$$

$$\frac{\partial x_2}{\partial t} = -u\frac{\partial x_2}{\partial Z} - r \tag{8-22b}$$

Suitable δ- and ϵ-bounds may be found for the set of equations by any of the several CPP techniques already presented. Such regions will include as an origin in the CPP the steady-state profile of the original TRAM, rather than that of the pseudo-PFTR, but this fact does not subvert the development. Secondly, note that the effect of the diffusion terms is such that were they alone to determine behavior, transients from any initial condition would never exceed the largest values on the initial temperature and concentration profiles. Another way to state this fact is to reiterate that diffusive behavior is always in the direction of reducing differences. Contrary developments can of course occur in response to flow, reaction, or heat transfer, but these terms have all been separated out as the pseudo-PFTR equations (8-22). This diffusive bound is illustrated on the CPP in Figure 8-13 for two hypothetical initial profiles. In each case the transients would be confined within the appropriate rectangle, were diffusion the only mechanism of change.

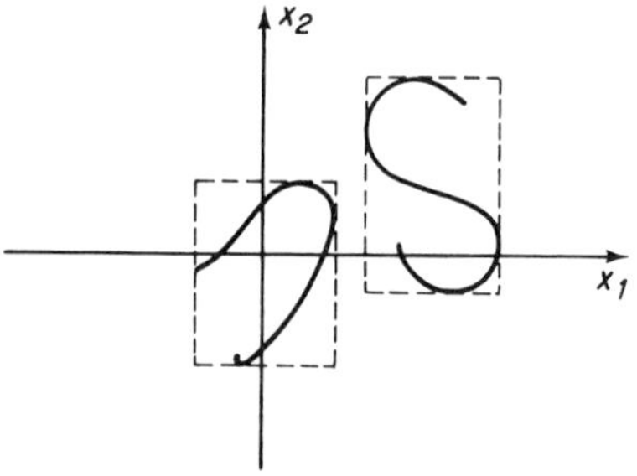

Figure 8-13. Rectangular diffusive bounds for two hypothetical initial profiles.

If the rectangular diffusive bound is entirely within a δ-region for the pseudo-PFTR, it is plausible that the transients that develop from the initial condition under study will never exceed the ϵ-bound for the pseudo-PFTR. The plausibility rests on the proposition that the addition of diffusion to a reactor system will tend to dampen its responses rather than aggravate them; but it must be emphasized that any idea of superposition is not rigorous for nonlinear systems and this proposal is only a conjecture on what might be a useful design procedure. Moreover it must be added that the difficulty is not merely that a proof is lacking. It is rather that the terms of the conjecture are so broad that they cannot possibly cover all possible situations.

In spite of this Wang has demonstrated by a number of numerical tests that the transients of a great many systems can be bounded in this way, particularly if the heats of reaction are low to moderate. Two of his numerical examples are summarized in Figures 8-14 and 8-15. In the first of these, the TRAM parameters are:

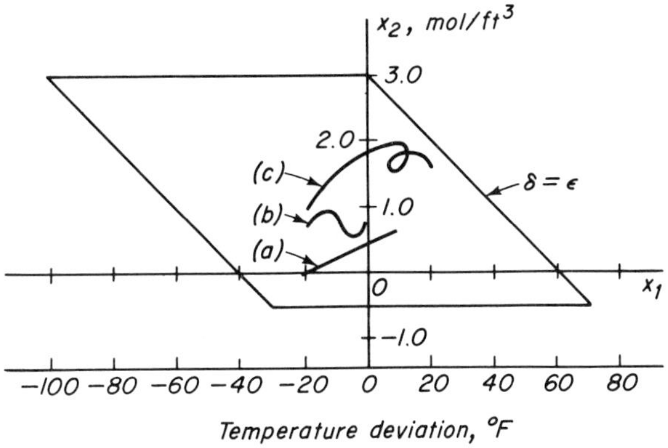

Figure 8-14a. A $\delta = \epsilon$ region for a TRAM showing three initial condition curves.

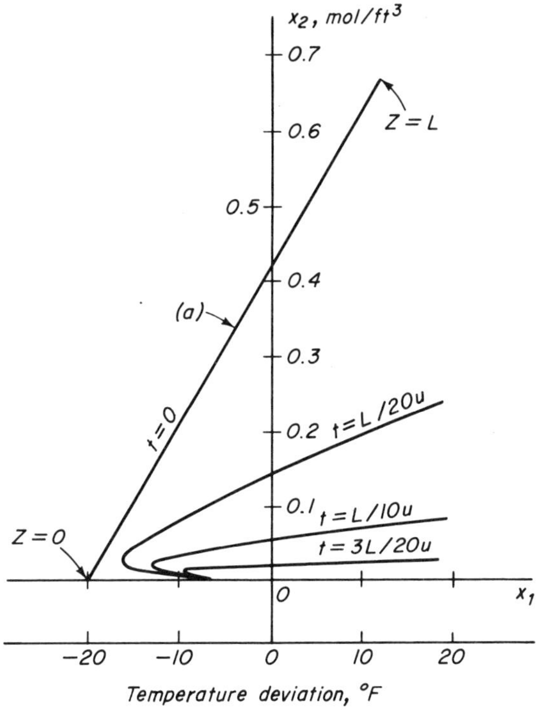

Figure 8-14b. Projections of TRAM transients from initial condition curve (a) of Figure 8-14a.

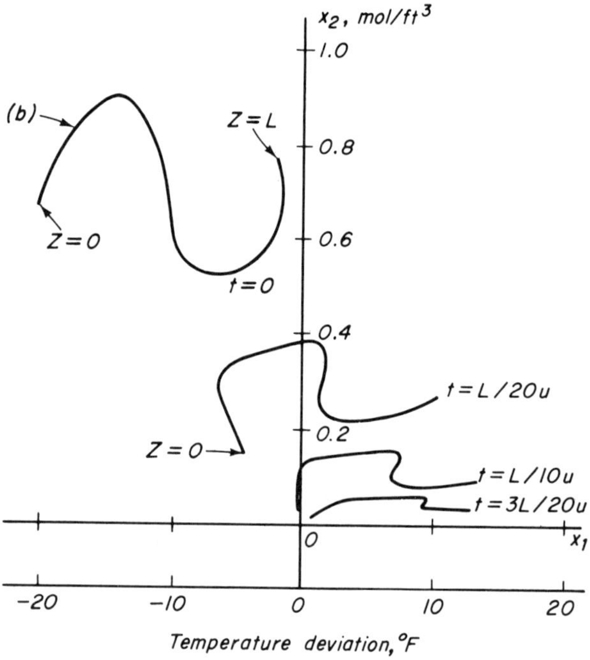

Figure 8-14c. Projections of TRAM transients from initial condition curve (b) of Figure 8-14a.

$$k_0 = 5 \times 10^7 \text{ hr}^{-1}$$
$$\Delta H = 1{,}000 \text{ Btu/lb-mol}$$
$$Q = 1.1 \times 10^4 \,^\circ\text{R}$$
$$C_p = 50 \text{ Btu/ft}^3 \,^\circ\text{R}$$
$$\alpha = 200 \text{ ft}^2/\text{hr}$$
$$D = 100 \text{ ft}^2/\text{hr}$$
$$L = 2 \text{ ft}$$
$$T_0 = 1{,}250^\circ\text{R}$$
$$C_0 = 0.7 \text{ lb-mol/ft}^3$$
$$u = 1{,}000 \text{ ft/hr}$$

The system is assumed to be adiabatic and to follow the usual first-order kinetics. The relatively shallow $r = 0$ lines for this case make it possible to sketch a simple $\delta = \epsilon$ parallelogram as in Figure 8-14a. The evolution in time of a particular initial disturbance is shown by Figures 8-14b and c. Parallel results are in Figures 8-15a, b, and c for a case using a ten-fold higher heat of reaction:

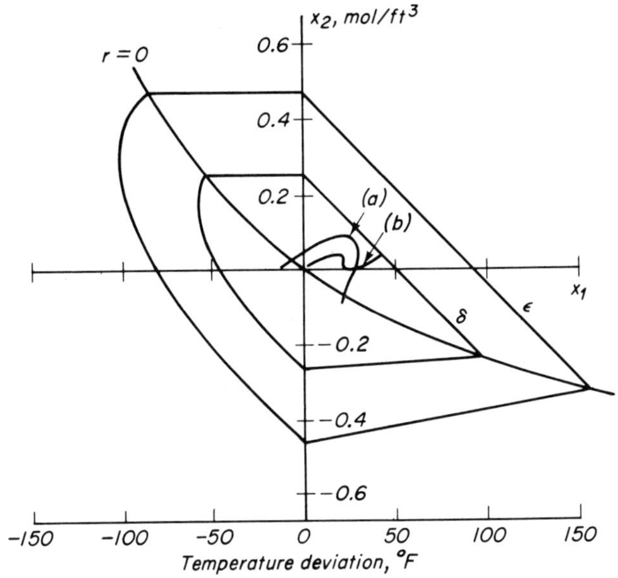

Figure 8-15a. A δ-region and ϵ-region for a TRAM, showing two initial condition curves.

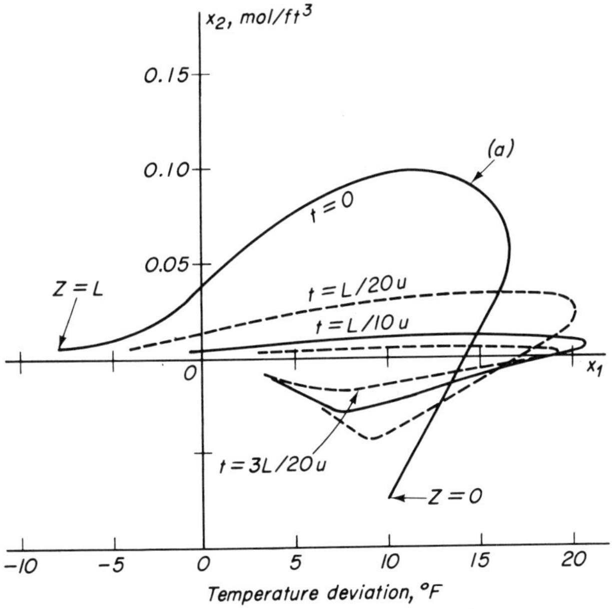

Figure 8-15b. Projections of TRAM transients from initial condition curve (a) of Figure 8-15a.

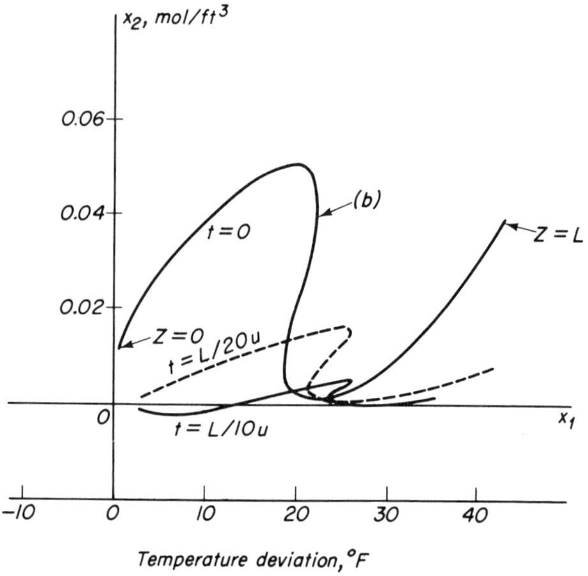

Figure 8-15c. Projections of TRAM transients from initial condition curve (b) of Figure 8-15a.

$$\Delta H = 10{,}000 \text{ Btu/lb-mol}$$

$$Q = 1.2 \times 10^4 \text{°R}$$

The diagram differs in the relatively more involved computation of suitable δ- and ϵ-contours but the results again support Wang's conjecture.

8-6. Weighted Residual Methods

In the previous chapter the modified collocation technique was used to reduce a distributed parameter model to a set of ordinary differential equations. The latter turned out to be linear but this property was the result of the earlier linearization, not at all inherent in the collocation itself. If instead of the linearized equations (7-58), the more general equations (7-13) are reduced by substitution of (7-45) one obtains

$$\frac{dx_1(\mathbf{z})}{dt} = \frac{\alpha}{L^2}\mathbf{N}_2\mathbf{N}_0^{-1}x_1(\mathbf{z}) + r(\mathbf{z}) \tag{8-23a}$$

$$\frac{dx_2(\mathbf{z})}{dt} = \frac{D}{L^2}\mathbf{N}_2\mathbf{N}_0^{-1}x_2(\mathbf{z}) - r(\mathbf{z}) \tag{8-23b}$$

where $r(\mathbf{z})$ is the diagonal matrix

$$r(\mathbf{z}) = r[x_1(\mathbf{z}), x_2(\mathbf{z})] = \begin{bmatrix} r(z_1) & 0 & \cdots & 0 \\ 0 & r(z_2) & \cdots & 0 \\ \cdots & & & \\ \cdots & & & \\ 0 & 0 & \cdots & r(z_n) \end{bmatrix} \tag{8-24}$$

Put into the even more concise notation of equation (7-65b) this relationship reduces to

$$\dot{\mathbf{x}} = \mathbf{f}(\mathbf{x})$$

suggesting that various lumped-parameter arguments may be useful. In particular one looks to the Liapunov analysis that was applicable to nonlinear systems of moderate dimensionality to yield regions of asymptotic stability.

Because modified collocation retains the state variables of the original distributed problem, any region of stability found for equation (8-23) is already expressed in terms of concentration and temperature deviations at the collocation points. If $n = 2$, for example, an RAS for (8-23) is a region in the four-dimensional space of

$$\mathbf{x}(\mathbf{z}) = \begin{bmatrix} x_1(z_1) \\ x_1(z_2) \\ x_2(z_1) \\ x_2(z_2) \end{bmatrix} = \begin{bmatrix} \eta(z_1) - \eta_s(z_1) \\ \eta(z_2) - \eta_s(z_2) \\ C(z_1) - C_s(z_1) \\ C(z_2) - C_s(z_2) \end{bmatrix} \tag{8-25}$$

It remains to interpret such a region in terms of the profiles of the original problem.

The defining equations (7-53) for the trial solution become at any given time:

$$x_1(z) = x_1(1) + \sum_{i=1}^{n} a_i \Phi_i(z) \tag{8-26a}$$

$$x_2(z) = x_2(1) + \sum_{i=1}^{n} b_i \Phi_i(z) \tag{8-26b}$$

From these the approximate profiles can be computed if the a_i and b_i coefficients are known. It is shown in the reference paper (and repeated in Appendix B) that these constants depend only on the choice of collocation points and the $\mathbf{x}(\mathbf{z})$ at these points, as follows:

$$a_i = \frac{1}{K_{i-1}} \mathbf{N}_3 \mathbf{N}_0^{-1} [x_1(\mathbf{z}) - x_1(1)] P_{i-1}(\mathbf{z}^2) \tag{8-27a}$$

$$b_i = \frac{1}{K_{i-1}} \mathbf{N}_3 \mathbf{N}_0^{-1} [x_2(\mathbf{z}) - x_2(1)] P_{i-1}(\mathbf{z}^2) \tag{8-27b}$$

where

$$\mathbf{N}_3 = \begin{bmatrix} 1 & \frac{1}{3} & \frac{1}{5} & \cdots & \frac{1}{2n+1} \end{bmatrix} \tag{8-28}$$

and the K_i, $\mathbf{N}_0$ and P_{i-1} are as previously defined by equations (7-54), (7-43b) and (7-52), respectively. Numerical values of $\mathbf{N}_3\mathbf{N}_0^{-1}$ computed by Villadsen and Stewart for various n are to be found in Appendix C. Using the $n = 2$ illustration again, equations (8-26) may be expanded to:

$$x_1(z) = x_1(1) + a_1(1 - z^2) + a_2(1 - z^2)(1 - 5z^2) \tag{8-29a}$$

$$x_2(z) = x_2(1) + b_1(1 - z^2) + b_2(1 - z^2)(1 - 5z^2) \tag{8-29b}$$

With the last collocation point placed at $z_3 = 1$, and the boundary condition $x_1(z) = x_2(1) = 0$ the first two coefficients are

$$a_1 = \frac{1}{K_0}\mathbf{N}_3\mathbf{N}_0^{-1}\begin{bmatrix} x_1(z_1) \\ x_1(z_2) \\ 0 \end{bmatrix} \tag{8-30a}$$

$$a_2 = \frac{1}{K_1}\mathbf{N}_3\mathbf{N}_0^{-1}\begin{bmatrix} (1 - 5z_1^2)x_1(z_1) \\ (1 - 5z_2^2)x_1(z_2) \\ 0 \end{bmatrix} \tag{8-30b}$$

and the b_i coefficients are similarly expressed in terms of x_2. From Appendix C:

$$\mathbf{N}_3\mathbf{N}_0^{-1} = [0.555 \quad 0.378 \quad 0.0667] \tag{8-31}$$

but it may be noted that the last element is irrelevant for this problem because of the zeros in equations (8-30).

The reader should satisfy himself that every point of the $\mathbf{x(z)}$ space generates a unique set of $x_1(z)$ and $x_2(z)$ profiles. This is sometimes called a *mapping*. For the $n = 2$ case detailed above, z_1 and z_2 are the roots of $P_2(z^2) = 0$, and only the four components of vector (8-25) are needed to fix the a_i, b_i, and the two profiles. As time progresses, changes in $\mathbf{x(z)}$ generate a trajectory as in any lumped-parameter model. Accordingly, the profiles change shape and position. Since every trajectory in an RAS must ultimately go to the origin $\mathbf{x(z)} = \mathbf{0}$, the corresponding disturbance profiles must decay to the distributed steady state: $x_1(z) = x_2(z) = 0$.

But it is possible to draw a stronger conclusion, for since there is a point to line correspondence, the RAS of $\mathbf{x(z)}$ space maps into a related region in the distributed-parameter space. To see why this is so, consider first all the points on a radial line from the $\mathbf{x(z)}$ origin to the boundary of an RAS. Along this line the ratios

$$\frac{x_1(z_1)}{x_1(z_2)}, \frac{x_1(z_1)}{x_1(z_3)}, \ldots, \frac{x_1(z_1)}{x_1(z_n)} \tag{8-32a}$$

and

$$\frac{x_2(z_1)}{x_2(z_2)}, \frac{x_2(z_1)}{x_2(z_3)}, \ldots, \frac{x_2(z_1)}{x_2(z_n)} \tag{8-32b}$$

are each constant and their contributions to equations (8-30) show the coefficients a_i and b_i to be proportional to the distance from the origin in $\mathbf{x}(\mathbf{z})$. Further, since equations (8-29) make the ordinates of the corresponding $x_1(z)$ and $x_2(z)$ profiles linear in a_i and b_i they must have the same property. These relationships are shown schematically in Figure 8-16, from which it

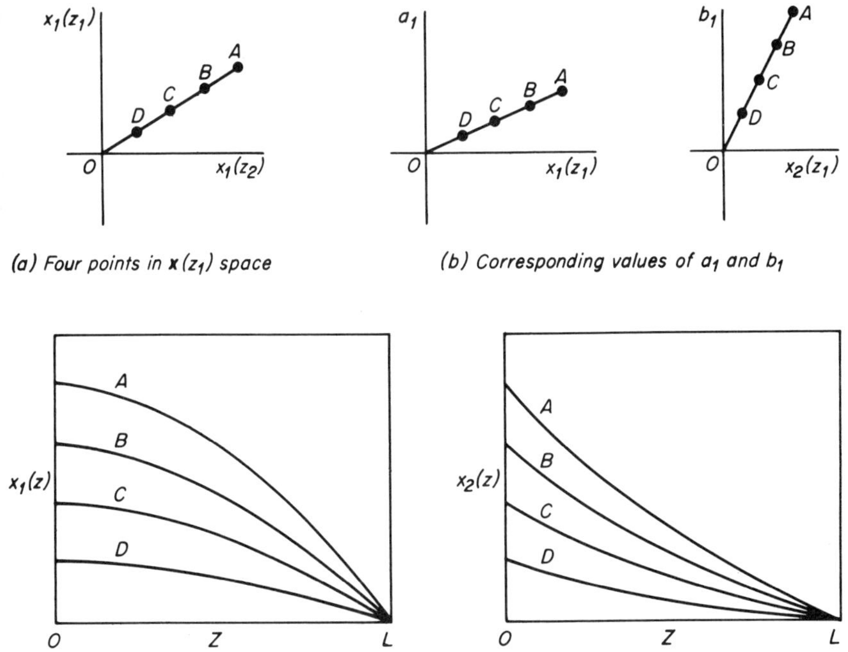

Figure 8-16. Relationships in mapping from $x(z)$ space into state-function space (schematic).

should be noted that the entire set of profiles corresponding to a given radial line in the LPS are of identical shape, differing only in their ordinate scale. One essentially maps a LPS radius into a DPS region by locating the curve for the point at the end of the radius and filling in the entire region between that curve and the steady-state line.

A second radial line produces a second DPS region, typically overlapping the first region to a considerable extent. A third radial line leads to yet another partly overlapping region, and so forth. Carried to its logical limit, the family of all radial lines maps into a relatively larger DPS region consisting of the contributions of the individual overlapping parts. This is called a *union.*

Actual trajectories in $\mathbf{x}(\mathbf{z})$ would of course not generally follow radial paths to the origin, but any path at all must pass through a sequence of points each of which is on some radius. If in addition any trajectory is constrained to lie within a region in $\mathbf{x}(\mathbf{z})$, its mapping in distributed space can never escape the union of the mappings of the possible finite radial lines. This is then a region of stability in the distributed space. To keep clear the distinction between the regions of stability in lumped-parameter space and this newer extension, the latter will be called a *Union of Stability*, or when appropriate a *Union of Asymptotic Stability* (UAS).

It is to be expected that the UAS will be conservative if the corresponding RAS is based on a Liapunov analysis, but any other method may be used to determine the bounded region in $\mathbf{x}(\mathbf{z})$ space. The mapping arguments that connect the RAS and UAS are not connected to the method used in finding the former. One reservation ought to be expressed, however, since it has implicitly been assumed that a disturbance under study would be adequately approximated by the trial solution. It appears that a relatively wide class of disturbances can be included under this umbrella, especially if the degree of approximation n is adjustable; nevertheless, one can imagine impulse or other peculiar forms that might cause difficulties.

EXAMPLE 8-4. Find a Union of Asymptotic Stability for the uncoupled catalyst particle model using the three steady-state case presented with Figure 6-10.

Solution: Working from equation (7-21), modified collocation produces the set of ordinary differential equations (8-23a). The subscript on x may be dropped, since temperature deviation is the only state variable. To determine an RAS one can examine the simplest Liapunov function

$$v = ||\,x(\mathbf{z})\,||^2 = [x(\mathbf{z})]^T\,[x(\mathbf{z})]$$

for which

$$\dot{v} = 2[x(\mathbf{z})]^T\,\mathbf{f} = 2[x(\mathbf{z})]^T\left[\frac{\alpha}{L^2}\mathbf{N}_2\mathbf{N}_0^{-1}x(\mathbf{z}) + r(\mathbf{z})\right]$$

A numerical study is needed to find the largest region $v < K$ within which $\dot{v} < 0$. Such calculations have been made by McGowin (1971), who used the Berger and Lapidus technique that was described in Chapter 5. For $n = 2$ the largest RAS for the low temperature steady state was found to be within the circle:

$$v = [x(z_1)]^2 + [x(z_2)]^2 = 0.0635\left(\frac{T_0C_p}{\Delta H}\right)^2$$

To find the corresponding UAS it is necessary to determine bounding temperature profiles by mapping points on the circumference of this circle onto the $x(\mathbf{z})$ plane. The coordinates of points on a circle are:

$$x(z_1) = \rho \cos \theta$$
$$x(z_2) = \rho \sin \theta$$

where the radius $\rho = 0.252\ (T_0 C_p/\Delta H)$, and θ is the angular parameter for the points chosen. Substitution into equations (8-27a) shows that the coefficients are also sinusoids in θ:

$$a_1 = \frac{3}{2}\rho[0.555 \cos \theta + 0.378 \sin \theta]$$

$$a_2 = \frac{21}{16}\rho[0.328 \cos \theta - 0.535 \sin \theta]$$

Numerical results for evenly spaced choices of θ from zero to π may be tabulated as:

θ	$\left[\frac{a_1}{(T_0 C_p/\Delta H)}\right]$	$\left[\frac{a_2}{(T_0 C_p/\Delta H)}\right]$
0	0.210	0.108
$\pi/6$	0.252	0.0056
$\pi/3$	0.228	−0.0982
$\pi/2$	0.142	−0.177
$2\pi/3$	0.018	−0.206
$5\pi/6$	−0.111	−0.182
π	−0.210	−0.108

from which $x(z)$ profiles may be computed via equation (8-29a). At $\theta = 0$ for example:

$$x(z) = \frac{T_0 C_p}{\Delta H}(1 - z^2)\,[0.210 + 0.108\,(1 - 5z^2)]$$

This curve and the other six are sketched in Figure 8-17a, labeled according to the θ used. It is not necessary to do any further computations for θ from π to 2π because the results differ only in sign from those shown and form a mirror image on Figure 8-17a. The UAS is the darkened region formed between the various curves and the horizontal (temperature steady state) axis. The same curves and some intermediate ones are also given in Figure 8-17b on the coordinates of Figure 6-10, where it is convenient to superimpose two steady states. It is interesting to note that the UAS includes the unstable steady state and also some profiles which cross this steady state. Additional profiles of a variety of shapes were studied by McGowin (1969) to demonstrate their stability with respect to the low or high temperature stable steady states. He was able in this manner to identify broad regions of stability such as are shown in Figure 8-18 for $n = 3$. The upper portion of the diagram is cut abruptly at the level of $(T/T_0) = 1.7$ because this is the maximum attainable adiabatic temperature. The regions cover most of the area between the profile of the intermediate steady state and the respective steady-state profiles under examination, a result in agreement with that of Luss and Lee (1968) obtained independently by another method.

McGowin's method is not restricted to an uncoupled single differential equation. It can for example be applied to the coupled catalyst particle

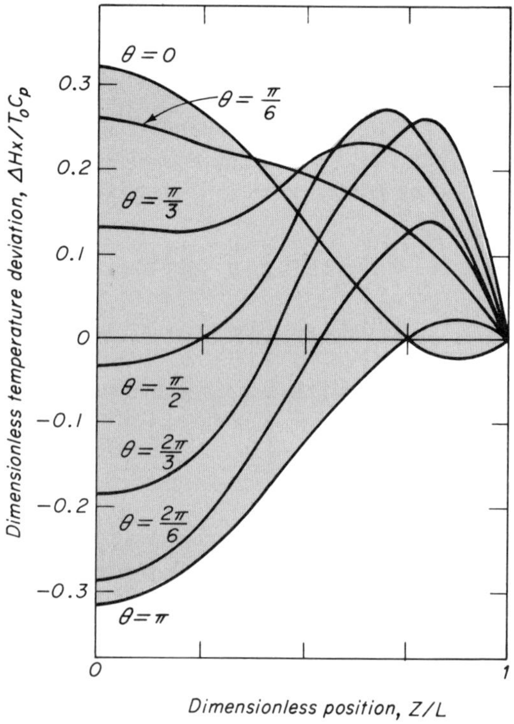

Figure 8-17a. A family of profiles and the corresponding UAS for Example 8-4.

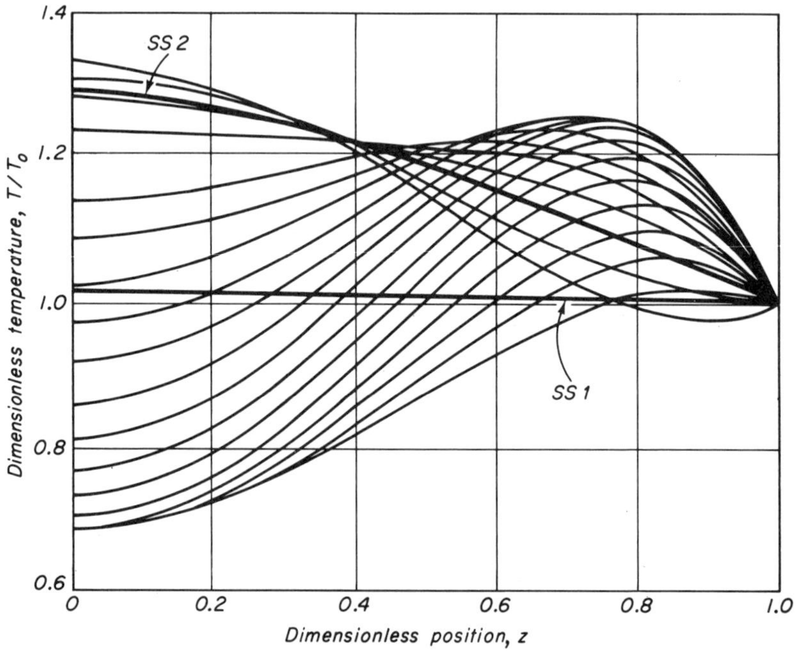

Figure 8-17b. The UAS of Figure 8-17a on the coordinates of Figure 6-10.

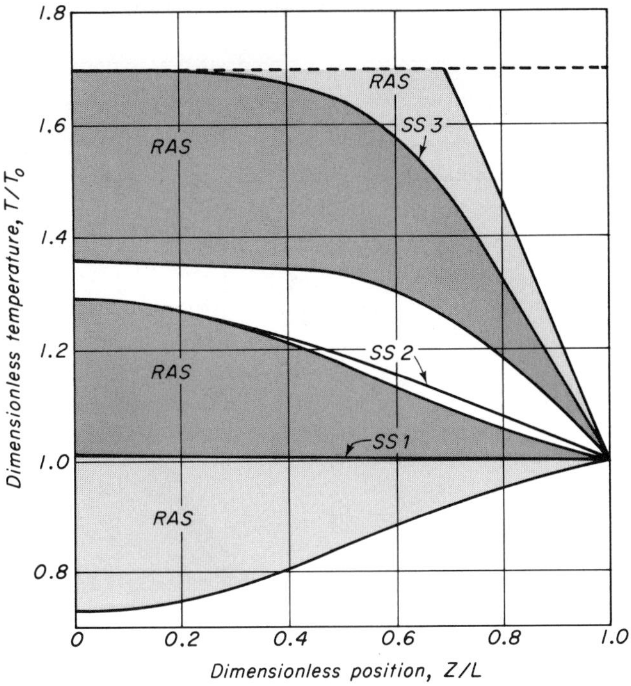

Figure 8-18. Regions of stability for catalyst particle (restricted case).

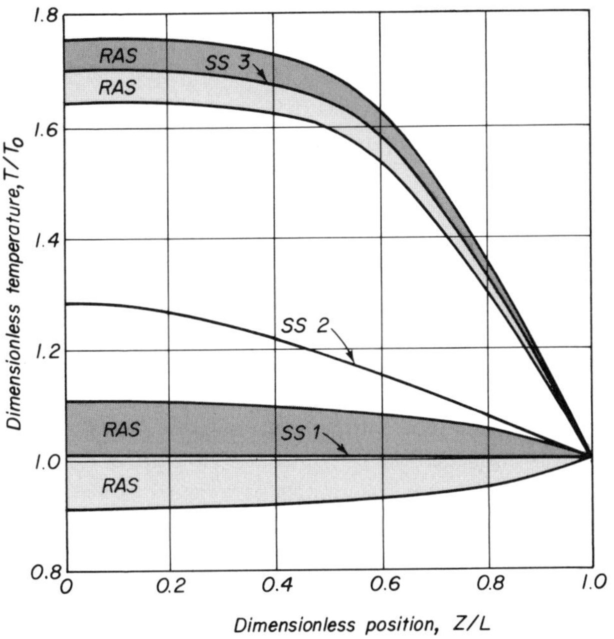

Figure 8-19a. Temperature regions of stability for catalyst particle (unrestricted case).

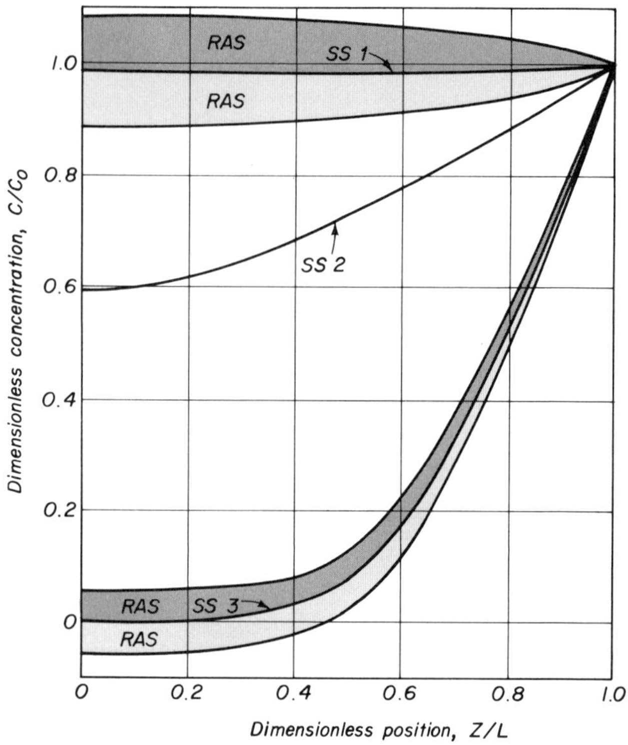

Figure 8-19b. Concentration regions of stability for catalyst particle (unrestricted case).

equations (7-11) which reduce by the substitutions of deviation variables (7-12) to:

$$\frac{\partial x_1}{\partial t} = \alpha\frac{\partial^2 x_1}{\partial Z^2} + r$$

$$\frac{\partial x_2}{\partial t} = D\frac{\partial^2 x_2}{\partial Z^2} - r$$

The collocation technique will now call forth equations (8-23). The details of solution are identical, but because the system dimensionality has doubled, it is to be expected that the computation will produce regions of greater conservatism. This expectation is confirmed in Figures 8-19 which report results obtained for the same system and parameters as were used in Figure 8-18; the new analysis is more general, however, in that the disturbances are not restricted in the way assumed in Section 7-3. Accordingly, the concentration-position Figure 8-19b is needed to complement the temperature-position Figure 8-19a. These regions of asymptotic stability as well as hose in Figure 8-18 are applicable only to concentration and temperature disturbances of opposite sign that are completely contained between their espective RAS boundary profiles and steady-state profiles. Regions computed for concentration and temperature disturbances of the same sign were smaller than those found for disturbances of the opposite sign. This s consistent with the tendency for large positive concentration and temperature disturbances to shift the system to another steady state.

3-7. Regions via a Maximum Principle

A major contribution to the understanding of catalyst particle transients was made by Luss and Lee (1968) by their recognition of the applicability of a theorem presented by Illin, Kalashnikov, and Oleinik (1962). This so-called *maximum principle* provides a constraint on the trajectories of a wide class of parabolic partial differential equations. Among the problems of concern in this book the theorem is particularly useful in considering equation (7-21), repeated here without the superfluous subscript:

$$\frac{\partial x}{\partial t} = \alpha \frac{\partial^2 x}{\partial Z^2} + r \tag{8-33}$$

A preliminary step is essential, however, because the theorem applies only to linear equations. With this restriction in mind, equation (6-86) provides that

$$\left.\frac{dR}{d\eta}\right|_{\eta *} = \frac{R(\eta) - R(\eta_s)}{\eta - \eta_s} = \frac{r(x, Z)}{x} \tag{8-34}$$

where some steady-state profile has been chosen as η_1 and the unnecessary subscript has been deleted from η_2, any other profile. Substitution of (8-34) into (8-33) gives

$$\frac{\partial x}{\partial t} = \alpha \frac{\partial^2 x}{\partial Z^2} + \left.\frac{dR}{d\eta}\right|_{\eta *} x \tag{8-35a}$$

The maximum principle guarantees that such a linear equation will have a solution that is nonnegative for all time and all axial-position values, provided only that the coefficients are real and finite, that $x(Z, t)$ is continuous over $0 \leq Z \leq L$, and that

$$x(L, t) \geq 0 \tag{8-35b}$$

and

$$x(Z, 0) \geq 0 \tag{8-35c}$$

Sign reversals on inequalities (8-35b and c) produce a nonpositive solution for $x(Z, t)$. Since condition (8-35b) is always met by the boundary condition (7-14b), it may be concluded that any initial disturbance that is entirely on one side of a steady state (condition 8-35c) remains on that side for all time. Moreover, since the derivation calls only for *some* steady state, this sweeping conclusion is as valid for systems with multiple steady states as for those where the steady state is unique. This means that the trajectory that develops from any initial profile lying between two steady states will never escape that region. In effect an intermediate unstable steady state divides the family of all one-sided profiles into two groups depending on their position relative to that steady state. Any initially positive deviation will evolve to the high temperature stable steady state, leaving any negative deviation to move to the remaining low temperature steady state. This is the basis for the regions RAS I and RAS II shown in Figure 8-20 for the same example as in Figure 8-18. A similar argument can be developed to bound the trajectories from initial profiles either below or above all three steady states, but it is simpler to take advantage of the physical-chemical limitations. Any feasible trajectory

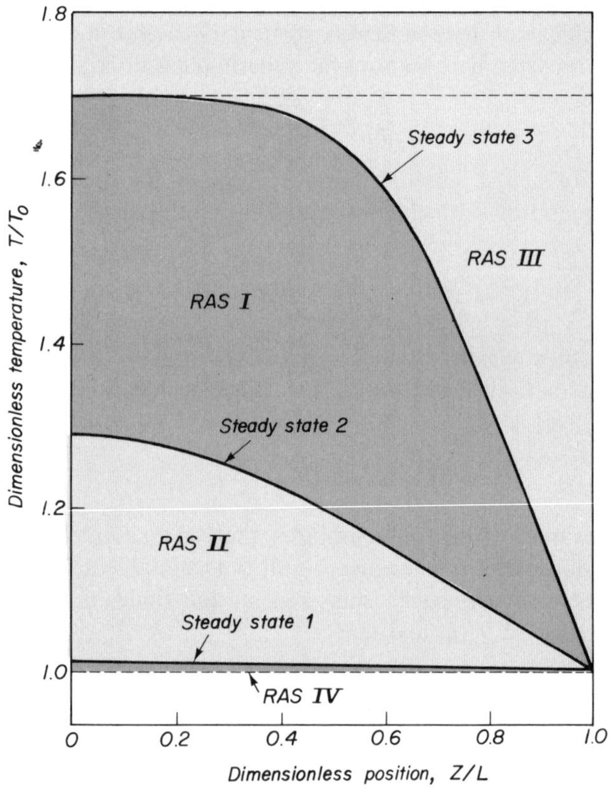

Figure 8-20. The steady-state profiles of Figure 6-10 with regions of asymptotic stability obtained via the maximum principle (after Luss and Lee, 1968).

cannot fall to temperatures below the surface condition T_0, nor can it rise above the adiabatic temperature T_{adi}, in the first instance because T_0 is the effective sink temperature and in the latter because no heat generation is possible other than that originating in the exothermic reaction [compare to equations (6-78) and (6-79)]. The consequence of this line of reasoning is to identify two more regions of stability wherein trajectories are trapped between stable steady states and the physical bounds T_0 and T_{adi}, respectively. These are identified as RAS III and RAS IV in Figure 8-20, regions which incidentally would occupy the entire plane in a system with a unique steady state.

It should be appreciated that the foregoing reasoning rests entirely on the maximum principle and the form of equation (8-33) and its boundary conditions. It did not require any particular restriction on R or dR/dT, other than continuity. The result is obviously superior to that developed in Section 8-6, both in its generality and its avoidance of detailed computation. On the other hand, the maximum principle is applicable only to single equation models, whereas the method of weighted residuals was also extended to more complex models.

For some purposes it is desirable to be able to demonstrate that subregions of an RAS are also regions of asymptotic stability. This may be achieved in the framework of the same theorem by defining $\dot{x} = (\partial x/\partial t)$ and differentiating equation (8-35a) to obtain

$$\frac{\partial \dot{x}}{\partial t} = \alpha\frac{\partial^2 \dot{x}}{\partial Z^2} + \left.\frac{dR}{d\eta}\right|_{\eta_*} \dot{x} \tag{8-36a}$$

Since boundary condition (7-14b) keeps the surface temperature constant:

$$\dot{x}(L, t) = 0 \tag{8-36b}$$

Comparison of equation sets (8-35) and (8-36) shows that the latter will also be subject to the maximum principle theorem for any initial disturbance that satisfies

$$\dot{x}(Z, 0) \geq 0 \tag{8-36c}$$

Thus, for example, if the disturbance profile is constant or linear in Z:

$$\frac{\partial^2 x}{\partial Z^2}(Z, 0) = 0 \tag{8-37}$$

and from equation (8-33)

$$(\partial x/\partial t) = \dot{x}(Z, 0) = r \tag{8-38}$$

Examination of Figure 7-4 serves as a reminder that

$$(dR/dT) < 0, \qquad 0 \leq Z \leq L \tag{8-39}$$

implies:

$$r < 0, \qquad x > 0 \tag{8-40a}$$

$$r > 0, \qquad x < 0 \tag{8-40b}$$

Using these inequalities in equation (8-38) leads to the result that any initial disturbance in this class will produce a transient for which

$$\dot{x}(Z, t) \leq 0 \tag{8-41}$$

for all time, or

$$\dot{x}(Z, t) \geq 0 \tag{8-42}$$

depending on whether the disturbance is positive or negative relative to the steady state. Either way the trajectory movement will converge uniformly and without oscillation to the steady state of reference, but can never pass that bound by virtue of equations (8-35). The region between any disturbance of the constant or linear class and the corresponding steady state is thus found to be a region of stability, provided only that $(dR/dT) < 0$ over the entire steady-state profile. This is of course the same condition that previously guaranteed *local* stability and uniqueness.

It should be emphasized that the concept of a subregion RAS as used here refers to any initial condition of the required linear class, but not to just any arbitrary profile as was the case for any of the four regions labeled I to IV in Figure 8-20. In this regard the conclusion resembles that for the UAS in Section 8-6.

None of the developments in this section can directly be applied to disturbances that initially fall on both sides of a steady state. Luss and Lee have observed however that transients in this category commonly pass into a single stability region before much time has passed. Since their subsequent fate is then sealed, it is convenient to depend on this fact to shorten such numerical computations. Several examples are to be found in the reference cited.

8-8. Regions and Liapunov Functionals

The analogies between Liapunov functions and functionals are strong enough to raise the question whether the latter can be used for distributed-parameter systems to formulate regions of stability in the state-function space. That this was feasible in the realm of lumped-parameter systems was a direct consequence of the fact that a closed contour in state space was uniquely associated with every value of the Liapunov function. In addition to local stability the very existence of a family of Liapunov contours guaranteed that trajectories would not leave certain regions in the state space. To test this notion for state-function space consider the simple functional equation (7-80) and note that the value of v is the area under the squared-profile-curve. Since there are an infinite number of $x(z)$ profiles that all produce the same area it is clear that there is no simple way to relate a given $v = K$ to a unique region in (x, z) space. Alternatively, the same conclusion can be arrived at by another argument. Suppose that it were established for a particular system that all the conditions (4-11) were satisfied for the v-functional equation (7-80). Would this imply that an upper bound could be found for $x(z)$? It would not, for since v can be interpreted as an area, its value can decrease even though the ordinate values of $x(z)$ increase over a part of the range in z. Such a counter-example is shown schematically in Figure 8-21, where v

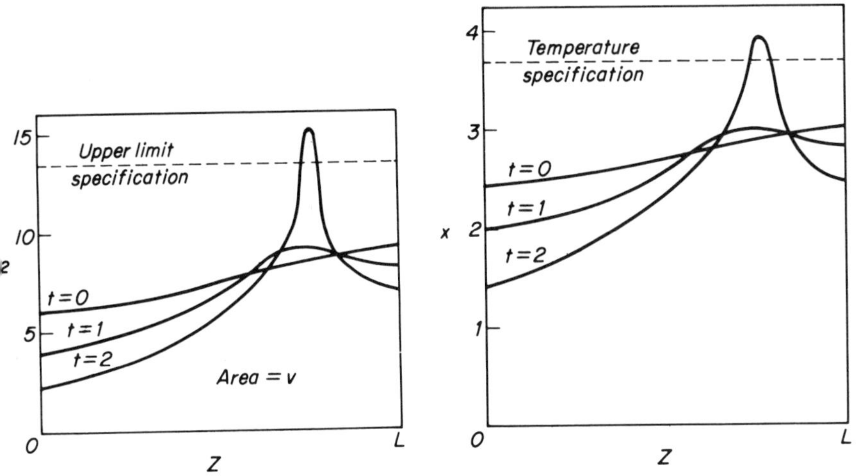

Figure 8-21. A case where v decreases with time, but local temperature rises above its permissible limit (schematic).

decreases monotonically even while local values of x exceed the maximum permissible value. The same reasoning applies to the more complicated v-functionals such as equation (7-79) which provide for a vector of state variable profiles.

These considerations reenforce the idea first presented in Chapter 5 to the effect that Liapunov stability is not necessarily coincident with good engineering design, but rather that practical stability must be the criterion. The two approaches can be brought a bit closer together by using for the Liapunov functional a suggestion due to Wei (1965):

$$v = \sup_{0 \le z \le 1} [\| x(z) \|] \tag{8-43}$$

where *Sup* refers to the smallest upper bound and can most easily be associated with some design specification or physical limitation on the system dependent variable. Though conceptually pleasing, this formulation is not of great engineering utility because it is virtually impossible to compute for even relatively simple problems, and there is no direct way to obtain $\dot{v}$ for this functional. These difficulties not withstanding, some progress in reactor problems can be made with Liapunov functionals.

Consider for a given system and a chosen v-functional that it is possible to demonstrate that within the span of *all physically possible* disturbance profiles

$$v[\mathbf{x}(z, t)] > 0 \tag{8-44}$$

$$\dot{v}[\mathbf{x}(z, t)] < 0 \tag{8-45}$$

Then, because condition (8-45) is true for *all* states of the system that could arise in practice, it must follow that the steady state under study is asymptotically stable. This encompasses more than just local stability, however, for the conditions guarantee decay to the steady state from any feasible initial condition. Moreover, the same reasoning leads to the conclusion that the given steady state must be the only stable one, for were this not the case at least some initial profile would move away from this steady state (increase the norm) and produce a violation of condition (8-45). The application of these ideas to a specific problem will depend on analysis or computation to establish inequalities (8-44) and (8-45). As usual, the first of these is easily taken care of by the way in which v is chosen, and the development hangs on the $\dot{v}$ condition.

An analytic result that fixes the sign-definiteness of $\dot{v}$ is usually unobtainable, and a numerical search over the entire space of feasible profiles is exhausting, if it is to be exhaustive. In view of this, Wang (1966) has suggested that one of the well-established optimization techniques be used to determine the maximum value of $\dot{v}$ in the region of interest. If this maximum is negative, condition (8-45) is proved, since $\dot{v}$ must always be less than its maximum value.

EXAMPLE 8-5. Use the Liapunov functional equation (7-79) to confirm the findings of Amundson and Raymond (1965) that the steady state of the catalyst particle is unique and stable for the usual kinetics and the parameter values:

$$\begin{aligned}
\alpha &= 1.2 \times 10^{-3}\ \text{cm}^2/\text{sec} \\
D &= 0.0615\ \text{cm}^2/\text{sec} \\
C_0 &= 9.76 \times 10^{-5}\ \text{mol/cm}^3 \\
T_0 &= 500^\circ\text{K} \\
Q &= 8{,}750^\circ\text{K} \\
k_0 &= 1.92 \times 10^7\ \text{sec}^{-1} \\
\Delta H &= 5{,}000\ \text{cal/mol} \\
C_p &= 0.25\ \text{cal/}^\circ\text{C cm}^3 \\
L &= 0.1\ \text{cm}
\end{aligned}$$

Solution: For the functional that is suggested

$$v = \int_0^1 (x_1^2 + x_2^2)\, dz$$

By differentiation and substitution from equations (7-13) and (7-23):

$$\dot{v} = 2\int_0^1 \left[\frac{\alpha x_1}{L^2}\frac{\partial^2 x_1}{\partial z^2} + \frac{D x_2}{L^2}\frac{\partial^2 x_2}{\partial z^2} + r(x_1 - x_2)\right] dz \tag{8-46}$$

Integration by parts gives

$$\dot{v} = \frac{2}{L^2}\left[\alpha x_1 \frac{\partial x_1}{\partial z} + D x_2 \frac{\partial x_2}{\partial z}\right]_0^1 + 2\int_0^1 \left[-\frac{\alpha}{L^2}\left(\frac{\partial x_1}{\partial z}\right)^2 - \frac{D}{L^2}\left(\frac{\partial x_2}{\partial z}\right)^2 + r(x_1 - x_2)\right] dz$$

which with boundary conditions (7-14) reduces to

$$\dot{v} = 2\int_0^1 \left[-\frac{\alpha}{L^2}\left(\frac{\partial x_1}{\partial z}\right)^2 - \frac{D}{L^2}\left(\frac{\partial x_2}{\partial z}\right)^2 + r(x_1 - x_2)\right] dz \tag{8-47}$$

Ranges of concentration and temperature are needed within which to seek sign constraints on $\dot{v}$. Since the reaction is irreversible

$$0 \leq C \leq C_0$$

or

$$-C_s \leq x_2 \leq C_0 - C_s$$

The highest adiabatic temperature is a convenient upper bound and a lower limit can be set (arbitrarily) where no appreciable reaction occurs. In the reference cited, Wang reported that the maximum $\dot{v}$ over the range

$$0.01 \leq x_1 \leq 0.03 \text{ mol/cm}^3$$

was negative (except for $\dot{v} = 0$ at the steady state) as found by a dynamic programming maximization. The conditions (8-44) and (8-45) are both satisfied for all feasible profiles in T and C, and the steady state is unique and stable.

8-9. Feedforward Control

Several of the viewpoints developed in this chapter lend themselves naturally to control applications, especially to so-called feedforward schemes. In control parlance the term feedforward is used for a regulator arrangement in which corrective action is taken in response to a disturbance *before* the input change has propagated through the system to affect the variables of particular interest. This is in contrast to feedback control which acts only after an error is detected in system output. The relative merits of each approach are discussed elsewhere [Perlmutter (1965)], but are not of concern in the following.

What is of interest may be introduced most simply by an example. Consider any (δ, ϵ) pair of regions developed by one of the techniques of this chapter. These have been called regions of stability to mean that disturbances not exceeding the δ-boundary at reactor inlet are known to be confined at reactor exit to be within the previously specified ϵ-region. Whenever rapid detection and manipulation are available to the engineer, it is obvious that a very simple and direct control can be put together by preheating (or pre-cooling) the fresh feed and/or by premixing a supplemental stream such as to compensate for any inlet distrubances that would otherwise exceed the limits of the δ-region. In circumstances where such direct manipulation of the state variables is not feasible, other feedforward alternatives are possible, and suggest themselves from the system equations.

Drawing upon the PFTR model for a specific analysis, note that the flow rate does not directly affect either the steady state solutions of equations (6-23) or the transients described by equations (8-3), except that it alters the residence time of the fluid in the reactor. Wang (1967) has therefore argued that one can describe such a system by a variable-length steady state curve such as is shown in Figure 8-22. A point on the τ-axis is the residence time

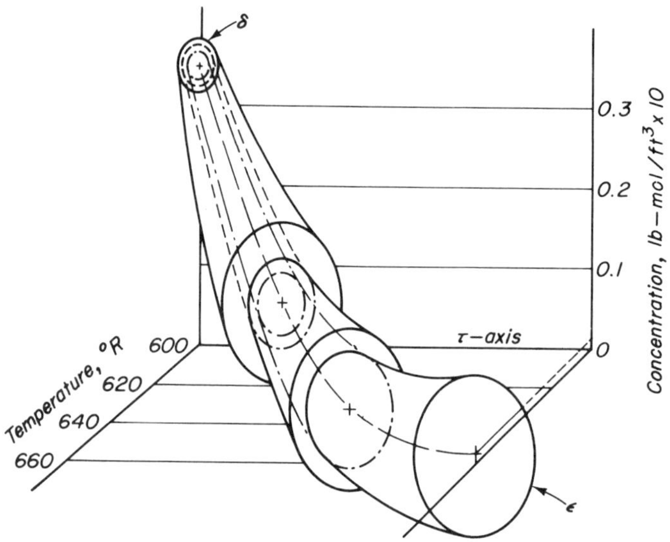

Figure 8-22. The family of δ, ϵ, generated by varying the flow rate.

at the reference flow rate, and corresponds to a particular (δ, ϵ) pair. When the flow rate changes, the new residence time will correspond to a new δ constraint for the same given ϵ specification. It is apparent that larger inlet disturbances can be tolerated by the same reactor with larger flow rate. In a control application, the flow rate could be changed in response to some disturbance in such a manner that at the set flow rate, the δ corresponding to the measured disturbance matches the desired ϵ. In effect, the flow rate is caused to respond to inlet concentration or temperature fluctuations by pushing the fluid out of the reactor before the transients can exceed the ϵ specification.

A similar argument can be developed for compensatory changes of the heat transfer coefficient by manipulation of coolant flow rate. For this case the logic is the same, but the numerical analysis more difficult because any change in h will affect both the steady state curve and the disturbance transients. Wang (1967) has illustrated this technique on a numerical example, but the increase in computational detail does not make this an attractive design method.

REFERENCES

AMUNDSON, N. R. and L. R. RAYMOND, *AIChE Journal*, **14**, 339 (1965).

BIRD, R. B., W. E. STEWART, and E. N. LIGHTFOOT, *Transport Phenomena*, New York: Wiley (1960) p. 73.

ILLIN, A. M., A. S. KALASHNIKOV, and O. A. OLEINIK, *Russian Math. Surveys*, **17**, No. 3, 1 (1962).

LASALLE, J. P. and S. LEFSCHETZ, *Stability by Liapunov's Direct Method*, Academic Press, New York (1961).

LUSS, DAN and J. C. M. LEE,, *Chem. Eng. Sci.*, **23**, 1237 (1968).

MCGOWIN, C. R., Ph. D. Thesis, Univ. of Pennsylvania, Philadelphia (1969).

MCGOWIN, C. R. and D. D. PERLMUTTER, *Chem. Eng. Sci.*, **26**, 275 (1971).

PERLMUTTER, D. D., *Introduction to Chemical Process Control*, New York: Wiley (1965).

WANG, F. S., Ph. D. Thesis, Univ. of Illinois, Urbana (1966).

WANG, F. S. and D. D. PERLMUTTER, *Proc. Joint Auto, Control Conf.* (June, 1967) p. 107.

WANG, F. S. and D. D. PERLMUTTER, *AIChE Journal*, **14**, 335 (1968a).

WANG, F. S. and D. D. PERLMUTTER, *Ibid*, 934 (1968b).

WANG, F. S. and D. D. PERLMUTTER, *Can. Journal Chem. Eng.*, **49**, 274 (1971).

WEI, JAMES, *Chem. Eng. Sci.*, **20**, 729 (1965).

WEISS, L. and E. F. INFANTE, *Proc. Nat'l Acad. Sci.*, **54**, No. 1, 44 (1965).

The Effects of Recycle

9

9-1. Recycle Equations

The reader with a chemical engineering background will need no convincing of the importance of recycle in reactor design. Those who are not familiar with typical process flow patterns may find it adequate justification to note that overall conversion can be greatly increased by repeated passes through a reactor of unreacted feed material. Because this makes it economically feasible to consider relatively lower conversions per pass, recycle can also be used as an adjunct to various control schemes.

The modeling of a recycle system differs in only one essential from the variety of models already considered. The new aspect of the problem is shown schematically in Figure 9-1, from which it is apparent that mixing relations are needed to relate the reactor feed conditions to those at exit and those of the fresh make-up stream. Assuming that there is no significant chemical

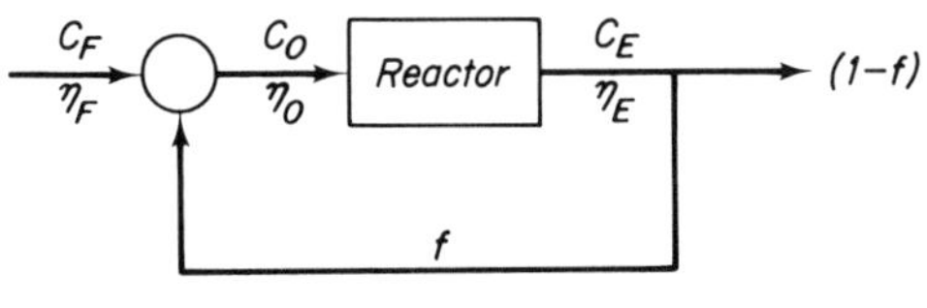

Figure 9-1. Schematic diagram of a reactor with recycle.

reaction, heat loss, or time delay in the recycle line, mass and energy conservation at the mixer dictate:

$$C_0 = (1 - f)C_F + fC_E \tag{9-1a}$$

$$\eta_0 = (1 - f)\eta_F + f\eta_E \tag{9-1b}$$

where f is the fraction of the reactor effluent which is recycled, and the subscripts 0, F, and E refer to reactor inlet, fresh feed, and reactor effluent conditions, respectively. In the event that the reactor exit conditions have a radial distribution as for the TRRM, the effluent values should be interpreted as radial averages:

$$C_E = 2\int_0^{L_0} C(L, Y)\, Y\, dY \tag{9-2a}$$

$$\eta_E = 2\int_0^{L_0} \eta(L, Y)\, Y\, dY \tag{9-2b}$$

Whenever it is convenient to do so, a reduced space variable may be defined as

$$y = \frac{Y}{L_0} \tag{9-3}$$

In combination with the usual deviation variables, direct substitution gives in place of equations (9-2):

$$x_{1E} = 2\int_0^1 x_1(1, y)\, y\, dy \tag{9-4}$$

$$x_{2E} = 2\int_0^1 x_2(1, y)\, y\, dy \tag{9-5}$$

9-2. Isothermal and Adiabatic PFTR-Recycle Systems

In Sections 6-3 and 6-5 it was emphasized that the PFTR cannot show multiple steady states, but that such behavior can occur when dispersive effects are introduced. If recycle is thought of as a form of backmixing, it should be expected that the PFTR-recycle combination can also in some cases have several solutions for a given feed condition. To test this notion consider the steady-state equation (6-21a) and suppose the reactor is isothermal with kinetics following equation (2-6). Then

$$u\frac{dC}{dZ} = \frac{-kC}{(1 + KC)^2} \tag{9-6}$$

and direct integration from the reactor inlet condition

$$C = C_0, \qquad Z = 0 \tag{9-7a}$$

to the effluent

$$C = C_E, \qquad Z = L \tag{9-7b}$$

gives

$$\ln\left(\frac{C_0}{C_E}\right) + 2K(C_0 - C_E) + \frac{K^2}{2}(C_0^2 - C_E^2) = \frac{kL}{u} \tag{9-8}$$

For selected values of the parameter K and the Damkohler number (kL/u), this relationship predicts the reactor effluent concentration as a function of the inlet level, but makes no provision for the recycle. The latter must be specified as equation (9-1a) which serves the same purpose of linking the input and effluent values, but based on the mass balance restrictions external to the reactor. The recycle system will exhibit multiple steady states if for a fixed fresh feed, the simultaneous equations (9-1a) and (9-8) are satisfied for more than single values of C_0 and C_E. Clearly this may depend on the recycle ratio in a particular problem.

A graphical solution is presented as Figure 9-2 which was computed for the choices:

$$K = 10 \text{ ft}^3/\text{mol}$$
$$(kL/u) = 20$$
$$C_F = 1.5 \text{ mol/ft}^3$$

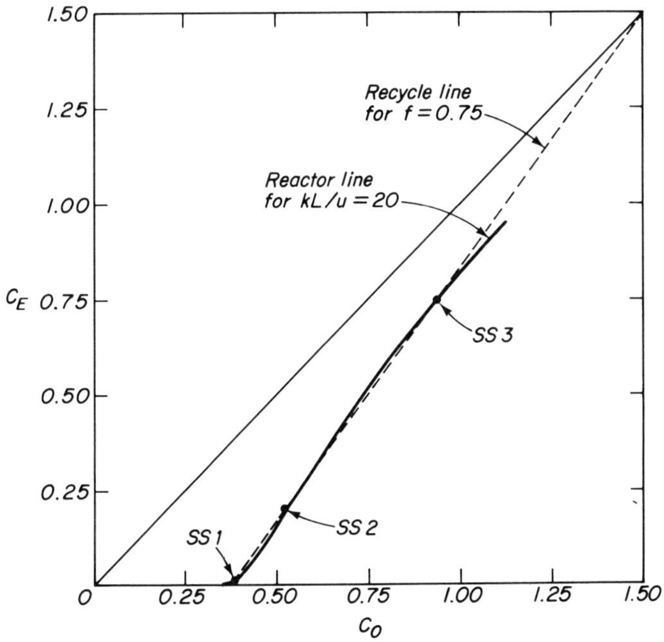

Figure 9-2. Multiple steady states for an isothermal PFTR–recycle system.

On these coordinates, the recycle condition (9-1a) is a straight line of slope $(1/f)$ which passes through the point $C_E = C_0 = C_F$. It should be noted that for these constants, the recycle line can only intersect the nonlinear reaction condition (9-8) at more than a single point over a rather narrow range of the recycle ratio:

$$0.74 < f < 0.77$$

For the designer such a result is of special concern not so much for the fact that multiple solutions are possible as for the indication that the system is extremely sensitive to relatively small changes in recycle fraction f, flow rate u, or rate constant k (in effect, temperature), especially in the ordinarily most desirable circumstance of high conversion. A system designed to yield $C_E < 0.01$ mol/ft³ by operation at $f = 0.78$ would produce $C_E = 0.82$ at $f = 0.73$. This would mean a drop in conversion from $> 99\%$ to 45% following a decrease in recycle fraction of only about 6%. This is of course a design to be avoided if at all possible. One method of insuring that no multiple steady states are possible in such an isothermal system is to choose the slope of the recycle line to be greater than that of the reaction line over its entire range:

$$\frac{1}{f} > \mathrm{Sup}\left[\frac{dC_E}{dC_0}\right] \tag{9-9}$$

but such a design will frequently lead to unnecessarily low conversions.

The treatment of PFTR-recycle problems by simultaneous solutions of reaction and recycle equations can equally well be developed for adiabatic reactors; it was in fact such an application that was used by Luss and Amundson (1967) in the paper that first presented this approach. By starting with the adiabatic subcase of equation (6-21b):

$$u\frac{d\eta}{dZ} = R(C, \eta) \tag{9-10}$$

Substitution of equation (6-25) reduces the system to a single dependent variable:

$$u\frac{d\eta}{dZ} = R(C_0, \eta_0, \eta) \tag{9-11}$$

which can be integrated (analytically or numerically) between the reactor entrance

$$\eta = \eta_0, \qquad Z = 0 \tag{9-12a}$$

and exit

$$\eta = \eta_E, \qquad Z = L \tag{9-12b}$$

The resulting reaction line, η_E versus η_0, is superimposed on the linear recycle condition (9-1b) to search for possible multiple intersections. Again a uniqueness condition can be written in terms of the slopes of the recycle and reaction lines:

$$\frac{1}{f} > \text{Sup}\left[\frac{d\eta_E}{d\eta_0}\right]$$

If the common first-order, Arrhenius kinetics are assumed, equation (9-11) integrates to

$$\int_{\eta_0}^{\eta_E} \frac{d\eta}{(\eta_0 + C_0 - \eta)\exp(-QC_p/\Delta H\eta)} = \frac{k_0 L}{u} \tag{9-13}$$

A sample result of Luss and Amundson is shown as Figure 9-3 computed for:

$$\frac{k_0 L}{u} = 5 \times 10^{13}$$

$$\eta_F = 2.5 \text{ mol/ft}^3$$

$$\frac{C_p Q}{\Delta H \eta_F} = 35$$

$$\eta_0 + C_0 = 3 \text{ mol/ft}^3$$

It may be noted that the reaction line approaches the adiabatic asymptote, whereas the isothermal case has no such natural limit. Triple steady states exist for

$$0.43 \leq f \leq 0.60$$

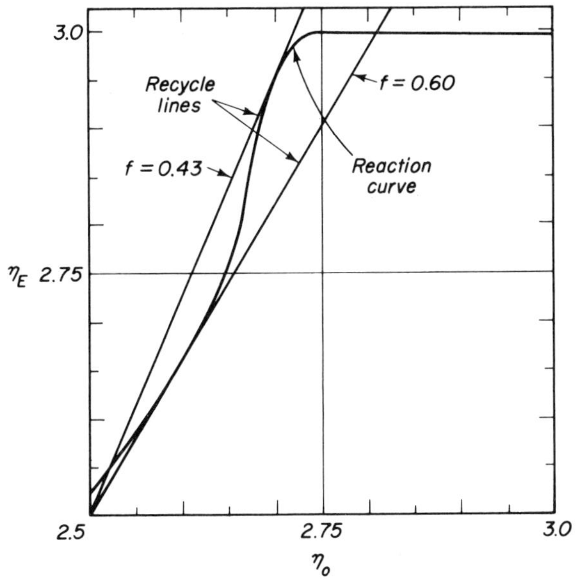

Figure 9-3. Multiple steady states for an adiabatic PFTR–recycle system (Luss and Amundson, 1967).

A very thorough experimental study by Root and Schmitz (1969, 1970) confirmed that adiabatic loop reactors show the predicted multiplicity of steady states. Experimental and computationally predicted outlet temperatures were in good agreement over a wide range of recycle ratios, even to the extent of hysteresis effects dependent on the direction of changes.

EXAMPLE 9-1. Are multiple steady states possible in the isothermal PFTR-recycle system for kinetics of the form:

$$\text{(a)}\quad R = kC^n$$

$$\text{(b)}\quad R = \frac{kC}{K + C} \quad \text{(Michaelis–Menten)}$$

Solution: The isothermal PFTR equation

$$\int_{C_E}^{C_0} \frac{1}{R}\,dC = \frac{L}{u}$$

is applicable to both parts of the problem. Noting that the left side integral is the area under the $(1/R)$ versus C curve permits one to conclude that for fixed reactor length and flow velocity, changes in reactor entrance condition are compensated for by response in effluent condition in such a way as to keep the area constant. In the symbols of Figure 9-4:

$$\frac{dC_E}{R_E} = \frac{dC_0}{R_0}$$

or

$$\frac{dC_E}{dC_0} = \frac{R_E}{R_0} \tag{9-14}$$

Part (a): For this kinetic equation:

$$\frac{dC_E}{dC_0} = \frac{kC_E^n}{kC_0^n} = \left[\frac{C_E}{C_0}\right]^n$$

But since $C_E < C_0$:

$$\text{Sup}\left[\frac{dC_E}{dC_0}\right] \leq 1$$

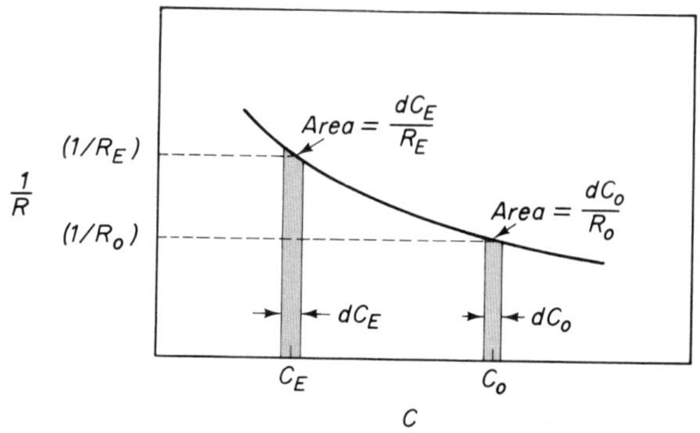

Figure 9-4. Hypothetical $1/R$ versus C curve, showing justification of equation (9-14).

for all reasonable values of n. Finally, because the fraction recycled must be less than unity:

$$\frac{1}{f} > 1$$

and the uniqueness condition (9-9) will be satisfied. It should be added, however, that this conclusion is only justified for constant density systems. Sano (1971) has computed some counter-examples for isothermal gas reactions with changing number of moles.

Part (b): For the Michaelis–Menten kinetics:

$$\frac{dC_E}{dC_0} = \frac{C_E(K + C_0)}{C_0(K + C_E)} = \frac{\dfrac{K}{C_0} + 1}{\dfrac{K}{C_E} + 1}$$

Using again that $C_E < C_0$, gives

$$\text{Sup}\left[\frac{dC_E}{dC_0}\right] \leq 1$$

and satisfies the same uniqueness condition as in part (a).

9-3. Finite-Difference Models for the PFTR-Recycle System

For the analysis of some problems, it is advantageous to emphasize the iterative quality that arises from recycle. Plug-flow models will yield inlet-outlet relationships which for the kth pass of a fluid element may be written in the form:

$$\eta_E^{(k)} = F_1[C_0^{(k)}, \eta_0^{(k)}]$$
$$C_E^{(k)} = F_2[C_0^{(k)}, \eta_0^{(k)}]$$

or

$$\eta_E^{(k+1)} = F_1[C_0^{(k+1)}, \eta_0^{(k+1)}] \tag{9-15a}$$
$$C_E^{(k+1)} = F_2[C_0^{(k+1)}, \eta_0^{(k+1)}] \tag{9-15b}$$

but as shown by Figure 9-1, a fraction of the effluent of the kth pass recycles to become part of the feed for the $(k + 1)$th pass through the reactor. Then equations (9-1) can be written with a discrete index:

$$C_0^{(k+1)} = (1 - f)C_F + fC_E^{(k)} \tag{9-16a}$$
$$\eta_0^{(k+1)} = (1 - f)\eta_F + f\eta_E^{(k)} \tag{9-16b}$$

Combining equations (9-15) and (9-16) produces a set of nonlinear difference equations with the fresh feed condition and fraction recycled as parameters:

$$\eta_E^{(k+1)} = F_1[C_E^{(k)}, \eta_E^{(k)}] \tag{9-17a}$$
$$C_E^{(k+1)} = F_2[C_E^{(k)}, \eta_E^{(k)}] \tag{9-17b}$$

Reilly and Schmitz (1966) have referred to books by Bellman (1949) and Goldberg (1958) to justify the linearization of such a system and to establish local stability conditions. In terms of deviation variables, the linearization of equations (9-17) gives

$$\mathbf{x}_E^{(k+1)} = \mathbf{J}_s \mathbf{x}_E^{(k)} \tag{9-18}$$

where

$$\mathbf{x}_E^{(k)} = \begin{bmatrix} x_{1E}^{(k)} \\ x_{2E}^{(k)} \end{bmatrix} = \begin{bmatrix} \eta_E^{(k)} - \eta_{Es} \\ C_E^{(k)} - C_{Es} \end{bmatrix} \tag{9-19}$$

and the Jacobian matrix is

$$\mathbf{J}_s = \begin{bmatrix} \frac{\partial F_1}{\partial \eta_E} & \frac{\partial F_1}{\partial C_E} \\ \frac{\partial F_2}{\partial \eta_E} & \frac{\partial F_2}{\partial C_E} \end{bmatrix}_s = \begin{bmatrix} \frac{\partial F_1}{\partial \eta_o} & \frac{\partial F_1}{\partial C_o} \\ \frac{\partial F_2}{\partial \eta_o} & \frac{\partial F_2}{\partial C_o} \end{bmatrix}_s \begin{bmatrix} \frac{\partial \eta_o}{\partial \eta_E} & \frac{\partial \eta_o}{\partial C_E} \\ \frac{\partial C_o}{\partial \eta_E} & \frac{\partial C_o}{\partial C_E} \end{bmatrix}_s \tag{9-20a}$$

Differentiation of the equations (9-1) identifies the last of these matrices as $f\mathbf{I}$ and

$$\mathbf{J}_s = f \begin{bmatrix} \frac{\partial F_1}{\partial \eta_0} & \frac{\partial F_1}{\partial C_0} \\ \frac{\partial F_2}{\partial \eta_0} & \frac{\partial F_2}{\partial C_0} \end{bmatrix} \tag{9-20b}$$

The asymptotic stability of the finite difference equations (9-18) depends on the necessary and sufficient condition that each eigenvalue of $\mathbf{J}_s$ must be smaller than unity in magnitude. As in the case of comparable differential equation models, the asymptotic stability of the linearized equations assures us that the original nonlinear system is locally stable, but note that the eigenvalue criterion for difference equations refers to the magnitudes of the eigenvalues, not to their sign. The linearization is inconclusive if it produces eigenvalues of marginal value for this test.

A stability test based on the eigenvalues of matrix (9-20b) is straightforward when the F_1 and F_2 are available, but these functions will only arise analytically in exceptional circumstances, since it requires a closed-form integration of simultaneous differential equations such as (6-21). If as is commonly the case, numerical integration is required, it is more convenient to express the desired $\mathbf{J}_s$ in terms of the f_1 and f_2 functions of equations (6-27). To this end Reilly and Schmitz (1966) defined the matrix variable

$$\mathbf{M}(Z) = \begin{bmatrix} \frac{\partial \eta}{\partial \eta_0} & \frac{\partial \eta}{\partial C_0} \\ \frac{\partial C}{\partial \eta_0} & \frac{\partial C}{\partial C_0} \end{bmatrix} \tag{9-21}$$

which reduces to the identity matrix at reactor entrance:

$$\mathbf{M}(0) = \mathbf{I} \tag{9-22}$$

and to

$$\mathbf{M}(L) = \begin{bmatrix} \dfrac{\partial \eta_E}{\partial \eta_0} & \dfrac{\partial \eta_E}{\partial C_0} \\[2ex] \dfrac{\partial C_E}{\partial \eta_0} & \dfrac{\partial C_E}{\partial C_0} \end{bmatrix} = \begin{bmatrix} \dfrac{\partial F_1}{\partial \eta_0} & \dfrac{\partial F_1}{\partial C_0} \\[2ex] \dfrac{\partial F_2}{\partial \eta_0} & \dfrac{\partial F_2}{\partial C_0} \end{bmatrix} \tag{9-23}$$

at reactor exit. Combining equations (9-20b) and (9-23) produces the alternative form

$$\mathbf{J}_s = f\mathbf{M}(L) \tag{9-24}$$

which serves to recast the problem as one dependent on a solution for $\mathbf{M}(Z)$. This step is accomplished by noting that

$$\frac{d\mathbf{M}}{dZ} = \begin{bmatrix} \dfrac{\partial f_1}{\partial \eta_0} & \dfrac{\partial f_1}{\partial C_0} \\[2ex] \dfrac{\partial f_2}{\partial \eta_0} & \dfrac{\partial f_2}{\partial C_0} \end{bmatrix} \tag{9-25a}$$

which may also be written as the auxiliary matrix equation

$$\frac{d\mathbf{M}}{dZ} = \begin{bmatrix} \dfrac{\partial f_1}{\partial \eta} & \dfrac{\partial f_1}{\partial C} \\[2ex] \dfrac{\partial f_2}{\partial \eta} & \dfrac{\partial f_2}{\partial C} \end{bmatrix} \mathbf{M} = \mathbf{A}(Z)\mathbf{M} \tag{9-25b}$$

Substituting the f_1 and f_2 functions of equations (6-21) gives for the coefficient matrix:

$$\mathbf{A}(Z) = \begin{bmatrix} \dfrac{1}{u}\dfrac{\partial R}{\partial \eta} - \dfrac{h}{uC_p} & \dfrac{1}{u}\dfrac{\partial R}{\partial C} \\[2ex] -\dfrac{1}{u}\dfrac{\partial R}{\partial \eta} & -\dfrac{1}{u}\dfrac{\partial R}{\partial C} \end{bmatrix} \tag{9-26}$$

The stability analysis of a given steady state requires by this approach that equation (9-25) be integrated from initial condition (9-22), using evaluations of $\mathbf{A}(Z)$ according to an equation such as (9-26). The exit condition on $\mathbf{M}$ is then used in equation (9-24). Specific kinetics reduce this to algebraic form. If for example equation (1-6a) is applicable:

$$\mathbf{A}(Z) = \left[\begin{array}{c|c} \dfrac{\Delta HQ}{uC_p}\dfrac{R}{T^2} - \dfrac{h}{uC_p} & \dfrac{nR}{uC} \\[2ex] \hline -\dfrac{\Delta HQ}{uC_p}\dfrac{R}{T^2} & -\dfrac{nR}{uC} \end{array}\right] \tag{9-27}$$

In the reference cited, Reilly and Schmitz make a most important observation on numerical computations: if one uses an iterative Newton–Raphson technique to find the steady states of a PFTR-recycle system, the matrix $\mathbf{J}_s$ will be available after the final iteration. It is then a simple matter to check the necessary eigenvalues for system stability. Results of a number of numerical tests are reproduced as Figure 9-5. It is worthy of notice in particular that *unique and unstable* steady states can occur over a range of feed conditions if $(hL/uC_p) > 0.79$. The same authors [Reilly and Schmitz (1967)] have thoroughly explored the implications of this and other related findings. They developed a (η_E, C_E) phase plane on which they were able to present a wide variety of behaviors—some similar to the conventional phase-plane expectations, and some of different character. Among other things they summarized the types of local responses that are associated with the signs, magnitudes and real or complex nature of the system eigenvalues. Oscillatory behavior can arise from either complex eigenvalues or from systems in which both eigenvalues are negative. They report that only positive values were found among the several thousand eigenvalues in their PFTR-recycle study.

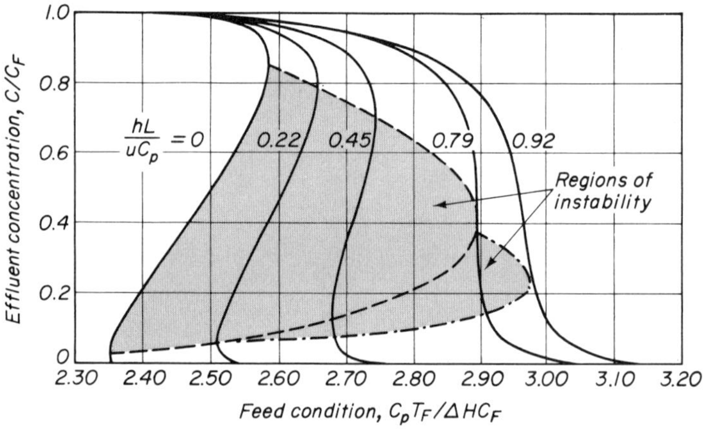

Figure 9-5. Steady-state operating curves for PFTR–recycle system with $f = 0.5$ (Reilly and Schmitz, 1966).

These results have been extended [Paraja and Reilly (1969)] to also include possible heat exchange or mixing in the recycle line. The effect is often to add stability to marginal systems.

EXAMPLE 9-2. Test for stability each of the three steady states shown in Figure 9-2 for $f = 0.75$.

Solution: Because the system under consideration is isothermal the vector $\mathbf{x}_E$ reduces to a scalar x for concentration deviation and

$$\mathbf{J}_s = f\frac{dF_2}{dC_0} = f\frac{dC_E}{dC_0}$$

From the arguments developed in Example 9-1:

$$\frac{dC_E}{dC_0} = \frac{R_E}{R_0} = \frac{C_E}{C_0}\left[\frac{1 + KC_0}{1 + KC_E}\right]^2$$

and the single eigenvalue of the scalar $\mathbf{J}_s$ is

$$\lambda = f\frac{C_E}{C_0}\left[\frac{1 + KC_0}{1 + KC_E}\right]^2 \tag{9-28}$$

The three steady states may be read from Figure 9-2:

Steady State	C_0	C_E
1	0.38	0.01
2	0.52	0.20
3	0.94	0.75

With $K = 10$ ft³/mol and $f = 0.75$, the respective eigenvalues may be tabulated and conclusions based on whether or not $|\lambda| < 1$:

Steady State	λ	Conclusion
1	0.37	Stable
2	1.24	Unstable
3	0.89	Stable

EXAMPLE 9-3. Apply the approach developed by Reilly and Schmitz to the isothermal irreversible reaction sequence

$$A \xrightarrow{R_1} B \xrightarrow{R_2} P$$

Demonstrate that the PFTR-recycle system cannot have an unstable steady state when both reactions are first-order:

$$R_1 = k_1 C_A$$
$$R_2 = k_2 C_B$$

Solution: The reactor steady-state profiles for the PFTR will be determined by

$$\frac{dC_A}{dZ} = -\frac{1}{u}R_1 \qquad = f_1 = -\frac{k_1}{u}C_A \tag{9-29a}$$

$$\frac{dC_B}{dZ} = \frac{1}{u}(R_1 - R_2) = f_2 = \frac{k_1}{u}C_A - \frac{k_2}{u}C_B \tag{9-29b}$$

and the coefficient matrix of the auxiliary equation will be

$$\mathbf{A} = \begin{bmatrix} \dfrac{\partial f_1}{\partial C_A} & \dfrac{\partial f_1}{\partial C_B} \\[2ex] \dfrac{\partial f_2}{\partial C_A} & \dfrac{\partial f_2}{\partial C_B} \end{bmatrix} = \frac{1}{u}\begin{bmatrix} -k_1 & 0 \\ k_1 & -k_2 \end{bmatrix}$$

Since in this case the $\mathbf{A}$ matrix is constant, equation (9-25) has a closed-form analytic solution:

$$\mathbf{M}(Z) = \left[\begin{array}{c|c} e^{-k_1 Z/u} & 0 \\ \hline \dfrac{k_1}{k_2 - k_1}(e^{-k_1 Z/u} - e^{-k_2 Z/u}) & e^{-k_2 Z/u} \end{array}\right] \tag{9-30}$$

and via equation (9-24):

$$\mathbf{J}_s = f\left[\begin{array}{c|c} e^{-k_1L/u} & 0 \\ \hline \dfrac{k_1}{k_2 - k_1}(e^{-k_1L/u} - e^{-k_2L/u}) & e^{-k_2L/u} \end{array}\right]$$

The eigenvalues of this matrix are:

$$\lambda_1 = fe^{-k_1L/u}$$

$$\lambda_2 = fe^{-k_2L/u}$$

which are both positive and smaller than one in magnitude for all feasible values of recycle fraction and Damkohler numbers. As a result any steady state of this system is stable.

9-4. The TRAM-Recycle System

The analysis of the local stability character of a steady state in a TRAM-recycle system can most readily be approached by the modified collocation method. Starting from the linearized and normalized equations (7-72), the substitutions of equations (7-42) and (7-45) yield:

$$\frac{d\mathbf{x}_1}{dt} = \frac{\alpha}{L^2}\mathbf{N}_2\mathbf{N}_0^{-1}\mathbf{x}_1 - \frac{u}{L}\mathbf{N}_1\mathbf{N}_0^{-1}\mathbf{x}_1 + \left(\mathbf{R}_\eta - \frac{h}{C_p}\right)\mathbf{x}_1 + \mathbf{R}_c\mathbf{x}_2 \tag{9-31a}$$

$$\frac{d\mathbf{x}_2}{dt} = \frac{D}{L^2}\mathbf{N}_2\mathbf{N}_0^{-1}\mathbf{x}_2 - \frac{u}{L}\mathbf{N}_1\mathbf{N}_0^{-1}\mathbf{x}_2 - \mathbf{R}_\eta\mathbf{x}_1 - \mathbf{R}_c\mathbf{x}_2 \tag{9-31b}$$

which is again the standard linear form with

$$\mathbf{A} = \left[\begin{array}{c|c} \dfrac{\alpha}{L^2}\mathbf{N}_2\mathbf{N}_0^{-1} - \dfrac{u}{L}\mathbf{N}_1\mathbf{N}_0^{-1} + \left(\mathbf{R}_\eta - \dfrac{h}{C_p}\right) & \mathbf{R}_c \\ \hline -\mathbf{R}_\eta & \dfrac{D}{L^2}\mathbf{N}_2\mathbf{N}_0^{-1} - \dfrac{u}{L}\mathbf{N}_1\mathbf{N}_0^{-1} - \mathbf{R}_c \end{array}\right] \tag{9-32}$$

These equations are in essence merely the extension to a two-equation system of the development in Section 7-6. It should here be recalled that somewhat arbitrary choices of trial solution form and collocation points must be made in the next steps. Satisfactory results for this problem were obtained by McGowin (1971) who chose the trial forms:

$$x_1(z, t) = \sum_{i=1}^{n} a_i(t)\, z^{i-1} \tag{9-33a}$$

$$x_2(z, t) = \sum_{i=1}^{n} b_i(t)\, z^{i-1} \tag{9-33b}$$

for which:

$$\mathbf{N}_0 = \begin{bmatrix} 1 & z_1 & z_1^2 & \cdots & z_1^{n-1} \\ 1 & z_2 & z_2^2 & \cdots & z_2^{n-1} \\ \cdots \\ \cdots \\ 1 & z_n & z_n^2 & \cdots & z_n^{n-1} \end{bmatrix} \tag{9-34a}$$

$$\mathbf{N}_1 = \begin{bmatrix} 0 & 1 & 2z_1 & 3z_1^2 & \cdots & (n-1)z_1^{n-2} \\ 0 & 1 & 2z_2 & 3z_2^2 & \cdots & (n-1)z_2^{n-2} \\ \cdots \\ \cdots \\ 0 & 1 & 2z_n & 3z_n^2 & \cdots & (n-1)z_n^{n-2} \end{bmatrix} \tag{9-34b}$$

$$\mathbf{N}_2 = \begin{bmatrix} 0 & 0 & 2 & 6z_1 & \cdots & (n-1)(n-2)z_1^{n-3} \\ 0 & 0 & 2 & 6z_2 & \cdots & (n-1)(n-2)z_2^{n-3} \\ \cdots \\ \cdots \\ 0 & 0 & 2 & 6z_n & \cdots & (n-1)(n-2)z_n^{n-3} \end{bmatrix} \tag{9-34c}$$

The appropriate boundary conditions must be examined as part of the process of chosing the z_j collocation points. In this problem the TRAM boundary equations (1-9) are applicable. The condition at $Z = L$ remains unchanged but the inlet conditions must be modified to account for the recycle. By combining recycle conditions (9-1) with boundary equations (1-9):

$$D\frac{\partial C}{\partial Z} = u[C - (1-f)C_F - fC_E] \tag{9-35a}$$

$$\alpha\frac{\partial \eta}{\partial Z} = u[\eta - (1-f)\eta_F - f\eta_E] \tag{9-35b}$$

In terms of deviation variables this becomes:

$$\frac{\partial x_1}{\partial z} = \frac{uL}{\alpha}[x_1(0) - fx_1(1)] \tag{9-36a}$$

$$\frac{\partial x_2}{\partial z} = \frac{uL}{D}[x_2(0) - fx_2(1)] \tag{9-36b}$$

Because the trial functions (9-33) do not necessarily satisfy either conditions (7-73c) or conditions (9-36) it is essential to impose these conditions on the reduced collocation equations by choosing two extra collocation points at the ends of the interval of interest: $z_{n+1} = 1$ and $z_{n+2} = 0$. At the reactor exit the boundary condition may be expressed as the $(n+1)$th component of equation set (7-42):

$$\left[\frac{\partial x_1}{\partial z}\right]_{n+1} = [\mathbf{N}_1\mathbf{N}_0^{-1}]_{n+1}\, x_1(\mathbf{z}) = 0 \tag{9-37a}$$

$$\left[\frac{\partial x_2}{\partial z}\right]_{n+1} = [\mathbf{N}_1\mathbf{N}_0^{-1}]_{n+1}\, x_2(\mathbf{z}) = 0 \tag{9-37b}$$

A similar reduction at reactor entrance gives:

$$\begin{aligned}\left[\frac{\partial x_1}{\partial z}\right]_{n+2} &= [\mathbf{N}_1\mathbf{N}_0^{-1}]_{n+2}\, x_1(\mathbf{z}) \\ &= \frac{uL}{\alpha}[x_1(z_{n+2}) - f x_1(z_{n+1})]\end{aligned} \tag{9-38a}$$

$$\begin{aligned}\left[\frac{\partial x_2}{\partial z}\right]_{n+2} &= [\mathbf{N}_1\mathbf{N}_0^{-1}]_{n+2}\, x_2(\mathbf{z}) \\ &= \frac{uL}{D}[x_2(z_{n+2}) - f x_2(z_{n+1})]\end{aligned} \tag{9-38b}$$

Although it would appear that the matrix **A** has been augmented by the two boundary conditions to form a $(2n+2)$ dimensional problem, McGowin has pointed out that the two auxiliary collocation points can be eliminated by simultaneous solution of equations (9-37) and (9-38) to express all variables as functions evaluated at the n points only.

By using the parameters chosen by Reilly and Schmitz (1966) for their PFTR-recycle study and adding suitable values of Peclet numbers, McGowin was able to extend the earlier results to survey the behavior of TRAM-recycle systems. He determined local stability behavior for a variety of steady states by computing the dominant eigenvalue of **A** for a range of increasing degree

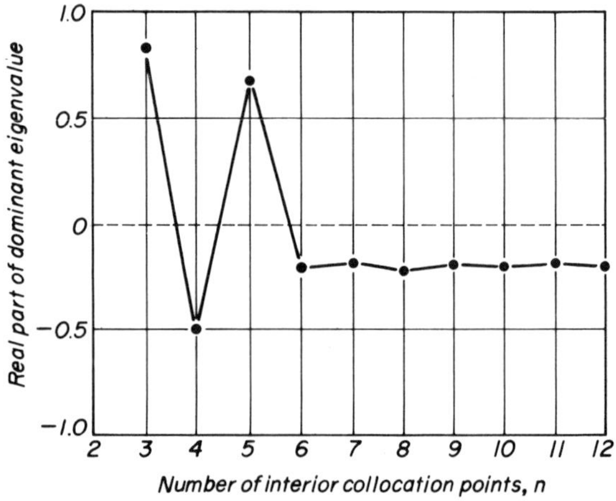

Figure 9-6. Convergence pattern of dominant eigenvalue estimates using collocation method.

of approximation n. A typical trend is shown in Figure 9-6 from which it is apparent that the result converges in a damped oscillatory manner. In this example the amplitude of the oscillations becomes less than the magnitude of the eigenvalue for $n \geq 6$ and the sign is determined. For some other steady states for this problem, the stability (sign of the dominant eigenvalue) was not definitely established until $n = 8$.

The results of McGowin's study may be summarized with reference to Figures 9-7. As the Peclet numbers for heat and mass approach zero, the

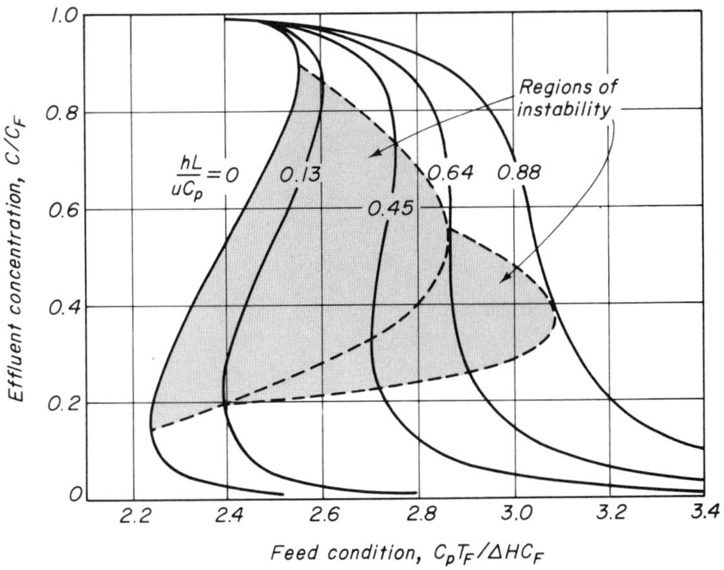

Figure 9-7a. Steady-state operating curves for TRAM–recycle system with $f = 0.5$, $uL/D = uL/\alpha = 0$.

TRAM-recycle system becomes increasingly like the CSTR-recycle system in character. Thus, except for numerical values, Figures 4-4 and 9-7a describe virtually the same reactor. At the other extreme of Peclet numbers, the TRAM approaches the PFTR. This can be seen already at $uL/D = uL/\alpha = 100$ in Figure 9-7b, which is hardly distinguishable from the Reilly and Schmitz PFTR results of Figure 9-5. Between these limits of axial dispersion, smooth transitions occur in both operating curves and instability regions. These are illustrated in Figures 9-7c and d for two different levels of heat transfer coefficient.

EXAMPLE 9-4. Use $n = 1$ and $z_1 = (\frac{1}{2})$ to demonstrate that the conditions (9-37) and (9-38) give x_1 and x_2 at the boundaries in terms of x_1 and x_2 at the internal collocation point.

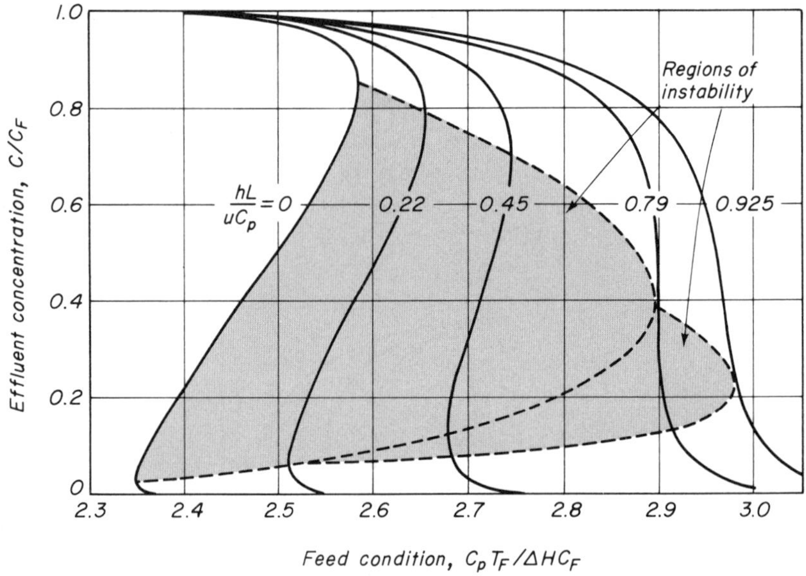

Figure 9-7b. Steady-state operating curves for TRAM–recycle system with $f = 0.5$, $uL/D = uL/\alpha = 100$.

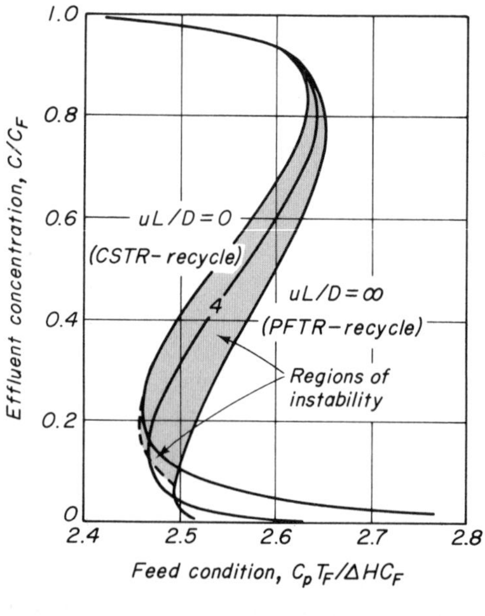

Figure 9-7c. Steady-state operating curves for TRAM–recycle system with $f = 0.5$, $uL/D = uL/\alpha$, $hL/uC_p = 0.2$.

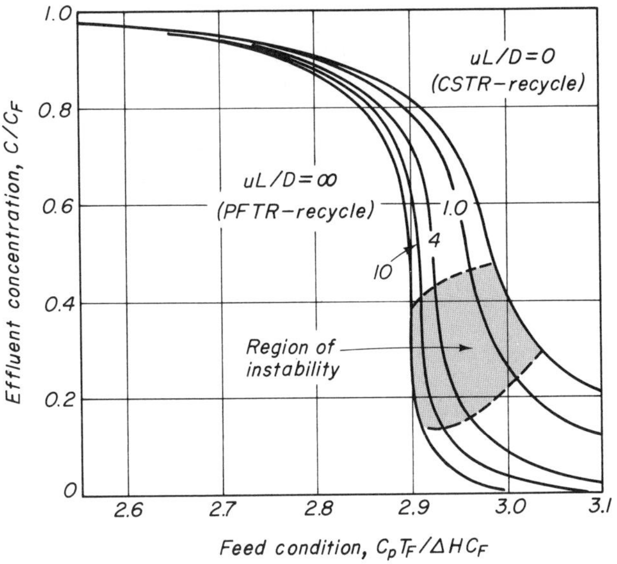

Figure 9-7d. Steady-state operating curves for TRAM–recycle system $f = 0.5$, $uL/D = uL/\alpha$, $hL/uC_p = 0.79$.

Solution: For the case of $n = 1$, $z_1 = (\frac{1}{2})$, equations (9-34) may be evaluated as:

$$\mathbf{N}_0 = \begin{bmatrix} 1 & z_1 & z_1^2 \\ 1 & 1 & 1 \\ 1 & 0 & 0 \end{bmatrix} = \begin{bmatrix} 1 & \frac{1}{2} & \frac{1}{4} \\ 1 & 1 & 1 \\ 1 & 0 & 0 \end{bmatrix}$$

$$\mathbf{N}_1 = \begin{bmatrix} 0 & 1 & 2z_1 \\ 0 & 1 & 2 \\ 0 & 1 & 0 \end{bmatrix} = \begin{bmatrix} 0 & 1 & 1 \\ 0 & 1 & 2 \\ 0 & 1 & 0 \end{bmatrix}$$

$$\mathrm{N}_1\mathrm{N}_0^{-1} = \begin{bmatrix} 0 & 1 & -1 \\ -4 & 3 & 1 \\ 4 & -1 & -3 \end{bmatrix}$$

and equations (9-37) and (9-38) are

$$-4x_1(\tfrac{1}{2}) + 3x_1(1) + x_1(0) = 0$$

and

$$4x_1(\tfrac{1}{2}) - x_1(1) - 3x_1(0) = \frac{uL}{\alpha}[x_1(0) - fx_1(1)]$$

Simultaneous solution gives:

$$x_1(0) = \left[\frac{4\left(2 + f\frac{uL}{\alpha}\right)}{8 + (3 + f)\frac{uL}{\alpha}}\right] x_1(\tfrac{1}{2}) \tag{9-39a}$$

$$x_1(1) = \left[\frac{4\left(2 + \frac{uL}{\alpha}\right)}{8 + (3 + f)\frac{uL}{\alpha}}\right] x_1(\tfrac{1}{2}) \tag{9-39b}$$

and similar results for x_2 with the parameter D replacing α.

EXAMPLE 9-5. Schmeal and Amundson (1966) studied the TRAM-recycle system for the case of isothermal first-order kinetics. With such simplification the system is linear and they were able to identify the characteristics of the eigenvalues associated with the response to a step input. Confirm Schmeal and Amundson's finding of complex eigenvalues by using the modified collocation approach of this section, placing the z_j at:

$$\begin{aligned} z_1 &= \tfrac{1}{3} \\ z_2 &= \tfrac{2}{3} \\ z_3 &= 0 \qquad \text{(inlet to reactor)} \\ z_4 &= 1 \qquad \text{(exit of reactor)} \end{aligned}$$

For numerical work assume $(D/uL) = 0.1$ and that $f = 0.61$.

Solution: Adopting trial solution (9-33a) leads to evaluation of the matrices (9-34) at the given collocation points:

$$\mathbf{N}_0 = \begin{bmatrix} 1 & \frac{1}{3} & \frac{1}{9} & \frac{1}{27} \\ 1 & \frac{2}{3} & \frac{4}{9} & \frac{8}{27} \\ 1 & 0 & 0 & 0 \\ 1 & 1 & 1 & 1 \end{bmatrix}$$

$$\mathbf{N}_1 = \begin{bmatrix} 0 & 1 & \frac{2}{3} & \frac{1}{3} \\ 0 & 1 & \frac{4}{3} & \frac{4}{3} \\ 0 & 1 & 0 & 0 \\ 0 & 1 & 2 & 3 \end{bmatrix}$$

$$\mathbf{N}_2 = \begin{bmatrix} 0 & 0 & 2 & 2 \\ 0 & 0 & 2 & 4 \\ 0 & 0 & 2 & 0 \\ 0 & 0 & 2 & 6 \end{bmatrix}$$

Furthermore:

$$\mathbf{N}_1\mathbf{N}_0^{-1} = \frac{1}{2}\begin{bmatrix} -3 & 6 & -2 & -1 \\ -42 & 3 & 1 & 2 \\ 18 & -9 & -11 & 2 \\ 9 & -18 & -2 & 11 \end{bmatrix}$$

$$\mathbf{N}_2\mathbf{N}_0^{-1} = 9\begin{bmatrix} -2 & 1 & 1 & 0 \\ 1 & -2 & 0 & 1 \\ -5 & 4 & 2 & -1 \\ 4 & -5 & -1 & 2 \end{bmatrix}$$

Equations (9-31) reduce in this case to

$$\frac{d\mathbf{x}}{dt} = \frac{D}{L^2}\mathbf{N}_2\mathbf{N}_0^{-1}\mathbf{x} - \frac{u}{L}\mathbf{N}_1\mathbf{N}_0^{-1}\mathbf{x} - \mathbf{kx} \tag{9-40}$$

where the vector $\mathbf{x}$ is

$$\mathbf{x} = \begin{bmatrix} x(\frac{1}{3}) \\ x(\frac{2}{3}) \\ x(0) \\ x(1) \end{bmatrix}$$

It was shown in Example 9-4 that boundary conditions (7-73c) and (9-36) can serve to eliminate $x(0)$ and $x(1)$ in terms of the interior collocation values. In particular for the problem at hand:

$$x(0) = 0.226x(\tfrac{1}{3}) + 0.503x(\tfrac{2}{3})$$
$$x(1) = -0.78x(\tfrac{1}{3}) + 1.73x(\tfrac{2}{3})$$

This reduces the **A** matrix for this problem to one of only second order with elements:

$$a_{11} = k\left[\frac{u}{kL}\left(1.33 - 15.9\frac{D}{uL}\right) - 1\right]$$

$$a_{12} = k\left[\frac{u}{kL}\left(-1.63 + 13.5\frac{D}{uL}\right)\right]$$

$$a_{21} = k\left[\frac{u}{kL}\left(21.7 + 1.98\frac{D}{uL}\right)\right]$$

$$a_{22} = k\left[\frac{u}{kL}\left(-3.48 - 2.43\frac{D}{uL}\right) - 1\right]$$

For any reasonable values of the Damkohler group, $(u/kL) \ll 1$, and with the given Peclet number matrix **A** may be estimated as

$$\mathbf{A} = k\begin{bmatrix} -1 & -0.28\left(\frac{u}{kL}\right) \\ 21.9\left(\frac{u}{kL}\right) & -1 \end{bmatrix}$$

and the eigenvalues are from equation (3-40):

$$\lambda = k\left[-1 \pm 2.47\frac{u}{kL}j\right]$$

The system is locally stable, but shows oscillatory responses as expected. Since the equations are all linear in this case, the same behavior pattern would be predicted for large or small disturbances.

9-5. The TRRM-Recycle System

Because no axial dispersion occurs in the TRRM, each plug of fluid moves axially and behaves independently of every other plug in spite of the radial changes. As a result the technique of Reilly and Schmitz is as applicable to the TRRM-recycle system as to the PFTR. Using the difference equation viewpoint leads directly to equations (9-18) and (9-20), but each variable that has radial variation must be integrated in this direction as by equations (9-2). The details of such computations have been presented by McGowin (1971b) for the example that earlier produced the steady state profiles of Figures 6-2.

Paralleling the development of Section 9-3 a matrix is defined as

$$\mathbf{M}(Y, Z) = \begin{bmatrix} \dfrac{\partial\eta}{\partial\eta_0} & \dfrac{\partial\eta}{\partial C_0} \\ \dfrac{\partial C}{\partial\eta_0} & \dfrac{\partial C}{\partial C_0} \end{bmatrix} \tag{9-41}$$

which reduces at the reactor entrance location to

$$\mathbf{M}(Y, 0) = \mathbf{I} \tag{9-42}$$

In comparison with equations (9-21) and (9-22) there appears to be little difference, however because the TRRM has radial variation and boundary conditions (1-11):

$$\frac{\partial\mathbf{M}}{\partial Y}(0, Z) = \mathbf{0} \tag{9-43}$$

and

$$\frac{\partial\mathbf{M}}{\partial Y}(L_0, Z) = \begin{bmatrix} -\dfrac{U_r}{\alpha C_p} & 0 \\ 0 & 0 \end{bmatrix}\mathbf{M}(L_0, Z) \tag{9-44}$$

For the evaluation of $\mathbf{J}_s$ it needs to be observed that the radial average $\mathbf{M}$ at any axial position is

$$\bar{\mathbf{M}}(L) = \begin{bmatrix} \dfrac{\partial\eta_E}{\partial\eta_0} & \dfrac{\partial\eta_E}{\partial C_0} \\ \dfrac{\partial C_E}{\partial\eta_0} & \dfrac{\partial C_E}{\partial C_0} \end{bmatrix} = \begin{bmatrix} \dfrac{\partial F_1}{\partial\eta_0} & \dfrac{\partial F_1}{\partial C_0} \\ \dfrac{\partial F_2}{\partial\eta_0} & \dfrac{\partial F_2}{\partial C_0} \end{bmatrix} \tag{9-45}$$

identically the function of equation (9-23), if η_E and C_E are understood to be radial averages as defined by equation (9-2). As a result of this the stability of a TRRM may be studied via the eigenvalues of

$$\mathbf{J}_s = f\bar{\mathbf{M}}(L) \tag{9-46}$$

An auxiliary equation is still needed for **M**. By following the procedure of McGinnis (1965), the required equation is

$$\frac{\partial \mathbf{M}}{\partial Z} = \begin{bmatrix} \frac{\alpha}{u} & 0 \\ 0 & \frac{D}{u} \end{bmatrix} \left[\frac{\partial^2 \mathbf{M}}{\partial Y^2} + \frac{1}{Y}\frac{\partial \mathbf{M}}{\partial Y}\right] + \frac{1}{u}\begin{bmatrix} -\frac{\partial R}{\partial \eta} & -\frac{\partial R}{\partial C} \\ \frac{\partial R}{\partial \eta} & \frac{\partial R}{\partial C} \end{bmatrix} \mathbf{M} \tag{9-47}$$

As before the solution to a particular problem requires integration of both the system equations and the auxiliary equations. This aspect of the problem is the same for the TRRM as it was for the PFTR problem, but since the radial variation has been included, the equations are considerably more difficult to handle. McGowin used the collocation method to get numerical axial profiles at a series of radial locations from the steady-state TRRM equations:

$$\frac{\partial C}{\partial Z} = \frac{D}{u}\left[\frac{\partial^2 C}{\partial Y^2} + \frac{1}{Y}\frac{\partial C}{\partial Y}\right] - \frac{R}{u} \tag{9-48a}$$

$$\frac{\partial \eta}{\partial Z} = \frac{\alpha}{u}\left[\frac{\partial^2 \eta}{\partial Y^2} + \frac{1}{Y}\frac{\partial \eta}{\partial Y}\right] + \frac{R}{u} \tag{9-48b}$$

and the associated auxiliary equations (9-47).

The numerical results are given in Figures 9-8 for the TRRM-recycle system. The first figure of this pair uses the Peclet number as parameter at constant level of heat transfer coefficient, whereas Figure 9-8b shows the effect of variation in $(U_r L/uC_p L_0)$ for a fixed value of $(uL_0/D) = (uL_0/\alpha) = 5$. The steady states' local stability character were determined by the eigenvalue criterion applied to matrix $\mathbf{J}_s$. These results are identified on both figures by the crosses and circles for local instability and stability, respectively.

The steady-state operating curves obtained for the TRRM-recycle system are generally of the form already familiar from the study of the other reactor models: the states are unique or triple, and when three steady states arise the low and high conversion states are typically locally stable with the intermediate state unstable. Important exceptions can also arise. That a unique steady state can be unstable is not surprising since the same behavior was found for the CSTR, PFTR-recycle, and TRAM systems. A typical result of this variety is shown in Figure 9-9. The numbered points show the successive states of the fluid each time it passes through the reactor entrance. The limit cycle is developed after approximately 40 passes around the recycle loop. It has a period of approximately 10 passes. The transient response was generated by recursively integrating the equations (9-48), computing the

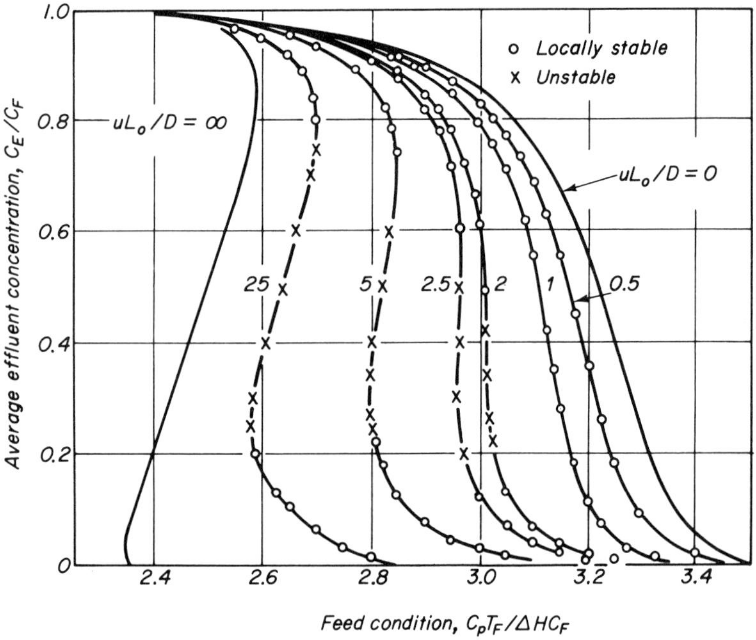

Figure 9-8a. Steady-state operating curves for TRRM–recycle system with $f = 0.5$, $uL_o/D = uL_o/\alpha$, $U_rL/uC_pL_0 = 0.79$.

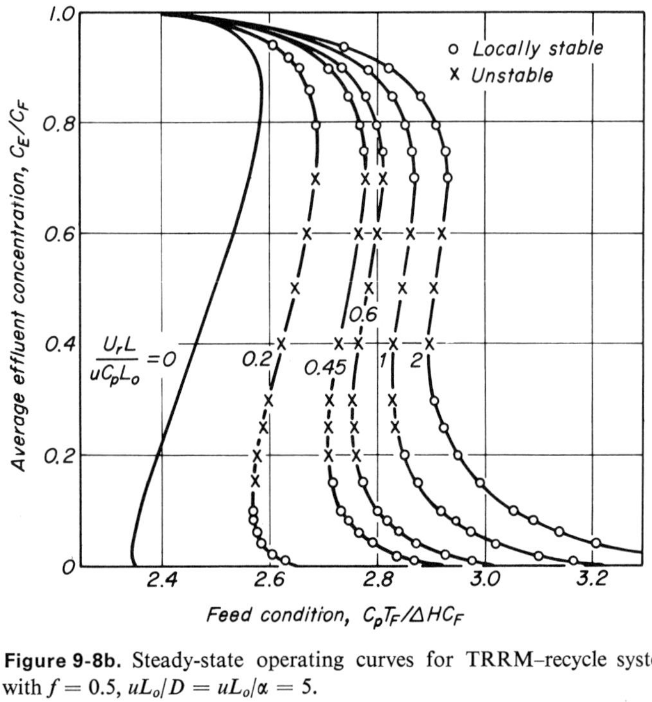

Figure 9-8b. Steady-state operating curves for TRRM–recycle system with $f = 0.5$, $uL_o/D = uL_o/\alpha = 5$.

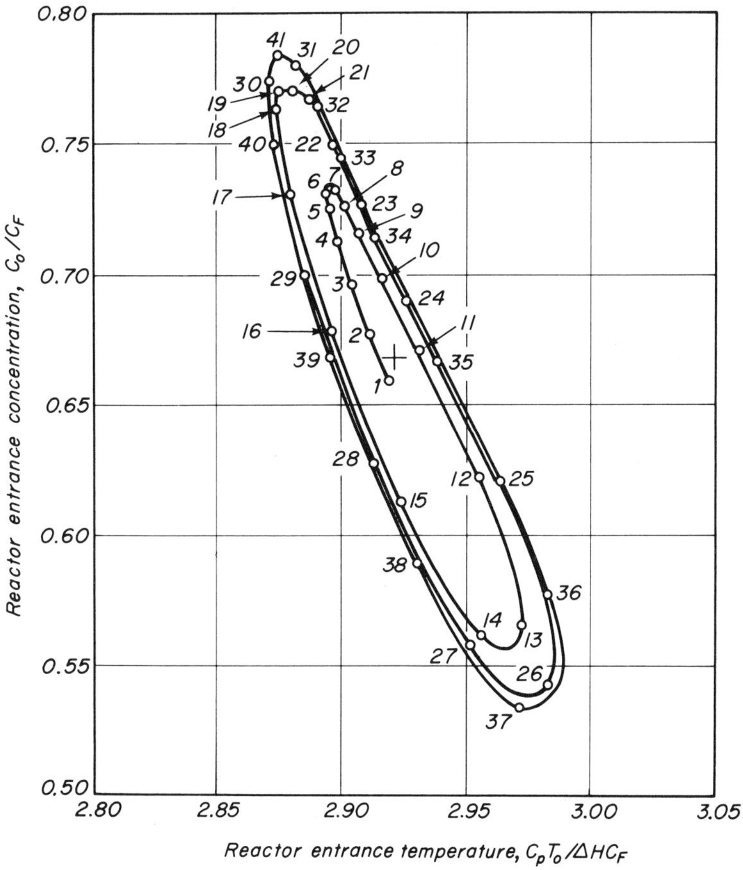

Figure 9-9. Discrete transient response and limit cycle for TRRM–recycle system with $uL_o/D = uL_o/\alpha = 2$, $U_rL/uC_pL_o = 0.79$, $f = 0.5$, $C_pT_F/\Delta HC_F = 3.015$.

radial average concentration and temperature at the reactor exit, and using the recycle boundary conditions (9-1) to compute updated entrance conditions.

Rather less to be expected is the finding that over the ranges of parameters

$$\frac{uL_0}{D} = \frac{uL_0}{\alpha} < 5 \qquad \text{at } (U_rL/uC_pL_0) = 0.79$$

or

$$\frac{U_rL}{uC_pL_0} > 0.45 \qquad \text{at } \frac{uL_0}{D} = \frac{uL_0}{\alpha} = 5$$

the high conversion steady state may be unstable. When this is the case for a system with triple steady state, only the low conversion state is physically attainable. The great majority of cases studied produced steady states associated with complex eigenvalues, indicating that the discrete transient response of the TRRM-recycle system usually follows an oscillatory pattern the signs of the reactor concentration and temperature disturbances alternate on successive passes around the recycle loop. In view of the Reilly and Schmitz (1967) results that always produced positive eigenvalues for the PFTR-recycle system, it is of interest also to note that some steady state of the TRRM-recycle system are associated with complex eigenvalues with negative real parts.

Comparisons among the operating curves of Figures 9-8 with those of Figures 9-7 show that the tubular reactor-recycle system is much more sensitive to changes in the radial Peclet numbers than to changes in the axial Peclet numbers. In most cases, the number of steady states that occur and the local stability character of the steady states remain the same in the TRAM-recycle system as axial mixing varies over a wide range, but in the TRRM-recycle system, relatively small changes in the radial Peclet numbers can cause the character of the operating condition to change. Figure 9-7d, for example, shows that for

$$C_p T_F/\Delta H C_F = 2.96$$
$$hL/uC_p = 0.79$$

the TRAM-recycle system possesses a unique unstable steady state (exhibits limit cycle behavior) for axial Peclet numbers between approximately 1.0 and 4. In contrast, Figure 9-8a indicates that the TRRM-recycle system can show either a unique low conversion stable steady state, or three steady state among which one is unstable, or a unique high conversion stable steady state —all within the narrow range of radial Peclet number between about 2 and 3.5. This sensitivity of the TRRM-recycle system to the radial Peclet numbers exists for the parameters chosen at

$$U_r L/uC_p L_0 = 0.79$$

and increases with increase in heat transfer coefficient.

9-6. Regions of Stability in Recycle Systems

In considering regions of stability it is well to distinguish between what has been done in the realm of local stability studies and what is to be examined in this section. Appreciable areas of the diagrams such as Figures 9-7 (the steady-state operating curves) are labeled as Regions of Instability to emphasize the local behavior of those steady states, but those results do not deal at all with disturbances of relatively large magnitude. The responses to disturbances in the latter category can be studied by one of two approaches.

First there is the most obvious but usually time consuming method of numerical integration of the differential equations of interest. This is most useful if a particular initial condition is to be tested or a result of another method is to be checked, as for example if a limit cycle is suspected. As noted above, the collocation method is attractive for many TRAM- or TRRM-recycle problems, but other numerical techniques are also available.

As a second alternative to the description of transients following large-magnitudes disturbances, Wang (1967) suggested a relatively simple corrollary of his composite phase-plane method presented in Chapter 8. This method has the merit of producing Regions of Practical Stability, which can be related to the engineering specifications on a system. It is a direct outgrowth of the feedforward control arguments in Section 8-8 that centered about a (δ, ϵ) pair and associated regions. The point was made in that discussion that appropriate preheating or premixing could provide control that would hold a reactor inlet condition to within a δ-region specification. Since the recycle stream in a reactor loop can be thought of as achieving precisely this effect, the design questions can be reformulated in this context. The object is to choose the various adjustable parameters such that the recycle mixing satisfies the (δ, ϵ) conditions.

To illustrate with more quantitative relationships, assume that the appropriate reactor inlet and effluent conditions are confined by:

$$x_{10}^2 + x_{20}^2 \leq \delta \tag{9-49a}$$

$$x_{1E}^2 + x_{2E}^2 \leq \epsilon \tag{9-49b}$$

and the fresh feed is characterized by

$$C_F^2 + \eta_F^2 = \delta_F \tag{9-50}$$

A safe but conservative way to satisfy (9-49a) is by

$$C_0^2 + \eta_0^2 \leq \delta \tag{9-51}$$

Substituting equations (9-1) into (9-51) gives

$$(1-f)^2(C_F^2 + \eta_F^2) + f^2(\eta_E^2 + C_E^2) + 2f(1-f)(\eta_E\eta_F + C_EC_F) \leq \delta \tag{9-52}$$

which reduces via (9-49) and (9-50) to

$$(1-f)^2\delta_F + f^2\epsilon + 2f(1-f)(\eta_E\eta_F + C_EC_F) \leq \delta \tag{9-53}$$

This inequality may be further reduced by noting that:

$$2\eta_E\eta_F \leq \eta_E^2 + \eta_F^2 \tag{9-54a}$$

$$2C_EC_F \leq C_E^2 + C_F^2 \tag{9-54b}$$

and the sum

$$2(\eta_E\eta_F + C_EC_F) \leq \epsilon + \delta_F \tag{9-55}$$

Substitution of inequality (9-55) into (9-53) gives

$$(1-f)\delta_F + f\epsilon \leq \delta \tag{9-56}$$

which may be treated as a design equation. A less conservative design is feasible if one avoids the chain of simplifying inequalities, and returns instead to equations (9-1) and (9-49).

It should be added in summary that any (δ, ϵ) regions developed via a composite phase-plane approach will be practical regions in the sense of being bounds corresponding to engineering specifications. Thus it is possible as it was also in the CSTR analysis to have a limit cycle (locally unstable steady state) within a region of stability. This particular point is discussed by Wang and Perlmutter (1968) in terms of a numerical example.

9-7. Internal versus External Recycle

Because the destabilizing effects of recycle are occasionally of major significance, it is most important to recognize that such feedback can occur in a variety of ways, either internal or external to a reactor. In fact the distinction between what is or is not considered a recycle system is often rather arbitrary, as much in the eye of the beholder as inherent in the system. To put it another way: one is not able to tell from the various equations of change alone whether recycle exists or not.

Several examples of reactor models that have feedback character due to internal mass or energy transfer have already been discussed in Chapter 6. The dispersion terms of the TRAM models have this property, as does the back-radiation included in the PFTR by Chen and Churchill (1970). To obtain a complete catalog of such possibilities is impractical and probably undesirable, but it is worthwhile to consider at least two other models with internal recycle in order to show the breadth of applicability of the viewpoint.

Having observed that the CSTR is not always an adequate model to represent mixing obtained in practice, Gall and Aris (1965) suggested a configuration consisting of combined tubular and stirred reactors, allowing for thermal or material feedback to different degrees. They used modified van Heerden diagrams to demonstrate that several familiar features can arise (steady state multiplicity, unstable steady states, extreme parametric sensitivity) as well as some novel interaction effects of considerable complexity. The recycle is entirely internal in the sense that an outsider would not find any evidence that the reactor product was being returned to the feed.

A related example may be found in the work of Iscol (1970), who modeled a Fluid Catalytic Cracker by four ordinary differential equations describing the mass and energy dynamics in reactor and regenerator, respectively. On superficial examination it is not obvious from the paired CSTR equations

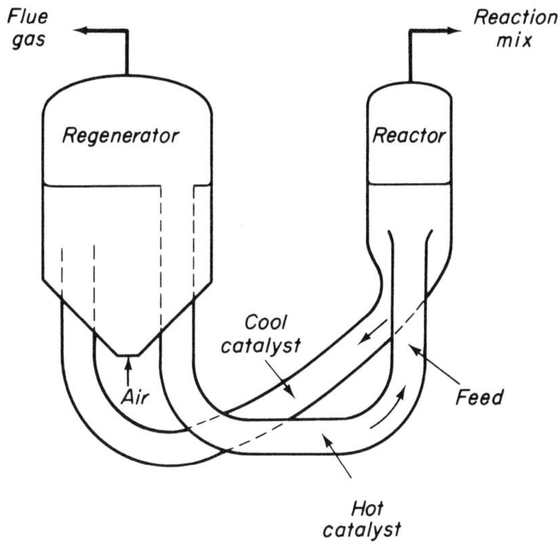

Figure 9-10. Basic structure of a Model IV catalytic cracker (after Iscol, 1970).

that a recycle exists, but of course the actual flow patterns as in diagrammatic Figure 9-10 show that each part provides and receives internal recycle from the other. Iscol carried his analysis through a detailed linearization and eigenvalue study to demonstrate that the system can be unstable with or without oscillations over a wide range of likely parameters. He suggests that his computational results may explain the well-known drifts and frequent adjustments that are typical in commercial cracking operation.

REFERENCES

BELLMAN, R., *A Survey of the Theory of the Boundedness, Stability and Asymptotic Behavior of Solutions of Differential and Difference Equations*, ONR, Washington, D. C. (1949) pp. 141–156.

CHEN, J. L. P., Ph. D. Thesis, Univ. of Pennsylvania, Philadelphia (1970).

GALL, C. E. and RUTHERFORD ARIS, *Can. Journal Chem. Eng.*, **43**, 16 (1965).

GOLDBERG, S., *Introduction to Difference Equations*, New York: Wiley (1958).

ISCOL, L., "The Dynamics and Stability of a Fluid Catalytic Cracker", *Proc. Joint Automatic Control Conf.* (1970).

LUSS, DAN and N. R. AMUNDSON, *AIChE Journal*, **13**, 279 (1967).

MCGINNIS, P. H., Jr., *Chem. Eng. Prog. Symp. Ser.*, No. 55, **61**, 2 (1959).

MCGOWIN, C. R., Ph. D. Thesis, Univ. of Pennsylvania, Philadelphia (1969).

McGowin, C. R. and D. D. Perlmutter, *Chem. Eng. Journal*, **2**, 125 (1971a).

McGowin, C. R. and D. D. Perlmutter, *AIChE Journal*, **17**, 842 (1971b).

Pareja, Guillermo and M. J. Reilly, *Ind. Eng. Chem. Fund.*, **8**, 442 (1969).

Reilly, M. J. and Schmitz, R. A., *AIChE Journal*, **12**, 153 (1966).

Reilly, M. J. and Schmitz, R. A., *Ibid.*, **13**, 519 (1967).

Root, R. B. and Schmitz, R. A., *AIChE Journal*, **15**, 670 (1969).

Root, R. B. and Schmitz, R. A., *Ibid.*, **16**, 356 (1970).

Sano, K., M. S. Thesis, Univ. of Pennsylvania, Philadelphia (1971).

Schmeal, W. R. and N. R. Amundson, *AIChE Journal*, **12**, 1202 (1966).

Wang, F. S. and D. D. Perlmutter, *Proc. Joint Automatic Control Conf.* (June, 1967), p. 107.

Wang, F. S. and D. D. Perlmutter, *AIChE Journal*, **14**, 976 (1968).

Appendix A: Rules for Transforming a Matrix into the Schwartz Form

The procedure detailed in this appendix consists of three steps, following in order the results of Ogata (1967), Routh (1884), and Chen and Chu (1966).

Step 1. Transformation from state-space to scalar form (after Ogata):

Assume that a linear system is described in state-space by a matrix of the form of equation (3-25b). The corresponding characteristic equation (3-28) may then be expanded to an nth degree polynomial in λ:

$$\alpha_0\lambda^n + \alpha_1\lambda^{n-1} + \alpha_2\lambda^{n-2} + \cdots + \alpha_{n-1}\lambda + \alpha_n = 0 \qquad \text{(A-1)}$$

But this is identical to the characteristic equation that would arise from Laplace transformation of the higher-order scalar equation

$$\alpha_0\frac{d^n x}{dt^n} + \alpha_1\frac{d^{n-1}x}{dt^{n-1}} + \alpha_2\frac{d^{n-2}x}{dt^{n-2}} + \cdots + \alpha_{n-1}\frac{dx}{dt} + \alpha_n x = 0 \qquad \text{(A-2)}$$

By dividing each term by α_0 and defining new coefficients by the ratios of the respective α's:

$$\frac{d^n x}{dt^n} + a_n\frac{d^{n-1}x}{dt^{n-1}} + a_{n-1}\frac{d^{n-2}x}{dt^{n-2}} + \cdots + a_2\frac{dx}{dt} + a_1 x = 0 \qquad \text{(A-3)}$$

This last equation may in turn also be put into an alternative matrix form by letting:

$$\begin{aligned} x_1 &= x \\ x_2 &= dx/dt \\ x_3 &= d^2x/dt^2 \\ &\cdot \\ &\cdot \\ &\cdot \\ x_n &= d^{n-1}x/dt^{n-1} \end{aligned} \tag{A-4}$$

Then $\dot{\mathbf{x}} = \mathbf{A}_n\mathbf{x}$ where

$$\mathbf{A}_n = \begin{bmatrix} 0 & 1 & 0 & \cdots & 0 & 0 \\ 0 & 0 & 1 & \cdots & 0 & 0 \\ 0 & 0 & 0 & \cdots & 0 & 0 \\ \cdots & & & & & \\ \cdots & & & & & \\ 0 & 0 & 0 & \cdots & 1 & 0 \\ 0 & 0 & 0 & \cdots & 0 & 1 \\ -a_1 & -a_2 & -a_3 & \cdots & -a_{n-1} & -a_n \end{bmatrix} \tag{A-5}$$

It may be noted in passing that a system described by characteristic equation (A-1) will be unstable if all the α_i coefficients are not of the same sign or if any $\alpha_i = 0$ [see for example, Dickson (1939)]. The converse is not necessarily true, however, and an equation with all $\alpha_i > 0$ must be examined further to draw stability conclusions.

Step 2. The Routh array:

Routh (1884) showed that a convenient algorithm can produce a necessary and sufficient stability test for equation (A-1). The method uses an array

$$\begin{matrix} C_{11} & C_{12} & C_{13} & \cdots \\ C_{21} & C_{22} & C_{23} & \cdots \\ C_{31} & C_{32} & \cdots & \\ C_{41} & C_{42} & \cdots & \\ C_{51} & \cdots & & \\ \cdots & & & \end{matrix} \tag{A-6}$$

for which the α_i supply the first two rows directly:

$$\begin{aligned} C_{11} &= \alpha_0 \\ C_{21} &= \alpha_1 \\ C_{12} &= \alpha_2 \\ C_{22} &= \alpha_3 \\ C_{13} &= \alpha_4, \text{ etc.} \end{aligned} \tag{A-7}$$

and the third and subsequent rows are computed from the 2×2 subdeterminants according to:

$$\begin{aligned}
C_{31} &= \frac{C_{21}C_{12} - C_{11}C_{22}}{C_{21}} \\
C_{32} &= \frac{C_{21}C_{13} - C_{11}C_{23}}{C_{21}} \\
C_{33} &= \frac{C_{21}C_{14} - C_{11}C_{24}}{C_{21}} \\
C_{41} &= \frac{C_{31}C_{22} - C_{21}C_{32}}{C_{31}} \\
C_{42} &= \frac{C_{31}C_{23} - C_{21}C_{33}}{C_{31}}, \quad \text{etc.}
\end{aligned} \tag{A-8}$$

Equation (A-1) models an asymptotically stable system if the coefficients α_i and the elements of the first row in the Routh array are all positive.

Step 3. Transformation from $\mathbf{A}_n$ *to* **B** (*after Chen and Chu*):

A matrix of the form (A-5) can always be related to an equivalent matrix **B** by the transformation equation (4-48). The needed matrix **T** has been tabulated in terms of the elements of the Routh array in the Chen and Chu reference cited:

$$\mathbf{T}_2 = \mathbf{I} \tag{A-9a}$$

$$\mathbf{T}_3 = \begin{bmatrix} 1 & 0 & 0 \\ 0 & 1 & 0 \\ (C_{22}/C_{21}) & 0 & 1 \end{bmatrix} \tag{A-9b}$$

$$\mathbf{T}_4 = \begin{bmatrix} 1 & 0 & 0 & 0 \\ 0 & 1 & 0 & 0 \\ (C_{32}/C_{31}) & 0 & 1 & 0 \\ 0 & (C_{22}/C_{21}) & 0 & 1 \end{bmatrix} \tag{A-9c}$$

$$\mathbf{T}_5 = \begin{bmatrix} 1 & 0 & 0 & 0 & 0 \\ 0 & 1 & 0 & 0 & 0 \\ (C_{42}/C_{41}) & 0 & 1 & 0 & 0 \\ 0 & (C_{32}/C_{31}) & 0 & 1 & 0 \\ (C_{23}/C_{21}) & 0 & (C_{22}/C_{21}) & 0 & 1 \end{bmatrix} \tag{A-9d}$$

The generalization of this pattern is also given in the reference paper.

The reader should note that computation of the **B** matrix by

$$\mathbf{B} = \mathbf{T}_n \mathbf{A}_n \mathbf{T}_n^{-1} \tag{A-10}$$

is only of interest if a Liapunov function is being sought. Otherwise, the question of stability is settled at an earlier stage, usually when the Routh array becomes available.

EXAMPLE A-1. Find **B**, given that

$$\mathbf{A} = \begin{bmatrix} 6 & -1 & -5 \\ 0 & 3 & 1 \\ 6 & 1 & 6 \end{bmatrix}$$

Solution: For the given matrix the characteristic equation is

$$\det(\lambda\mathbf{I} - \mathbf{A}) = \lambda^3 - 15\lambda^2 + 101\lambda - 186 = 0$$

and the Routh array is

$$\begin{array}{rr} 1 & 101 \\ -15 & -186 \\ 88.6 & \\ -186 & \end{array}$$

Substitution into equations (A-9b) and (A-5) gives

$$\mathbf{T}_3 = \begin{bmatrix} 1 & 0 & 0 \\ 0 & 1 & 0 \\ 12.4 & 0 & 1 \end{bmatrix}$$

and

$$\mathbf{A}_3 = \begin{bmatrix} 0 & 1 & 0 \\ 0 & 0 & 1 \\ 186 & -101 & 15 \end{bmatrix}$$

and via equation (A-10):

$$\mathbf{B} = \begin{bmatrix} 0 & 1 & 0 \\ -12.4 & 0 & 1 \\ 0 & -88.6 & 15 \end{bmatrix}$$

REFERENCES

CHEN, C. F. and H. CHU, *IEEE Tr. Auto. Control*, **AC-11**, No. 2, 303 (April, 1966).

DICKSON, L. E., *New First Course in the Theory of Equations*, New York: Wiley (1939).

OGATA, K., *State Space Analysis of Control Systems*, Englewood Cliffs, N. J.: Prentice-Hall (1967).

ROUTH, E. J., *A Treatise on the Dynamics of a System of Rigid Bodies*, 4 Ed., London: Macmillan (1884).

Appendix B: Derivation of Inequality (6-68)[1]

When particular kinetics make multiple solutions possible, whether they in fact occur or not will depend on the size of the catalyst particle. A system with unique steady state for a given size may exhibit multiplicity as the significant dimension is reduced. If this occurs, the point of transition is called a *bifurcation.* Important contributions in modern topology are concerned with justifying linearization at bifurcation points of partial differential equations. That is to say, one needs to know when the eigenvalues of linear approximations are faithful descriptions of local behavior in practical nonlinear problems. The work of Krasnoselskii (1964) justifies such a procedure in the case of the diffusion equation. Accordingly, it is proper to approximate the rate function for any nearby solution profile C_2 by a truncated expansion about C_1, the bifurcation solution:

$$R(C_2) = R(C_1) + \left.\frac{dR}{dC}\right|_{C_1}(C_2 - C_1) \tag{B-1}$$

Since equation (6-40) must be satisfied by any solution, equations (6-63) and (6-64) remain valid for the new definition of subscripts. Combination of equations (6-62), (B-1), and (6-64) gives

$$0 = D\frac{d^2\Phi}{dZ^2} - \left.\frac{dR}{dC}\right|_{C_1}\Phi \tag{B-2}$$

[1]After Luss (1968).

If equation (6-63a) is multiplied by Φ, equation (B-2) is multiplied by $(C_1 - C_0)$, and one product subtracted from the other:

$$\left[R(C_1) - (C_1 - C_0)\frac{dR}{dC}\bigg|_{C_1}\right]\Phi = D\left[\Phi\frac{d^2C_1}{dZ^2} - (C_1 - C_0)\frac{d^2\Phi}{dZ^2}\right] \tag{B-3}$$

Integration with respect to the distance variable will produce an integral on the right that is identically zero by Green's identity [Wylie (1966)]:

$$\int_0^L \left[R(C_1) - (C_1 - C_0)\frac{dR}{dC}\bigg|_{C_1}\right]\Phi dZ = 0 \tag{B-4}$$

Such an integral can only be zero if the integrand changes sign over the interval of integration. This can be denied as a possibility and uniqueness accordingly guaranteed if

$$R(C_1) - (C_1 - C_0)\frac{dR}{dC}\bigg|_{C_1} \geq 0 \tag{B-5}$$

or what is equivalent, if

$$(C_1 - C_0)\frac{d \ln R}{dC}\bigg|_{C_1} \leq 1 \tag{B-6}$$

Inequality (6-68) includes (B-6) and is thus sufficient to establish uniqueness.

REFERENCES

KRASNOSELSKII, M. A., *Topological Methods in the Theory of Nonlinear Integral Equations*, New York: Pergamon Press (1964).

LUSS, DAN, *Chem. Eng. Sci.*, **23**, 1249 (1968).

WYLIE, C. R. JR., *Advanced Engineering Mathematics*, New York: McGraw-Hill, Inc. (1966) p. 575.

Appendix C: Collocation Constants

TABLE 1
COLLOCATION CONSTANTS[1]

n	$\mathbf{z}$	$\mathbf{N}_1\mathbf{N}_0^{-1}$	$\mathbf{N}_2\mathbf{N}_0^{-1}$	$[\mathbf{N}_3\mathbf{N}_0^{-1}]^T$
1	$\begin{bmatrix} 0.44721 \\ 1 \end{bmatrix}$	$\begin{bmatrix} -1.1180 & 1.1180 \\ -2.5000 & 2.5000 \end{bmatrix}$	$\begin{bmatrix} -2.5 & 2.5 \\ -2.5 & 2.5 \end{bmatrix}$	$\begin{bmatrix} 0.83333 \\ 0.166667 \end{bmatrix}$
2	$\begin{bmatrix} 0.285232 \\ 0.765055 \\ 1 \end{bmatrix}$	$\begin{bmatrix} -1.752962 & 2.507614 & -0.754652 \\ -1.370599 & -0.653547 & 2.024146 \\ 1.791503 & -8.791503 & 7 \end{bmatrix}$	$\begin{bmatrix} -4.739870 & 5.677124 & -0.937254 \\ 8.322876 & -23.260130 & 14.937254 \\ 19.071891 & -47.071891 & 28 \end{bmatrix}$	$\begin{bmatrix} 0.554858 \\ 0.378475 \\ 0.066667 \end{bmatrix}$
3	$\begin{bmatrix} 0.20929922 \\ 0.59170018 \\ 0.87174015 \\ 1 \end{bmatrix}$	$\begin{bmatrix} -2.3889244 & 3.5135453 & -1.7401020 & 0.61548115 \\ -1.5027316 & -0.84502256 & 3.3435154 & -0.99576125 \\ 0.81782916 & -3.6741370 & -0.57356542 & 3.4298732 \\ -1.4877192 & 5.6276304 & -17.639911 & \frac{27}{2} \end{bmatrix}$	$\begin{bmatrix} -8.1079245 & 9.6033328 & -2.0342937 & 0.53888539 \\ 11.611597 & -27.294644 & 19.309454 & -3.6264063 \\ -3.9821807 & 31.261322 & -77.097432 & 49.81829 \\ -35.432632 & 122.93626 & -200.003632 & \frac{225}{2} \end{bmatrix}$	$\begin{bmatrix} 0.41245880 \\ 0.34112269 \\ 0.21070423 \\ \frac{1}{28} \end{bmatrix}$

[1] Villadsen, J. V. and W. E. Stewart, *Chem. Eng. Sci.*, **22**, 1483 (1967).

Appendix D: *Some Exercise Problems*

D-1. A chemical reactor system is modeled by

$$\frac{\partial C}{\partial t} = D\frac{\partial^2 C}{\partial Z^2} - u\frac{\partial C}{\partial Z} - kC$$

$$\frac{D}{u}\frac{\partial C}{\partial Z} = C - C_0, \qquad Z = 0$$

$$\frac{\partial C}{\partial Z} = 0, \qquad Z = L$$

What are the bases for this model? What laws of physics and chemistry? What do the symbols stand for? What assumptions are involved? (Ch.1).

D-2. Derive equations (1-14) through (1-17) from first principles, carefully defining the symbols and noting the major assumptions. (Ch. 1).

D-3. It is proposed to use existing idle equipment to process a new feed stream of reactant concentration $C_0 = 1$ g mol/lit and heat capacity $C_p = 1{,}000$ cal/lit°K. A CSTR is available that can supply heat transfer to the extent of $U = 30$ cal/sec°K for a flow rate $q = 0.03$ lit/sec. Laboratory kinetic measurements indicate that the reaction is second-order in the single reactant, exothermic, and follows Arrhenius temperature behavior:

$$n \;\; = 2$$
$$\Delta H = 79{,}200 \text{ cal/g-mol}$$
$$Q \;\; = 12{,}500°\text{K}$$

Preliminary calculations show a steady state at:

$$T_s \;\; = 300°\text{K}$$
$$C_s \;\; = 0.5 \text{ g-mol/lit}$$

(a) It is proposed to avoid steady-state multiplicity by preheating the feed stream. Over what range of T_0 can you guarantee uniqueness? Will the system necessarily show multiple steady states with a lower temperature? Explain.

(b) How does your estimate in part (a) compare with the value of T_0 that corresponds to the calculated unstable steady state? Assume that coolant is available at any convenient temperature. (Ch. 2).

D-4. When a new catalyst was tried in the chain-reaction oxidation described by Matsuuro and Kato, it was found that the kinetics changed to replace equations (1-21) with:

$$R_1 = k_1 C_B^{1/2} - k_2 C_A^2$$
$$R_2 = k_2 C_A^2 - k_3 C_B$$

Recall that the reaction is to be run in an isothermal CSTR with neither A nor B in the feed stream. Can this new system have multiple steady states? If not, explain why not. If so, derive a uniqueness criterion in terms of the system parameters. (Ch. 2).

D-5. The solution to a certain linear problem includes

$$\mathbf{X}(t) = \begin{bmatrix} e^{-t}\cos 3t & e^{-t}\sin 3t \\ -e^{-t}\sin 3t & e^{-t}\cos 3t \end{bmatrix}$$

Find the maximum (μ/c) for the bound

$$\|\mathbf{X}(t)\| \leq ce^{-\mu t}$$

for (a) the Euclidean norm, (b) the square norm. (Ch. 3).

D-6. Consider the linear system $\dot{\mathbf{x}} = \mathbf{Ax}$, where

$$\mathbf{A} = \begin{bmatrix} -5 & 1 & -3 \\ 0 & -2 & -4 \\ -4 & 0 & -4 \end{bmatrix}$$

(a) Does Gerschgorin's theorem apply? Explain your answer.

(b) Can you establish stability (or instability) by using the spherical Liapunov function

$$v = x_1^2 + x_2^2 + x_3^2$$

Support your answer with calculations.

(c) Find the system steady state (or steady states). Is it (are they) stable?

(d) If the given matrix **A** had been the result of a prior linearization of a nonlinear system, what can you conclude about the original nonlinear system? Explain why. (Ch. 4).

D-7. Analyze for local stability characteristics about *each* steady state (Ch. 4):

(a) $dx/dt = x - x^2$

(b) $du/dt = -k(u - u^3);\ \ k > 0$

D-8. Analyze for local stability by linearization:

$$\frac{d^2x}{dt^2} + a(x^2 - 1)\frac{dx}{dt} + x = 0$$

How is your answer affected by the sign of the coefficient a? [Hint: Reduce to a set of two equations by the substitution of $x = x_1$ and $(dx/dt) = x_2$] (Ch. 4).

D-9. With reference to problem D-3, is the given steady state stable? It is proposed that an experimental confirmation should be obtained to demonstrate that the unstable steady state does indeed exist. Is this a good idea? Explain. (Ch. 4).

D-10. With reference to Problem D-4:

(a) Is the steady state stable? If there are multiple steady states, which are stable? Justify your answer.

(b) Is total washout of free-radicals a problem in this modified system? Describe how you would estimate the holding time required for practical operation. (Ch. 4).

D-11. Koga and Humphrey (1967) have analyzed the dynamic behavior of a chemostat system by the simultaneous equations:

$$\frac{dx_1}{dt} = x_1\left[\mu\frac{x_2}{K + x_2} - D\right]$$

$$\frac{dx_2}{dt} = D(S_R - x_2) - \frac{\mu}{Y}\left(\frac{x_1 x_2}{K + x_2}\right)$$

where

x_1 = cell population

x_2 = substrate concentration

S_R = substrate input concentration, constant

D = dilution rate, constant

μ = specific growth rate, constant

K = a constant

Y = yield coefficient, a constant

Find the system steady states. Are they stable? Considering the tracking function

$$E = x_1 + Yx_2$$

what graphical conclusions can you justify on the $(x_1 x_2)$ plane? Can you find a region of stability? (Ch. 5).

D-12. With reference to Problem D-1, outline how you would answer the question of possible steady-state multiplicity. (Ch. 6).

D-13. Chiou and Cohen (1968) made a numerical study of the adiabatic TRAM for the reaction

$$3\,Na_2S_2O_3 + 5\,H_2O_2 \longrightarrow \text{Products}$$

The kinetics for this reaction follow the rate equation

$$R = k_0 C_1 C_2 \exp(-Q/T)$$

where the subscripts 1 and 2 refer to the thiosulfate and peroxide reactants, respectively. The thermodynamics and kinetic constants

$$\Delta H = 138 \text{ kcal/g-mol of thiosulfate (exothermic)}$$

$$Q = 8{,}630°\text{K}$$

are applicable for feed concentrations in the range

$$C_{20} \geq 4C_{10}$$

(a) Are uniqueness criteria (6-95) applicable to this system?
(b) How must condition (6-99) be modified to account for the kinetics of the thiosulfate-peroxide reaction?
(c) Assuming reasonable values for feed temperature and aqueous heat capacity, estimate reactant feed concentrations for which you could assure a unique steady state.
(d) In the reference cited, Chiou and Cohen demonstrated that steady-state multiplicity, does occur for

$$T_0 = 273°\text{K}$$

$$C_{10} = 0.7 \text{ g-mol/lit}$$

$$C_{20} = 2.8 \text{ g-mol/lit}$$

$$\alpha = 4.8 \text{ cm}^2/\text{sec}$$

$$L = 42 \text{ cm}$$

$$u = 28.6 \text{ cm/sec}$$

Is this result consistent with your findings?
(e) Describe qualitatively how you would develop criterion (6-95b) for this system to account for flow rate affects.

D-14. Referring again to Problem D-1, outline how you would test for local stability of a steady state by using the modified collocation method with the trial solution

$$x(z, t) = \sum_{i=1}^{n} a_i(t) z^{i-1}$$

where x is the concentration-deviation variable and z is normalized distance. (Ch. 7).

D-15. Consider an isothermal PFTR-recycle system with irreversible kinetics of the Michaelis–Menten form:

$$R = \frac{kC}{K + C}$$

where C is the reactant concentration and k and K are available constants.

(a) Can the system have multiple steady states? Does the answer depend on k and K?

(b) How would you determine whether or not a particular steady state of this system is stable? (Ch. 9).

REFERENCES

CHIOU, J. K. S. and W. C. COHEN, Paper presented at AIChE Meeting, St. Louis, Mo., Feb., 1968. To appear in *AIChE Journal* (1972).

KOGA, S. and A. E. HUMPHREY, *Biotech. and Bioengineering*, **9**, 375 (1967).

Author Index

Subject Index

A

B

C

V

W

Z